THE ANALYSIS OF AIR POLLUTANTS

W. LEITHE

Head of the Main Laboratory of

Österrichischen Stickstoffwerke A.G., Linz/Donau

Translated by R. KONDOR

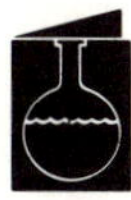

ANN ARBOR SCIENCE PUBLISHERS

ANN ARBOR • 1971

ANN ARBOR SCIENCE PUBLISHERS, INC.
Drawer No. 1425, Ann Arbor, Michigan 48106

Library of Congress Catalog Card Number 70-120994
ISBN 250 39910 5

This book is a translation of
DIE ANALYSE DER LUFT UND IHRER VERUNREINIGUNGEN
IN DER FREIEN ATMOSPHÄRE UND AM ARBEITSPLATZ
Wissenschaftliche Verlagsgesellschaft mbH
Stuttgart 1968

First English Language Printing 1970
Second Printing 1971

Printed in U.S.A.

Preface

Preventing the pollution of the most important part of our environment, the air we breathe, has become a complex and extremely serious problem, because of increases in population density, traffic and industrialization. A large section of the public and specialists in radically different professions — chemists, physicians, engineers, biologists and jurists — are concerned with this problem. But before discussing pollution, and taking the necessary measures, we must have an accurate knowledge of the types and concentrations of the pollutants from various sources. It is the task of physical and chemical analysts to gather evidence of these pollutants at the work site, in industry and trade. The results of air analysis must be known to protect the workers from injury and the products from contamination.

Every chemist, whether in research, industry, or in public institutions, will have to deal with air pollution problems and must be informed about the required techniques.

Information on air analysis is found throughout the literature. Some important methods of investigation are included in the Directives of the Working Group "Prevention of Air Pollution" (Reinhaltung der Luft) in the VDI (Verein deutscher Ingenieure) and the ICI Manual. Collections of methodological directives have been published by international organizations such as the OECD (Organization for Economic Cooperation and Development) and IUPAC (International Union of Pure and Applied Chemistry). The standard formulas of US and British trade organizations are also available. Many books on gas analysis contain sections on air, and papers on air analysis are found in numerous periodicals.

It is an exciting task, for this author, to survey a field of analytical practice which has such great contemporary interest. The task is all the more relevant since the analytical methods, and even the sanitation and legislative angles in many cases are still under discussion. Thus, textbooks in this field cannot present established facts and must include personal opinions to move discussion forward. However, analysis of air pollution cannot be delayed until the theory is worked out since solutions are urgently needed now all over the world.

The author's experience as chief chemist of a large chemical plant on the fringe of a densely populated residential area has given him comprehensive insight into the air pollution problem. He lectures at the University of Vienna and is active in international committees for preventing air pollution.

This book contains detailed instructions for practical analysis, adapting several methods to characteristic problems. These instructions were taken from German, British and American literature. The book is not intended as an introduction to air analysis and is more for the laboratory than the office. Readers should have a basic knowledge of general analytical practice since conventional methods of titration, preparation of titers, colorimetric measurements, etc., are not described in detail.*

Wherever he works, the analyst must not restrict his activity to receiving samples, analyzing them, and handing over results. Thus a book on air analysis dealing only with the theory and practice of analytical methods would be of little value. The analyst participates in preliminary discussions with physicians, engineers, and sometimes jurists, considering the necessity for and the aim of the air analysis. After the purpose of the analysis has been established, the analyst plans the experiments, selecting the suitable method. The analyst has problems in common with the authority ordering the analysis: sanitary survey with the physician; eliminating proven emissions and their consequences with the engineer, and the legal problems with the jurist in civil service.

To facilitate discussions on problems bordering on the analyst's field, and thus to ensure the success of his work with the physician, engineer or jurist, the first chapter of the book contains an introduction to relevant sections of sanitation, technology and legislation. The origin and extent of air pollution, combined with meteorological factors, are weighed against the damage it does to man and plants. From these facts the necessary legal limits (MIK- and MAK-values) are derived. Also, a survey of basic technical possibilities for preventing air pollution is presented.

A general chapter on air analysis gives a comprehensive survey on how to plan the experiments and sampling procedures. We distinguish between long-term procedures and individual determinations. The latter can be carried out either as orientative rapid tests with monitoring instruments or as accurate determinations of the gas levels. In accordance with the modern trend toward continuous recording, we discuss the development of the appropriate instruments first generally; gas chromatography is treated in a separate section. Some sections explain the testing of new analytical methods and the preparation of test mixtures from gases.

In the chapter "Special Topics" we shall discuss first the pollutants in the form of dust and smoke, frequently a no man's land between physics and chemistry, and the numerous analytical methods applied. This chapter includes a short introduction

* The book "Analytische Chemie in der Industriellen Praxis", Frankfurt, 1964, written by the present author, deals with general problems of analytical practice.

to investigating radioactive pollutants. The more important gaseous and vapor pollutants are then comprehensively dealt with. Sections on the more important pollutants discuss their analytical chemistry, their origin, propagation, hygiene and the present state of techniques for waste gas elimination. The analytical treatment should be comprehensive enough to supercede the original literature in most cases. The analytical part facilitates the selection of appropriate methods, it gives the concentration to be expected, the presence of other pollutants and the instrumental means available.

The author has utilized the large store of information and experience in this field available in his firm. He expresses his thanks to the management for their permission to publish these data and for their support. He also wishes to thank his co-workers who helped him in obtaining the necessary data. The author is grateful to Dr. Megay and Dr. Frenzel for information on radioactive pollutants.

Any suggestions and critical remarks from workers in the field of air analysis will be most welcome by the author. It is the author's hope that the book will be useful to many of his colleagues and that repeated analysis will show the right way to prevent air pollution in towns and industrial plants.

Linz, March 1968 *Wolfgang Leithe*

Table of Contents

1

General

1.1 Introduction

The air around us (the atmosphere) is the most important part of our natural environment. Terminate the supply of breathable air and death follows in short order. Under natural biological conditions, breathing is not at all hazardous, provided the air is of proper and uniform composition. Human activity, however, has polluted the air with biologically harmful substances and it is only in exceptional cases that this pollution is still insignificant. Recently, there have been occasions when the supply of satisfactory air for breathing has been actually endangered. Unfortunately, our senses cannot adequately evaluate the quality of the air.

Our sense of smell does not respond to all harmful air pollutants, such as, for example, carbon monoxide, carbon dioxide, nitrous and nitric oxides. Certain toxic substances, such as hydrocyanic acid, though perceived when present in very low amounts, do not induce defensive reaction mechanisms. Further, the irritative and harmful effects of some gases do not occur simultaneously; in the case of ammonia first the organ is irritated and the harmful effect occurs afterwards, while the opposite trend is observed for SO_2 and NO_2. Moreover, certain limits are imposed on the quantitative evaluation of air by smelling. Apart from the fact that the odor-sensitivity of humans to many air pollutants varies, one can get accustomed to some toxicants, such as hydrogen sulfide or nitric oxide. We have no sensory organ at all for radioactive pollutants, which have become increasingly important over the past decades.

These examples show that with the ever increasing use of toxic substances and the increase in air pollutants, both as far as their type and concentration are concerned, additional means for their evaluation had to be found. Initially, empirical measures were employed, such as the observation of a burning candle which indicated by flickering or extinguishing an excessive carbon monoxide level, or the behavior of song birds in hazardous places which indicated penetration of carbon monoxide.

Later, chemical analysis and physical measurements were used to detect and evaluate pollutants.

The tasks of the investigation of air can be roughly divided into two groups:

a) investigation of the free atmosphere in residential areas in the interest of the whole population, and of the air in agricultural regions in order to assess damages to plants and injuries inflicted upon animals;

b) inspection of work places in factories and workshops subjected to the hazards of waste gases as a means of personnel protection.

It is noteworthy that the air in living rooms, offices, etc., where the city inhabitant usually stays longer than in the open air, is rather infrequently analyzed; it is apparently assumed that the values measured in the open air are also indicative of the enclosed spaces. This assumption is by no means true; thus, for example, a certain measured sulfur dioxide level rapidly decreases in a room where lime-containing paints have been used. The same should apply to nitrogen dioxide, hydrogen fluoride and other gases which form acids with water. Furthermore, the conditions for the dust level in the outside air and the rooms adjacent thereto are by no means the same.

It is noteworthy that modern technology, though it has led to a considerable deterioration of the open air, has effected great improvement in the air quality of living rooms due to the fact that the conditions of lighting, heating and sewage removal have been bettered owing to the higher living standards of large strata of the population. Thus, for example, the influence of the soot originating from candles or oil lamps on the incidence of lung diseases can hardly be assessed today.

As in many other cases of analytical practice the results of air analysis led to numerous consequences in the fields of public sanitation, engineering, etc. On the other hand, the practical requirements imposed on sensitivity, specificity, time saving and frequency of air analysis increased and became more varied. This and the problems of chemical warfare in both world wars gave vital impetus to the systematic development of trace analysis and continuously operating automatic monitoring instruments. This interplay between analysis and practice is continuing at a rapid pace with the result that interesting and important innovations in methods of air analysis are being produced which lead to important new information and technical progress in air pollution prevention.

1.2 Definition of the terms "clean air", "air pollutants", "emissions", and "immissions"

"Clean" air, i.e., air occurring in areas sufficiently distant from places of human activity or other abnormal influences has the following composition in accordance with VDI-Richtlinie (Directive) 2104:

Air component	Vol. %	Air component	Vol. %
N_2	78.10	Krypton	0.0001
O_2	20.93	Neon	0.0018
Argon	0.93	Helium	0.0005
CO_2	0.03–0.04	Xenon	0.00001

The above directive states the hydrogen level as 0.01 ‰. However, according to recent, more accurate data, this level is considerably lower.

With regard to the water content of air (air humidity), see page 137.

The pollutants listed below are also present and occur as traces below 1 ppm in "clean" air (see, e.g., Junge [125]):

Air component	ppm (vol.)	Air component	ppm (vol.)
CO	$(1\text{–}20) \cdot 10^{-2}$	N_2O	0.25–0.6
Ozone	$(0.5) \cdot 10^{-2}$	$NO + NO_2$	$(0\text{–}3) \cdot 10^{-3}$
H_2	0.4–1	NH_3	$(0\text{–}2) \cdot 10^{-2}$
CH_4	1.2–1.5		

These trace substances partly originate from high air layers (O_3) and are partly due to decomposition and putrefaction processes (NH_3, CO, perhaps also N_2O) or weather conditions (NO_2).

Since the number of air components is large and since they are widely distributed in low concentrations (they are often found in uniform amounts independent of the sampling site), it is frequently difficult to define the limitations of the term "air pollutants" or "foreign substances in the air".

The "Technische Anleitung" (Technical Directives) TAL (see page 7) as well as VDI-Richtlinie 2104 define "air pollutants" as "solid, liquid or gaseous substances changing the natural composition of the atmosphere".

Emissions, according to TAL, are air pollutants which enter the atmosphere after leaving the plant. The same directive defines immissions as air pollutants which occur near the source (usually 1.5 m above ground level, the upper limit of vegetation or at a height of 1.5 m above the top of a building).

VDI-Richtlinie 2104 defines emissions as solid, liquid or gaseous substances of all types and origin escaping into the outside air, while immissions are here defined as the discharge of solid, liquid or gaseous pollutants which are permanently or temporarily near ground level.

In daily usage the term "pollution" has a derogatory connotation. Accordingly, in practice we shall speak only about "air pollutants" or "toxicants" when the substance in question occurs in disadvantageous or evidently harmful concentrations.

In this sense air pollution is defined as follows by the World Health Organization (WHO): "Air pollution occurs when one or several air pollutants are present in such amounts and for such a long period in the outside air that they are harmful to humans, animals, plants or property, contribute to damage or may impair the well-being or use of property to a measurable degree".

1.3 History of air pollution

Air pollution caused by human activity has always existed since fire was used. Further, official measures against strong nuisances of this type were introduced at an early stage. We mention, for example, the banishment of very malodorous trades to the periphery of towns and the prohibition against using "sea" coal for heating purposes in medieval London.

"Smoking stacks" were apparently considered during the earlier stages of the Industrial Revolution as the symbol of industrial activity and prosperity. However, in the second half of the last century the abundance of such stacks and the occurrence of highly repulsive waste gases in the early stages of chemical industry (especially hydrochloric acid and hydrogen sulfide from the Leblanc Soda Process) forced the authorities to take the appropriate steps. The first example of such legislation is the Alcali Act of 1864 in England.

The public and, consequently, the legislative authorities were later compelled, as a result of serious catastrophes, to set up increasingly stringent measures against air pollutions. These mishaps were not caused by the local discharge of concentrated poisonous gases in accidents but by the accumulation of waste gases from industry or heating (up to that time considered as harmless) due to very unfavorable meteorological conditions. The first large-scale catastrophe occurred in 1930 in the Meuse valley near Liège, when under the influence of a smoke blanket caused by an inversion (see page 17) the waste gases from the heavy industry accumulated to such a degree that within a few days several thousand people became sick with respiratory troubles and 60 persons met their death. Under similar conditions 17 persons died in Donora, USA in 1948. The smoke catastrophes in London were considerably larger; thus, in 1952 within 14 days the death toll following acute diseases of the respiratory tracts was about 4000 higher than during comparable intervals of other years. This occurrence was repeated in 1956 with about 1000 deaths.

The smog situation (smog is a fabricated word for smoke and fog) in the cities of Los Angeles and San Francisco, which were hitherto notorious for their agreeable weather, is not characterized by a higher mortality rate but by considerable damage to materials and vegetation. Investigation of this phenomenon by research chemists showed that automobile exhaust gases are mainly responsible and not the SO_2-

containing flue gases as in Europe. The olefins and the nitrogen oxides from these gases produced the various fog-forming respiratory irritants, the reactions being catalyzed by solar radiation. However, also in other places, the increasing nuisance caused by waste gases of all types, the damaging of forests, agricultural plants and thus cattle, as well as other objects, such as buildings, metals, textiles, etc., led to much indignation. The damages due to waste gases in the USA are estimated at about 7–10 billion dollars per year in addition to the harm to the health and well-being of the population which cannot be expressed in terms of money.

These catastrophes and inconveniences were widely noted by the public and publication media, and induced the authorities to support large-scale studies and to take preventive steps. First, large-scale research in the field of sanitation and particularly in the field of chemical analysis was conducted which, especially in the USA, was subsidized by public sources. Numerous new, adequately sensitive and rapid methods and instruments for analyzing toxicants were developed and applied. These served for collecting and evaluating comprehensive numerical data in order to establish in cooperation with physicians which qualitative or quantitative composition of the environmental air can be considered as harmless, or at which pollutant levels technical and legal countermeasures are required.

Even before large-scale attempts were made to prevent air pollution, the sanitary conditions at individual work sites were studied in highly industrialized states in order to protect workers against health hazards. The chemist's task was to develop analytical methods by means of which the pollutant concentration could be established and after that compared with the requirements set up by the occupational hygienist.

Germany was spared catastrophes of such an extent, probably because of the different climatic conditions. However, also in this country the air has considerably deteriorated due to industrial expansion, vehicular traffic and higher domestic fuel consumption. About 15 years ago intensive scientific and technical research and development widely supported by public and private sources was initiated. This not only led to stricter legislative measures but also to numerous improvements in technical processes intended to find a compromise reducing the air pollutants to bearable limits without imposing intolerable demands on industrial production or civil requirements.

1.4 Legislation and public measures

The framework of public-legal arguments pertaining to air pollution control reflects the attempts of society to reconcile two legal interests:

a) the right of each person to demand clean air and corresponding reparations for damages to health and property by foreign pollutants;

b) the justifiable requirements for heating, transportation and chances to earn money in industry.

At the present state of technology it is not yet possible to completely satisfy both requirements. Thus the authorities have to find a compromise acceptable to both parties.

In fact, the civil laws and trade regulations in force in the different states permit, in principle, the authorities to take steps against the originators of air pollutions leading to health and material damages or unacceptable nuisance, and to compel them to eliminate the inconvenience and pay for the damage. Plants to be erected require the permission of the Baurechtsämter (Board of Works) and the Gewerbepolizei (a police force controlling infringements by trade and industry). These authorities in conjunction with technical and medical experts examine the possibilities of damages caused by waste gases inside and outside the plant and prescribe the measures to be taken in granting the licenses.

The division of these official functions varies in accordance with the constitution and the legislature in the different states. Sometimes these functions are carried out by a special central authority (for example, in the United Kingdom by the Alkali Inspectorate) or by the different regional, district or local authorities. The problem of the hygienic or economical requirements is mostly left to the discretion of the authorities, who make the decisions, whenever possible, after hearing the opinions of experts; frequently experiences and corresponding decisions of other states are taken into account.

As the problems of air pollution became increasingly topical and very urgent, it was often necessary to supplement these older, mostly general, regulations with new and more stringent requirements, and to provide numerical data in order to reach, insofar as possible, a unified legal position. In the German Federal Republic this is based on the "Gesetz zur Änderung der Gewerbeordnung and Ergänzung des Bürgerlichen Gesetzbuches" ("Luftreinhaltegesetz") (Law for the Correction of Trade Regulations and Supplement to the Civil Code (Clean Air Maintenance Law)), which went into force on 1 June 1960. The frequently used para. 16 of the Trade Regulations controls the granting of licenses to erect plants, which may lead to inconveniences, risks and nuisances to local residents (the category of such plants was expanded by subsequent legal regulations, for example, those of 4 Aug. 1960). According to para. 25 of this law additional injunctions can be prescribed to licensed plants; these injunctions, however, must have bearing on the corresponding state of technology and be economically feasible for plants of this type. The "Clean Air Maintenance Law" also deals with measures to control emissions by qualified experts and the installation of suitable measuring devices. The plant in question is liable for the costs of such measures.

Another law which went into force on 17 May 1965, provides for the erection of control stations in areas in the German Federal Republic with increased accumulation of air pollutants. This permits the timely detection of hazardous air pollutant concentrations by continuous recording, so that the necessary measures can be taken.

These federal laws are supplemented by state laws for protection against immissions. In this context we mention the law of Nordrhein-Westfalen, the state exposed to the greatest hazards. Finally, we mention the regulations of the "Strassenverkehrsordnung" (Traffic Ordinance), according to which air pollution due to automobile exhaust gases must not exceed the level unavoidable in the present state of technology. Legal regulations in accordance with the US Standards are being drawn up for establishing the maximum permissible concentration.

In order that the most important problems of waste gases could be dealt with by the authorities in the most uniform manner, the competent federal ministers issued a "Technische Anleitung zur Reinhaltung der Luft" (Technical Directives for Clean Air Maintenance) ("TAL"). This serves as a general administrative prescription for plants subject to authorization, according to para. 16 of the Trade Regulations. It contains definitions, general principles for authorization and supplementary directives, restrictions on the emission of smoke and dust, immission limits for dusts, gases and vapors. Furthermore, it contains directives for determining the basic load of SO_2 and for calculating the minimum stack heights. In the minimum requirements for the individual types of plants the corresponding directives of the VDI (Union of German Engineers) are taken into account. This applies, for example, to heating plants, concrete plants, ironworks and to the permissible dust levels in these plants.

In dangerous situations the authorities have the right to undertake immediate steps to reduce immissions. The state Nordrhein-Westfalen has two stages of alert. Stage I is reached when the SO_2-level in the air is 2.5 mg/m^3 at several measuring stations and a further increase is to be expected because of meteorological conditions with poor diffusion. At this stage the competent authorities are warned and further measures are prepared. At alert stage II (SO_2-level above 5 mg/m^3) the vehicular traffic is restricted and fuels poor in sulfur (which have to be in store for this purpose) must be used.

A local agreement between the SO_2-emittent and the authority is illustrated by the example in Linz on the Danube. The Österreichische Stickstoffwerke in this city agreed to interrupt their production of sulfuric acid, when the SO_2-level recorded by the monitoring instruments in the residential area reaches a value of 4 mg SO_2/m^3 for more than 15 min. This situation, however, has not occurred so far.

Immediate measures to be taken when reaching the alert stages are also prescribed in other countries. Thus, for example, in Los Angeles three stages of public alert are specified when the toxicant concentrations listed in the table below are exceeded.

In the first stage preparative measures are taken only. When stage II or III is reached, restrictions on or interruption of certain plants and vehicular traffic must be implemented.

	Stage I, ppm	Stage II, ppm	Stage III, ppm
CO	100	200	300
Sulfur oxides	3	5	10
Nitrogen oxides	3	5	10
Ozone	0.5	1.0	1.5

In the German Federal Republic numerous institutions are dealing with the problem of air pollution control. Air analysis is currently being carried out in the "Institut für Wasser-, Boden- und Lufthygiene" (Institute for Water, Soil and Air Hygiene) at the Federal Health Institute in Berlin, the "Landesanstalt für Immissions- und Bodennutzungsschutz" (State Institute for Protection against Immission and for Soil Utilization) in Essen and Bochum, the "Institut für gewerbliche Wasserwirtschaft und Luftreinhaltung" (Institute for Industrial Water Economy and Clean Air Maintenance) in Cologne, several technical inspection associations, the different working groups of the VDI Committee "Reinhaltung der Luft" (Clean Air Maintenance) and numerous public and private institutions. The "Deutsche Forschungsgesellschaft" (German Research Association) has launched a comprehensive program for determining pollutants at numerous slightly and heavily polluted sites.

Also in other countries numerous public and private institutions, technical associations and university institutes are dealing with the analysis of air.* In the USA the problem of air pollution is studied very intensively both theoretically and experimentally owing to the high industrial production level and the excessive vehicular traffic on the one hand, and the very unfavorable meteorological conditions in some areas on the other, which lead to very peculiar and annoying pollution phenomena (see page 17).

Among the international corporations the European Council in Strassbourg, the OECD (Organization for Economic Cooperation and Development), the WHO (World Health Organization), the IUPAC (International Union of Pure and Applied Chemistry), the working group "Clean Air Maintenance" of the European Federation of Chemical Engineering and the International Union of Air Pollution Prevention as the head organization are dealing with the different problems of air pollution.

The study group CONCAWE (Conservation of Clean Air and Water, Western Europe), The Hague, established and maintained by the leading petroleum firms, is an early example of industrial cooperation in the field of air pollution control.

* A list of organizations and persons in the European industrial countries dealing with the problem of air pollution was published by the European Federation of Chemical Engineering.

1.5 Books and periodicals on analysis of air and prevention of air pollution

The book by M. B. Jacobs: "The Chemical Analysis of Air Pollutants", Interscience Publishers, New York–London, 1960 is a special monograph on the analysis of air.

Collections of analytical methods published by individual organizations include:

VDI Manual "Reinhaltung der Luft" (Clean Air Maintenance) (VDI-Verlag, Düsseldorf); it contains several VDI directives on air analysis.

ICI Manual "The Determination of Toxic Substances in Air", edited by N. W. Hanson, D. A. Reilly and H. E. Stagg, Heffer Publishing House, Cambridge, 1965. This mainly contains methods for analyzing the air at the work site. Similar problems are dealt with in the collection "Methods for the Determination of Toxic Substances in Air (edited by J. C. Gage, N. Strafford and R. Truhaut; Butterworths, London, 1962), published on behalf of the IUPAC, and in the Booklets 1–16 on "Methods for the Detection of Toxic Substances in Air", published by the British Factory Inspectorate.

The booklet "Methods of Measuring Air Pollution", Paris 1964, contains data on smoke, SO_2, SO_3, hydrocarbons and fluorides.

In the USA the following books were published on the analysis of air: "Manual of Analytical Methods Recommended for Sampling and Analysis of Atmospheric Contaminants", Cincinnati, 1957, published by the American Conference of Governmental Industrial Hygienists, and Volume 23 of the ASTM standards (Industrial Water and Atmospheric Analysis), Philadelphia, 1968.

The following books give general information on problems of air pollution control.

"Die Verunreinigung der Luft" (Air Pollution) (German translation of 14 lectures held at the International Conference of the WHO in 1957), Verlag Chemie, Weinheim.

H. Jung: "Luftverunreinigung und industrielle Staubbekämpfung" (Air Pollution and Industrial Dust Prevention), Akademieverlag, Berlin (East), 1965.

W. Knop and W. Teske: "Technik der Luftreinhaltung" (Techniques of Clean Air Maintenance), Krauskopf-Verlag, Mainz, 1965.

"Air Pollution", edited by A. C. Stern, 3 volumes, Academic Press, New York–London, Volume I, II, 1962; Volume III, 1968.

The book by F. Bayer and G. Wagner: "Gasanalyse" (Gas Analysis) (Volume 39 of the series "Die Chemische Analyse", 3rd Edition), Verlag F. Enke, Stuttgart, 1960, contains a few sections dealing with air analysis.

In the following a survey is given of existing periodicals with articles and reports on the analysis of air and its pollutants.

In Germany: "Staub, Reinhaltung der Luft". [Cover-to-cover translation into English currently published by Israel Program for Scientific Translations, Jerusalem. The English translation (starting with Vol. 25, 1965) is for sale in the area of the United States of America solely by the Clearinghouse for Federal Scientific and Technical Information, U.S. Department of Commerce, Springfield, Va. For sale outside the United States of America solely by VDI-Verlag GmbH, 4 Düsseldorf, Post Box 1139.] Numerous original articles, abstracts and summaries of lectures and reports held at conferences in the German Federal Republic, "Gesundheits-Ingenieur", "Wasser/Luft/Betrieb", "Zeitschrift für Analytische Chemie" with comprehensive reports on methods of air analysis.

Outside Germany: "International Journal of Air and Water Pollution" (since 1967: "Atmospheric Environment"), "American Industrial Association Journal" (USA), "Journal of the Air Pollution Control Association" (USA), "Gigiena and Sanitariya" (USSR) [cover-to-cover translation into English currently published by Israel Program for Scientific Translations, Jerusalem. The English translation, "Hygiene and Sanitation" (starting with Vol. 29, 1964) is available from the U.S. Department of Commerce, Clearinghouse for Federal Scientific and Technical Information, Springfield, Va.]

1.6 Survey of pollution sources and source control

To enable the reader to get a better understanding of the following sections we shall give a brief, up-to-date survey on the origin of air pollutants and modern means of preventing and controlling air pollution.

Three large source categories—domestic sources, automobiles and industrial sources—are to an approximately equal extent, responsible for the current state of air pollution, which varies, of course, with season and geographical location.

1.6.1 Domestic sources

Domestic sources considerably contribute to air pollution especially in cities during the heating season. This is mainly due to the use of coal stoves (which are frequently inadequate) with strong smoke and soot generation, and the SO_2-discharge, depending on the sulfur level in the fuel. The atmosphere of London is heavily polluted because of the traditional use of open fireplaces. As a result of the regulations under the Clean Air Act of 1956 (improvement in fireplace heating by using coal which generates little smoke, utilization of better stoves, transition to gas and oil, and long-distance heating) soot and smoke discharge was lowered to one third of the amount present in 1952 and damage to the respiratory organs decreased. However, an appreciable reduction in the total SO_2-concentration has so far not been observed.

Also in other places the flue gas discharge has already decreased or improvement is to be expected owing to the continuous technical advances in the heating of living rooms. The gradual transition from coal to coke heating in slow-combustion stoves is expected to lead to a decrease in SO_2 and smoke generation. Further, utilization of oil and gas, whereby smoke can be more easily suppressed, reduction in smoke and dust generation as well as in the discharge of uncombusted gaseous particles is probable. It depends on the SO_2-content in the fuel oil whether the discharge of SO_2 will be lower; we know, however, that in the coking of coal a considerable portion of the sulfur present in the coal is extracted, since it is partly processed together with weak ammonia water and partly removed from the gas by efficient desulfurizing processes.

Electrical heating, which due to high costs is being only slowly applied on a larger scale, is very suitable for preventing air pollution, especially when the current is generated by water power (or by nuclear energy in the future). Even in power plants, which burn coal or oil, the possibilities of reducing air pollution or removing the pollutants in a suitable way are better than for domestic stoves.

1.6.2 Air pollution caused by automobile exhaust gases

The exhaust gases from motor vehicles with gasoline and diesel engines are responsible to a considerable (sometimes even predominant) degree for the pollution of air, especially in cities. The toxicant level in the exhaust gases depends strongly on the driving conditions.

While the CO_2-level is not important as a toxicant, the actual air pollutants occurring in the exhaust gas of gasoline engines are carbon monoxide, gaseous paraffinic and olefinic hydrocarbons originating from cracking of the fuels, uncombusted fuel components; more highly condensed, especially polycyclic aromatics and soot, partial oxidation products such as aldehydes and, finally, nitrogen oxides — the product of nitrogen oxidation which is thermodynamically favored during explosion.

The table below lists new data on the toxicant levels in gasoline-engine exhaust.

Table 1.6.2
COMPOSITION OF AUTOMOBILE EXHAUST GASES
(see VDI-Richtlinie No. 2282)

	Idling	Full load	
		Low speed	High speed
Nitrogen oxides	0–50 ppm	1000 ppm	4000 ppm
CO_2	6.5–8 vol. %	7–11%	12–13%
H_2O	7–10 vol. %	9–11%	10–11%
O_2	1–1.5 vol. %	0.5–2%	0.1–0.4%
CO	3–10 vol. %	3–8%	1–5%
H_2	0.5–4 vol. %	0.2–1%	0.1–0.2%
Hydrocarbons	300–8000 ppm	200–500 ppm	100–300 ppm

Lead compounds (Pb) 60 mg/m^3, 3,4-Benzopyrene 0.001–0.01 mg/m^3

These exhaust gases, when mixed with the environmental air, especially in the presence of intense solar irradiation (see page 5), can participate in numerous, sometimes photochemically induced, reactions. Thus, for example, in the cities of California, where vehicular traffic is very dense, solar irradiation intense, and where the location in valleys favors inversion, the exhaust gases may lead to the simultaneous formation of peroxides and ozone caused by oxidation. These compounds produce smog which irritates the eye and damages plants and material.

Though in Europe such extreme annoyances and damages have not yet been observed, the CO-level may reach the corresponding MAK-values (p. 20), especially at street intersections during the busy traffic hours. The formation of nitrogen oxides is less troublesome, since they are generated during high-speed driving with full load which, of course, occurs only when the traffic is less dense. Little is known about the effect of uncombusted hydrocarbons. The polycyclic aromatics, especially 3,4-benzopyrene (benzo[a]pyrene), which are analyzed at numerous laboratories, are considered as potentially carcinogenic because of the skin cancer induced by continuous application of these compounds in animal experiments. It is probable, but by no means certain, that these compounds are responsible for the widespread increase in lung cancer (see also page 255). There is no doubt, however, that the dust cloud and mist which frequently hang above streets with dense vehicular traffic constitute a health hazard, since they absorb the curative short-wave components of sunlight.

Intensive research is being carried out to reduce the toxicant level in automobile exhaust both in the USA and in Europe. In the USA since 1968 this is all the more so for the reason that the permissible toxicant concentrations in the exhaust gases are to be reduced to the relatively low value of 1.5–2.3% CO (depending on the cubic motor capacity) and to 275–400 ppm hydrocarbons.

Two research trends exist:

a) completion of the combustion process in the cylinder, if necessary, by means of certain additives;

b) decrease in the amount of uncombusted components by afterburning of the exhaust gases after introducing secondary air, especially with the participation of catalysts after they leave the cylinder, as well as venting the exhaust gases of the crankcase into the intake manifold.

A permanent solution satisfying all those concerned has so far not been found. However, we should not be deterred by the higher costs involved in the purchase and operation of such units if we are to keep in pace with research in other fields of air pollution control.

1.6.3 Industrial pollutants

1.6.3.1 *Origin*

It is a known fact that air pollution is predominant in dense industrial areas (e.g., the Ruhr Valley). The problems of air pollution control in these areas are more involved, since economical aspects play here an important role.

Thermal power plants, especially those using coal, which discharge soot, ash and sulfur dioxide, metallurgical plants, which emit soot, dust, gaseous iron oxide, SO_2

and sometimes fluorides, and cement plants, because of their dust discharge are the most important sources of industrial air pollution. Plants of the heavy inorganic chemical industries emit various waste gases depending on their production program (SO_2, SiF_4, HF, NO, NO_2 and many others). Organochemical production plants (cellulose, oil refineries, etc.) emit malodorous waste gases. Further, all industrial enterprises are equipped with heating plants and frequently with power plants which also discharge waste gases.

In the chemical industry alone the sources of pollutants can be divided into several groups from the standpoint of production engineering:

1. Incomplete yields either because the chemical equilibrium of the process does not involve a 100% conversion of the reactants (e.g., production of sulfuric acid by the contact process) or because the yield of the final product is much less than theoretically expected (dust losses in the cement and coal industries, incomplete washing of the nitrogen oxides in the production of nitric acid by oxidation of NH_3).

2. Discharge of secondary components and impurities of the raw material into the atmosphere: fluorine in the form of HF and SiF_4 from crude phosphates, ores and ceramic raw materials, sulfur as SO_2 or H_2S from natural gas, petroleum, tar, coal and sulfide-containing ores, potassium compounds in the production of cement, arsenic and selenium from pyrites for the production of sulfuric acid.

3. Losses of auxiliary substances, for example, volatile organic solvents in many production processes, carbon disulfide in addition to H_2S in the rayon and rayon staple industry, nitrogen oxides in the production of sulfuric acid by the lead-chamber or Glover process, fluorine compounds in aluminum production.

4. Malodorous substances and oxidation products in the spent air from oxidation, heating or drying processes (for example, in the production of phthalic acid from naphthalene or xylene), in drying processes of the food, soap and glue industries, in roasting processes (coffee, malt). Mercaptans and H_2S are generated in the cooking process of sulfate pulp.

1.6.3.2 *Countermeasures*

Whereas the special recovery and cleaning possibilities for the individual air pollutants, including dust, will be treated in a special section, we give here a brief survey on some of the generally applicable measures for the elimination or cleaning of waste gases:

1. High stacks.
2. Flaring.
3. Catalytic-oxidative combustion.
4. Adsorption on activated carbon.
5. Washing out of soluble constituents.
6. Elimination of malodorous substances by adding chemicals.

The general measures for preventing or reducing the emission of toxicants include, of course, the continuous reminder to the personnel to conscientiously carry out all possible measures to ensure clean air maintenance.

1.6.3.2.1 High stacks

The purpose of high stacks is to remove undesirable waste gases from their source and the surrounding vegetation in such a way that it can be ensured that the waste gases are no longer harmful to humans or materials when the sufficiently diluted toxicants later arrive at the inhabited earth surface by diffusion or meteorological effects (katabatic wind, temperature inversion).

Data for calculating minimum stack heights are given in the TAL (see page 7) and in VDI-Richtlinie 2289 in the form of nomograms. The nomograms can be applied to gases or dusts (particle size $< 10\ \mu$) under the following conditions:

a) constant temperature and waste gas amount;

b) flat terrain without any buildings or large trees;

c) allowance for immissions already present in the surrounding air;

d) the mean value of these immissions over time must not exhibit any appreciable differences in the horizontal direction. The following parameters appear in the nomograms:

1. Inside diameter of the stack outlet.
2. Temperature of the waste gas at the stack outlet.
3. Total amount of the waste gases in m^3(STP)/hr.
4. Amount of toxicants by weight in kg/hr.
5. Concentration difference between the permissible immission limit and the value for the basic load, determined according to the method on page 22 (for SO_2 the maximum basic load is 0.35 mg/m^3).
6. A factor allowing for the immission period.
7. Mean wind velocity.

The ground-level concentration in the area affected by the smoke plume is proportional to the amount of substance and inversely proportional to the square of the stack height. Maximum ground-level concentrations occur at a distance of 10–20 horizontally plotted stack heights.

See also the booklet "The Calculation of Atmospheric Dispersion from a Stack" published by CONCAWE (see page 8).

1.6.3.2.2 Flaring of waste gases

Combustible waste gases should by no means be introduced uncombusted into the stack. If possible (provided the gas cannot be used as a raw material for chemical processing), its calorific value can be utilized, for example, by combustion of the gas underneath a steam boiler. In the event that this is impossible for technical or safety reasons (e.g., hazards of mixing with air), the gas can be provisionally flared at open tubes. As a safety precaution the flare tube outlet must be 4–10 m above the ground at a distance of at least 120 m from combustible substances. The flame should burn without noise and with low luminosity. Formation of soot can be prevented by blowing in air and/or steam.

1.6.3.2.3 Catalytic-oxidative combustion

If combustible toxicants are present in the waste gas in concentrations too low to permit flaring, they can in many cases be eliminated by catalytic afterburning at 200–500°C. When the required temperatures cannot be maintained by the heat of the oxidation reaction, either the upper temperature limit must be maintained by external heating or by supplying cheap fuel gases, or the lower limit maintained by heat removal. Noble metals on ceramic carriers and frequently oxides of base metals are used and marketed as catalysts. When selecting these catalysts, the presence of catalytic poisons must be taken into account. The throughput, as a rule, is 10,000–20,000 m^3 spent air per m^3 of catalyst per hour.

For further details on apparatus and plants for catalytic afterburning the reader is referred to the papers of Herrmann [105, 106].

1.6.3.2.4 Adsorption processes

Adsorption processes are especially advantageous, when the adsorbed substances can be used again after recovery of the adsorbent. Accordingly, this procedure was found to be convenient, for example, for the recovery of carbon disulfide in the production of rayon and rayon staple, and of solvents in numerous other industries. The most widely applied adsorbing agent is activated carbon. It is hydrophobic and therefore not deactivated by steam. Its adsorptive capacity increases with decreasing temperature, increasing molecular weight, C-content and boiling point of the substance to be adsorbed. Regeneration with steam is usually simple. In some cases chemical reactions occur on the very active surface, especially oxidation with air oxygen, which must be taken into account when selecting the recovery method. Thus, for example, hydrogen sulfide is adsorbed in the presence of oxygen as elemental sulfur.

The cleaning of air and waste gas is a discontinuous process; adsorption proceeds until the efficiency has decreased to a certain point. Then the pollutant is desorbed (in the reverse direction) and the recovery process ends by drying and cooling the adsorbing agent. Dust must be removed from the air or waste gas, since it reduces the activity of the adsorbent by clogging its pores.

Activated-carbon filters are frequently used to clean the air in air-conditioned rooms and in respirators.

1.6.3.2.5 Technical scrubbing of air and waste gases

In *analytical* determinations of air pollutants it is important that the toxicant is completely removed by the washing agent. In *technical* scrubbing of air and waste gases an additional requirement is the reuse of the solvent or adsorbent. It is usually not permitted to discharge the solvent containing the toxicant into the sewer or stream (even when the material in question occurs in a very dilute aqueous solution), since the requirements for clean water maintenance are just as strict as those for clean air maintenance. Therefore, the recovery procedure must simultaneously involve separation of the toxicant from the washing liquid. The separated toxicant can then be either utilized as such or otherwise rendered harmless.

In the technical procedures complete adsorption of the toxicant is not absolutely necessary; frequently, a scrubbing effect of about 80–90% is sufficient.

The absorption of the toxicant can sometimes be due to physical dissolution processes; the vapor pressure of the toxicant, though reduced, is not completely eliminated, hence separation is incomplete (e.g., benzene scrubbing of gases by means of oils). Chemical binding of the toxicant during the absorption process is more convenient. In subsequent recovery, for example, by heating it is mostly attempted to break these bonds. Thus weak bases (ethanolamines, amino acids, aromatic amines) in the cold absorb acidic toxicants (SO_2, H_2S) to form salts. In hot solution, on the other hand, the acidic gas is again released by hydrolysis.

The scrubbing process should take place in a small apparatus with a high throughput rate. The interface necessary for mass transfer should be created with the smallest possible energy expenditure. Both continuous and noncontinuous processes can be employed. The washing liquid flows in suitable scrubbers in reverse-flow direction to the gas stream to be scrubbed, while the interface is increased by means of packing bodies, for example, Raschig rings. It is also possible to atomize the liquid by two rotating shafts as done in Ströder washers. Frequently, an additional separation of the liquid mist formed is necessary.

1.6.3.2.6 Removal of malodorous substances by addition of chemicals

In numerous cases, especially for very highly diluted pollutants causing odor annoyance, attempts are made to convert these substances by chemical reactions into odorless compounds prior to being discharged into the air. This frequently involves difficulties, since the compounds used for this purpose (e.g., chlorine or oxygen) are themselves very strong air pollutants. Therefore, the compounds must be metered out very accurately and the reactions must proceed quantitatively. Otherwise an excess of these reagents will form which sometimes has a more harmful effect than the toxicant to be eliminated.

Odors in enclosed spaces which are very annoying even when their intensity is very low (e.g., lavatory odors), are sometimes screened by stronger, but subjectively less unpleasant, odors (e.g., p-dichlorobenzene). In the USA such "odor neutralizers" are sometimes recommended as additives to the gases before they are discharged into the open air.

1.7 Meteorological effects on the dispersion of air pollutants

Waste gases discharged from a stack, chimney or automobile exhaust pipe are diluted with the environmental air on their path to the site where they can effect damage. The dilution depends on the distance and the prevailing wind and weather conditions.

The extent to which the damages caused by waste gases can be reduced with increasing distance between emission source and residential area is a problem of plant siting. Thus, industry involving the emission of very malodorous substances (e.g., sulfate pulp) near dense residential areas will unavoidably lead to complaints.

The erection of high stacks is the most widely applied method of increasing the distance from the emission source. Such stacks have the additional advantage that for warm waste gases a suction effect is produced and the gases are mixed with environmental air upon leaving the stack. For further data details see page 14.

The so-called inversions constitute a special factor preventing the dispersion and adequate dilution of the toxicant in the atmosphere. Under normal conditions the air temperature decreases with increasing height. The warm ground air with a lower specific gravity and the colder, heavier air above it mutually diffuse so that both layers are thoroughly mixed. The opposite situation, i.e., when a warm layer occurs above a cold ground layer, preventing the ascent of the former, is called inversion. We are acquainted with the frequent short-term nocturnal inversions which develop due to convective cooling of the ground air and heat losses. They vanish, however, after sunrise when the soil is gradually warmed. On the other hand, inversions which develop in plains between mountains and river valleys in the cold season under persistent high-pressure atmospheric conditions last longer (sometimes several weeks). They are accompanied by dense and persistent fogs. Under such conditions the pollutants accumulated near the ground are not adequately diluted and may constitute, mainly because of the cumulative effect of SO_2 and soot in the effluents, a serious health hazard. Frequently, high stacks project above the inversion layer and thus above the danger zone. It thus stands to reason that the authorities in regions subjected to inversion hazards take into account the possibility of pollutant accumulation and prepare the necessary measures.

Precipitations clean the air very effectively. Analysis of the air during rain- or snowfall is therefore pointless.

The most widely differing reactions can take place between the pollutants in the atmosphere. This is mainly dependent on the weather situation (humidity, solar radiation). Under European conditions (SO_2-containing stack effluents together with soot and ash components) the formation of wet sulfuric acid surfaces on the soot and ash particles is especially serious.

The Los Angeles smog is caused by other factors (see page 11). It is formed from the olefins and nitrogen oxides of automobile exhaust due to the effect of oxygen subjected to intense solar irradiation. Radicals are temporarily formed together with ozone, resulting in pungent and eye-irritating aldehydes and peroxides, for example, peroxyacetyl nitrate CH_3COONO_2, which was synthesized under smog-simulated experimental conditions.

1.8 Hygiene of air pollutants

Pollutants are mainly taken up by the respiratory organs. The total volume of respiratory air is 6–10 m^3 per day, while 4–7 m^3 arrive at the alveoli. In a normal breath about 0.5 liter and in a deep breath about 1.5–2 liter air is inhaled. The air inhaled through the mouth or nose enters via the trachea and bronchia into the interior of the lung (alveolar region), where in the numerous vesicles of the lung the gas exchange with the blood and lymph takes place.

The pollutants are taken up in different ways, depending on the state of aggregation and solubility. Coarse dusts of 5–40 μ diameter are partially retained in the nose and mouth and partially in the trachea and bronchia. The bronchia are provided with a ciliated epithelium, whose cilia convey the dust particles adhering to the mucus into the larynx where they are either coughed out or swallowed into the stomach.

Dust particles and aerosols of 0.5–5 μ diameter arrive at the alveoli where they are deposited. They can be taken up by the phagocytes and transported by the blood or lymph stream. Fine dust (diameter $< 5\ \mu$) is partially exhaled.

Depending on their solubility in water, gaseous pollutants are partially taken up by the mucous membrane of the nose or mouth and partially by the bronchia. Gases, which are sparingly soluble in water, penetrate as far as the alveolar region. If a water-soluble gas is adsorbed by fine solid particles, for example, soot, it can penetrate into the inner regions. The same applies to toxicants present in the form of very fine droplets (aerosols).

In a simplified approach the toxic effect on a living organism, when the toxicant is stored in the organism without losing its activity, is expressed by the equation $k = ct$, where k is the occurrence of a certain observable effect, c is the concentration and t the exposure period of the organism to the toxicant. On the other hand, there frequently exist lower limit concentrations a, below which the toxicant has no or only a slight effect for arbitrarily long periods of exposure, because it is either not taken up at all or separated (or rendered harmless in some other way) by the defense mechanisms of the organism. In this case the equation becomes $(c-a) \cdot t = k$.

For the concentration and effect on the sense of smell see page 97.

An additional difficulty in practice is the fact that the effect of a toxicant depends on other variable conditions. Thus, a certain toxicant level may have different effects in dry and humid weather or the effects can be cumulative when several toxicants are simultaneously present.

The harmful effects of pollutants are in the first instance determined via experience with diseases contracted in the open polluted air or at the work site. Here, accurate data on the air composition are not available and the air must be analyzed after diagnosis. Experiments on humans are, of course, possible only in very rare cases but results of a large number of experiments with animals with accurately dosed toxicant concentrations are available. Unfortunately, the applicability of such results to humans is very dubious and inaccurate because both the qualitative and quantitative susceptibility of most animals toward pollutants differs from that of humans.

If the effects on humans are less extreme (annoyance without ensuing disease), the personal standpoint with regard to the pollution source is of importance. A person who is sharing the profits of a plant emitting waste gases will have a different attitude than a local resident who has to bear the unpleasant consequences. This, of course, makes the decision on the tolerability of a pollutant, which though annoying, does not constitute a direct health hazard, more difficult for the authorities.

1.8.1 Effects of air pollution on vegetation

Vegetation is subjected to a much higher extent to the hazards of air pollutions than the animal organism. While man and animal have adjusted their gas metabolism to oxygen, whose concentration in the inhaled air is about 21%, green plants assimilate carbon dioxide which is present in the air in much lower concentrations (0.03%). They thus come into much closer contact with harmful air components. All environmental conditions, enhancing the respiratory activity due to the opening of the stomata of the leaves (light, air humidity, heat), promote the toxic effects of pollutants.

On the other hand, relaxation periods occur during the natural time intervals of lesser assimilation (night, winter). Thus, the different climatological as well as ecological conditions must be taken into account when evaluating damage to plants. Furthermore, these conditions lead to additional difficulties in establishing the harmless limit concentrations. The correct diagnosis of chronic damage caused by waste gases is very difficult owing to the similarity of the symptoms, and requires skilled and experienced scientists. On the other hand, comprehensive experimental data are obtained in the field of plant damage.

Finally, the different plant species possess different susceptibilities toward the effects of waste gases. Whereas deciduous trees and larches, which annually shed their leaves (or needles) suffer less permanent damage; spruces, firs and Scots pines with perennial needles are especially subjected to the chronic effects of waste gases. It is important to note that trees, adequately supplied with nutrients, are considerably more resistant against waste gases than insufficiently nurtured plants. Fertilization of forests is therefore a suitable protective measure against damage caused by waste gas.

Thus, the biological measures to avoid or at least reduce the effects of pollutants include mainly the selection of comparatively resistant plants, since there is no necessity to cultivate the most sensitive species in regions with high pollutant concentrations.

Interesting experiments are currently being carried out in various areas with plant species subjected to damage by waste gases, for example, conifers, which were bred by selection of resistant strains.

1.8.2 MAK-values

In modern production techniques a large number of substances are used which penetrate in the form of gases, vapors or dusts into the atmosphere of the work area and thus present a health hazard to the workers. To prevent health injuries a special committee was selected by the German Research Association in 1955 for the task of testing substances injurious to health. This committee publishes a list of MAK-values (Maximale Arbeitsplatz-Konzentrationen—maximum concentrations at the work site) on the basis of known data on the toxic effects of a large number of substances used in production. This list is being continually corrected and supplemented according to the most recent research results. The last list in the appendix of this book (see page 269) was published in 1966. These MAK-values represent limit values for the average concentration of pollutants at the work place during an 8-hour working day and are expressed in cm^3/m^3 (ppm) (20°C, 760 mm Hg) for gases and vapors and in mg/m^3 air for suspended particles. These values apply to the pure substance in question and not to mixtures of toxicants, whereby the individual effects can be enhanced or attenuated. Such values cannot be used to deduce the effects of higher concentrations for shorter durations and cannot be compared with the MIK-values, since the latter also include other types of damage (e.g., to plants). Further, the MAK-values cannot serve as a criterion for actual or alleged damages.

Similar values are being published by the American Conference of Governmental Industrial Hygienists under the title "Threshold Limit Values". Lists of this kind are also published in the countries of Eastern Europe and the USSR.

1.8.3 MIK-values of the Verein Deutscher Ingenieure (VDI — Union of German Engineers)

The VDI Committee "Reinhaltung der Luft" proposed the so-called MIK-values (Maximale Immissions-Konzentrationen—maximum ground-level concentrations) as limits for various important pollutants near the ground. They are defined as those concentrations in the near-ground air layer, which, in accordance with our present experience, are in general harmless to humans, animals and plants when present only over a certain period and with a certain frequency. They should be understood as orientative values serving to control immissions; these values apply only within certain dispersion limits and as such do not constitute a standard for evaluating presumed immission damages.

Since especially the accurate and specific standard analytical methods require a larger air volume, for which the sampling time is longer, a measurement interval of 30 minutes was standardized. The concentration data thus represent a mean value over this period (half-hourly mean). The readings of recording instruments with a shorter operation time have to be averaged over the individual half-hourly periods. It is assumed that possible peak values within this half-hour period do not exceed the given limit. Hazards due to such short-time excess concentrations are not expected in the free atmosphere, provided the emission source is sufficiently separated (e.g., by high stacks) and the concentration of the toxicants in the waste gases emitted does not exceed certain limits (the so-called MEK-values).

The VDI distinguishes between:

1. The "sustained-effect maximum" (MIK_D-value)—highest permissible mean concentration in the given measurement interval (half-hourly mean).

2. As it was established that some plants tolerate short-term exposure to gases with toxicant levels somewhat exceeding the MIK_D-values, provided periods with lower toxicant concentrations follow, intermittent concentrations (which may at most be twice as high as the long-term duration concentrations), expressed as half-hourly mean values (MIK_K-values), are temporarily permitted; this limit must always be supplemented by its permissible frequency.

Thus, for example, for SO_2 a MIK_D-value of 0.5 mg SO_2/m^3 (0.2 ppm) and a MIK_K-value of 75 mg/m^3 (0.3 ppm) were proposed. The MIK_K-value must not occur more than once in 2 hours. Thus, within this period one of the half-hourly mean values may be 0.3 ppm, while the other three half-hourly means must not exceed 0.2 ppm.

By definition, the MIK-values are distinguished from the MAK-values as follows:

1. The MIK-values do not apply to 8-hour intervals but to permanent exposure.

2. They do not only take into account healthy adults but also babies, children and diseased persons.

3. They also take into account potential damage to plants and odorous annoyance.

The MIK-values of organic compounds are listed in accordance with VDI 2306 in Table II on page 267.

In the case of organic pollutants the MIK-values are extremely low when the substances have a strong and unpleasant odor (e.g., the MAK-value for triethylamine is 25 ppm, while the MIK_D-value is as low as 0.01 ppm). Unpleasant and repulsive odors are considered to be intolerable in the open air, even when they present no toxic hazards; in such cases the MIK-value is below the odor threshold. Within the plant, however, the authorities have no objections against annoying smells which do not constitute a health hazard, the more so since the plant management may compensate the workers for this annoyance, for example, by means of an additional allowance.

MIK-VALUES (VDI) AND MAXIMUM IMMISSION CONCENTRATIONS OF THE TECHNISCHE ANLEITUNG (TAL)

Toxi-cant	MIK_D ppm	mg/m³	MIK_K ppm	mg/m³		TAL ppm	mg/m³	
NO_2	0.5	1	1	2	(3 times daily)	1	2	(Once within 8 hr)
Cl_2	0.1	0.3	0.5	1.5	(3 times daily)	0.3	0.6	(Once within 8 hr)
H_2S	0.1	0.15	0.2	0.3	(3 times daily)	0.15	0.3	(Once within 8 hr)
SO_2	0.2	0.5	0.3	0.75	(Once within 2 hr)	0.4	0.75	(Once within 2 hr)
HNO_3	0.5	1.3	1	2.6	(Once within 2 hr)			
HCl	0.5	0.7	1	1.4	(Once within 2 hr)			

1.8.4 Immission limits of the "Technische Anleitung"

The MIK-values established by the VDI are recommendations for which the organization takes responsibility but which have no legal power. On the other hand, the "Technische Anleitung" is a federal administrative regulation on immission limits for dust and certain gaseous pollutants which must be adhered to by all concerned. While the MIK-values of the VDI apply "because of individual differences in constitution and environment within certain dispersion limits", the "Technische Anleitung" contains special instructions for the statistical processing of the immission limits with the specific purpose of obtaining the basic loads of small or large regions. The immission value of a region, for which a large number of spatially distributed individual measurements are available (see page 30), is given by the sum of the arithmetic mean of the individual values and the 97.5% confidence limit of the mean value or the 97.5% confidence limit of the individual values above the arithmetic mean. It is here assumed that the distribution of the individual values above the mean approximately corresponds to half a normal (Gaussian) distribution.

Thus, for SO_2

$$\overline{X} + \Delta\overline{X}_{(P=97.5\%)} \leqq 0.4 \text{ and } \overline{X} + \Delta X_{(P=97.5\%)} \leqq 0.75,$$

where $\Delta X_{(P=97.5\%)} = S_{X_i} > \overline{X} \cdot t$ is the 97.5% confidence limit of the individual values with one degree of freedom (here t is the factor of the Student distribution, which is about 2 for P = 97.5%); $\Delta\overline{X}_{(P=97.5\%)} = S_{X_i} > \overline{X} \cdot t/\sqrt{2z}$ is the 97.5% confidence limit of the arithmetic mean with one degree of freedom; $S_{X_i} > \overline{X}$ is the standard deviation of all the individual values above the arithmetic mean $\overline{X}$:

$$S_{X_i} = \pm\sqrt{\frac{2\Sigma(\overline{X}-X_i)^2}{2z-1}} \text{ for all } X_i > \overline{X}.$$

Here X_i is the individual value, $\overline{X}$ is the arithmetic mean of all individual values and z is the number of all individual values above the arithmetic mean.

The VDI Directive 2289 characterizes the basic load of SO_2 in a region by the "95%-value of the cumulative frequency of all SO_2-concentrations". Such data can also be found in other publications. They express the value which was just not attained by 95% of all determinations carried out. A 95% cumulative frequency of 0.5 mg SO_2/m^3 thus indicates that of all the values 95% are below 0.5 mg/m^3 and 5% above this value.

Maximum emission concentrations (MEK-values)

Limits are frequently set for the toxicant concentrations of directly emitted waste gases by the competent authorities and trade organization, for example, the VDI Commission "Reinhaltung der Luft." These limits are sometimes called the MEK-values. Not only the total amount of pollutants but also their concentration in the discharged waste gas is of special importance for assessing the risk of acute injuries caused by insufficient dilution. Such limits are not only prescribed by the authorities of the German Federal Republic but also by those concerned in other countries, for example, the British Alkali Inspectorate.

Though the analysis of the waste gases themselves is not the subject of this book, some relevant data will be reported later on in the chapter on special topics.

2

Methodology

2.1 Introduction

It is the task of the analytical chemist to select or modify the method to be applied so that it will satisfy the specific requirements of the analysis and make it possible to obtain the relevant information. The more varied the prerequisites and requirements of this complex field of tasks, the lesser the possibility to achieve the goal with a universal method, which may or may not be standardized. Increasing experience and research have resulted in an ever growing number of methodical variations which satisfy the specific requirements of the individual tasks. Further, when the problem becomes of material importance and the methodical prerequisites for instrumentation are given, the instrument designers also play a role.

The number of different methods and instruments thus available are so wide in scope that a comprehensive survey is practically impossible, and the analytical chemist has to select the appropriate procedure to satisfy his needs.

Air analysis is a very instructive example for such a widely developed complex field of tasks. There is probably no other branch of practical analysis where the prerequisites and requirements are so varied. The continuous attempts to prevent air pollution provide an additional stimulus to methodical improvements and advances in instrumentation in air analysis. The selection of the methods and instrumental means thus requires effort and experience from the practical analysis.

We now proceed with a discussion of some prerequisites and requirements of air analysis.

2.1.1 Concentration

2.1.1.1 *Concentration data*

The data on air pollutants are always referred to 1 m^3 of air. The amount of pollutants is either expressed in terms of weight (mg/m^3 or $\mu g/m^3$) or as the volume ratio in the gaseous state (ppm = parts per million = 1:10⁶; ppb = parts per billion = 1:10⁹, or sometimes also pphm = parts per 100 million). Whereas in Germany both modes of expression are used, in the USA air pollutants are mostly expressed in ppm. This has the advantage of being independent of pressure and temperature, while the ratio mg/m^3 is variable in accordance with the gas laws. Sometimes the pollutants are referred in Germany to 1 Nm^3 (m^3 of dry gas at NTP). However, the idea of expressing pollutants which normally occur in the liquid state (e.g., benzene) in ppm makes little sense as the toxic effect corresponds to the weight amount and not to the volume.

It is frequently necessary to convert ppm into mg/Nm^3 and vice versa. This is done in accordance with the following formulas:

$$\text{ppm} = \text{mg/Nm}^3 \cdot \frac{\text{mol. volume}}{\text{mol. weight}}\,; \quad \text{mg/Nm}^3 = \text{ppm.} \frac{\text{mol. weight}}{\text{mol. volume}}\,.$$

The molecular volume is usually taken to be 22.4 liter. This value corresponds better to the individual toxicant in large dilution than the values calculated from the individual gas densities, which are determined for the pure gas at normal pressure.

The conversion of ppm into mg/Nm^3 and vice versa for air components having molecular weights between 2 and 200 can be done using Table I (page 265).

2.1.1.2 *Concentration range*

Though in this text we shall not deal with waste gases containing toxicants in high concentrations, a large concentration range must also be considered for the analysis of air when allowance is made for the widely differing limit values as well as for the concentrations occurring in practice. While carbon dioxide occurs in concentrations varying between 200 and 2000 ppm or higher, hydrogen fluoride levels as low as 1 ppm are sometimes present and must be determined. Since the concentration range varies over 6 orders of magnitude, the required air sample volumes differ, and consequently, widely differing supply and processing equipment and analytical methods must be used. The demands imposed on the detection limits for the most harmful pollutants must be satisfied by the choice of a very sensitive method of determination or by a large air sample volume.

2.1.1.3 *Variations in concentration*

The toxicant concentrations are more or less variable with regard to degree, duration and frequency of emissions, and depend on the climatic conditions (wind, precipitations). The formulation of the problem completely determines whether numerous instantaneous values or only sporadic average values over shorter or longer time intervals are to be obtained. For the measurement of low toxicant concentrations in the free atmosphere half-hourly mean values* are frequently required, assuming that short-term penetrations of pollutants with acute toxic effect are not to be expected; this especially applies in the case of high stacks, whereby the emissions are diluted. At the work place, however, sometimes instantaneous toxicant concentrations have to be measured in order to help the physician assess possible acute injuries. On the other hand, longer sampling intervals must be applied to clarify whether the MAK-values are exceeded (see page 20). When attempting to solve the problem of chronic injuries, average samples taken over days, months or even years may be of interest. These requirements do not only affect the sampling procedure but also the selection of the most advantageous analytical method.

2.1.2 Specificity

In some cases no special specificity of the determination is required. An example of this is a work place which is only accessible to a limited amount of pollutants and a special toxicant has to be measured. Sometimes we are interested only in the sum of all combustible pollutants without inquiring into the identity of the individual substances. As a rule, however, especially in the free atmosphere of dense industrial areas or cities with heavy vehicular traffic, numerous interfering substances are present, so that the specificity of the analytical method applied must be high. The same holds true in cases where we have to check the data of physical measuring instruments (which frequently are not very specific themselves) by individual measurements. The interference of accompanying substances is frequently called the "*cross sensitivity*" and must, insofar as it is unavoidable, be allowed for.

2.1.3 Required accuracy

The "accuracy" of an air analysis consists of its "sensitivity", "reproducibility" and "correctness" (see also Kaiser and Specker [127]).

The decisive criterion of the *sensitivity* of an analytical technique is the smallest amount of the substance to be determined which can be detected with "*certainty*". This was referred to as the "sensitivity limit" by Feigl, and is expressed in μg.

The *limit concentration* of the toxicant in the air which can be determined by the technique applied is another parameter characterizing the sensitivity of a determination. It is related to the result of the analysis and is expressed in ppm or ppb. In physical procedures, where the air sample is directly measured without converting the toxicant into another phase, the limit concentration at the same time characterizes the sensitivity. In chemical procedures the air component can be extracted into a certain phase, for example, an absorbent. When the conditions of absorption are favorable and the sampling period is sufficiently long, the detection limit of the determination can be achieved by very low toxicant concentrations.

* According to Lahmann [146a] in immission measurements the results of half-hourly mean values do not differ from those of the 10-min mean values when the number of determinations is large.

In order to define the term "certainty", with which this lowest amount can be determined as the detection limit, we must start from the "blank" of the analytical method. This is the value obtained by the analytical method in question in the absence of the toxicant in the air. It is found by analyzing an air sample of the same volume, whereby the toxicant has been removed by a reliable method. Very often a "reagent blank" is sufficient. For example, in spectrophotometric determinations the extinction of the mixture of the reagent in the reference cell is used as the zero value.

When the blank is determined frequently, oscillating values are obtained due to random errors. We take the arithmetic mean and the standard deviation

$$s = \pm \sqrt{\frac{\Sigma (X_i - \overline{X})^2}{N-1}},$$

where X_i are the individual values, $\overline{X}$ is the arithmetic mean of all the individual values and N is the number of individual values. The arithmetic mean of the blanks is taken as the zero value. If in the analysis of an unknown air sample we obtain the value $\overline{X} + s$, then this value is higher than the zero level with a probability of 68%. If it is $\overline{X} + 2s$, the probability increases to 95%, while for $\overline{X} + 3s$, it increases to 99.7%. Thus, the substance effecting this probability increase is almost certainly present.

The standard deviation, calculated by the above formula, which, when expressed in %, is called the relative standard deviation, also characterizes the *reproducibility* and the *confidence limits* of a result or the concentration data derived therefrom, when the measurements are repeated under the same conditions. This reproducibility is for some methods (especially physical measurements) very good, i.e., better than 1% of the measured value. It, however, does not necessarily correspond to the true value of the toxicant concentration present. Furthermore, systematic errors caused by changes in the measurement conditions (temperature, pressure, fluctuations in the supply voltage, etc.) or by interfering substances may distort the results.

The *correctness* of an analytical procedure can either be checked by control determination with a reliable, though more complicated, standard procedure (which is not used for merely this reason in routine determination) or by test analysis of manufactured air mixtures of known composition. It must be noted, however, that it is difficult to prepare such mixtures of very low concentration and that their stability is often low (see page 47ff).

The requirements for the sensitivity and accuracy of an analytical procedure should not be carried too far in air analysis. The limit concentration should amount to 10% of the toxicant concentration which can be definitely established as causing serious damage. As far as the accuracy (correctness and reproducibility) is concerned, it should be borne in mind that also the hazardous concentrations are only rough orientative values. As a rule, a reproducibility with a standard deviation of ± 5%

of the measured value will be adequate. Thus, the normal accuracy of, for example, colorimetric, polarographic or gas-chromatographic methods which can be attained without any special precautions, is sufficient. In investigations determining whether MAK-values at the work place are exceeded or not, the less accurate gas monitoring devices ($\pm$ 10–20%) should be suitable.

2.1.4 Frequency of the analysis

Repetition and frequency of the analysis are dependent on the confidence limits within which the influence of wind, weather, location and frequency of emissions have to be determined. This also decides the problem of whether single or continuous determinations are required.

2.1.5 Analysis time

When discussing this problem, we should first consider those cases where instantaneous penetration of the toxicant and occurrence of the associated acute damages are to be expected; such cases must be immediately recorded by the analytical instrument and a warning mechanism must be actuated. Here, continuously recording automatic instruments are indispensable. These are set up at locations where such hazards are most likely to occur. In the event that a rapid ambulant analysis of several toxicants is necessary, a mobile measuring station is very convenient (page 46).

2.1.6 Costs of the analysis

There are sometimes analytical problems which do not concern injuries to health or health hazards but rather damages to property. In such cases it is sufficient to obtain results with simpler and cheaper methods which provide less information (e.g., the rag method). On the other hand, whenever there exists a definite health hazard, economical considerations at the expense of the reliability and the comprehensiveness of the information are out of place. Experience has shown that in regard to the importance of air pollution control the necessary means are readily available, so that the required analysis need not be jeopardized due to high costs.

2.2 Tasks and aims of air analysis as a basis for selecting the method and experimental conditions

(See also VDI-Richtlinie 2450)

Because of the many-sided importance of air and its pollutants for all biological and technical processes, the tasks of the air analysts differ over a very wide range.

The motive for air analysis is always to determine whether deviations from the natural composition occur. The nitrogen level is rarely interesting, while the oxygen level is determined only in exceptional cases, since both elements are abundantly present in almost constant concentrations in the atmosphere. Sometimes the analyst has to determine variations in the concentrations of natural low-concentration air components (e.g., determination of CO_2 in biological experiments, measurement of the humidity when studying weather or climate). However, the major portion of the analyses deals with foreign, usually undesirable substances in the air. These determinations are necessary for the following reasons.

I. To provide a basis for the evaluation of injuries to humans and damage to plants and materials, which have already occurred, are possible or are to be expected. These include, for example:

1. Individual, periodical or continuous air analysis to test the quality of the air at a given location with the aim of establishing whether the composition of the air is satisfactory or whether foreign substances are present in dangerous or even forbidden concentrations. The results may lead to the removal of the pollution source or compensation for damages.

2. Experiments to establish unknown pollution sources.

3. Long-duration investigations to determine the basic load of a larger region. On the basis of the results licenses are granted to or fines imposed upon the industrial enterprises in question. The results may also be decisive in recommending the cultivation of waste-gas-resistant plants.

4. Purity tests of the air before it is processed, for example, in oxygen factories.

5. Manufacture and testing of air mixtures used in behavioral experiments on man, animal, plant or material in the presence of certain pollutants.

II. To check the effectiveness of technical measures for air pollution prevention in order to avoid loss of valuable substances on the one hand, and to prevent damages and thus abide with the law, on the other. This also includes checking and supervision of measuring, monitoring and control instruments.

III. To determine substances, which are foreign to air but harmless, in order to establish events or activities which have taken place (e.g., the Alko (alcohol) test on the respiratory air of car drivers).

Thus, air analyses are recommended by the following institutions or persons:

1. chemical-technical plants,
2. chemical-technical research,
3. government offices,
4. physicians and hygienists,
5. other scientific workers (medical researchers, biologists, botanists, meteorologists, etc.),
6. the air analysts themselves, for example, in order to check known or to develop new analytical methods.

As far as the *environment*, whose air purity has to be determined by analysis, is concerned, we distinguish between three situations:

1. open air.
2. air in living rooms, offices, etc. (see page 2).
3. air in factories, workshops, etc.

2.3 General hints for selecting methods and equipment

The following section provides some hints for experiments and selecting methods for special tasks of air analysis.

1. *Determination of the ground-level pollutant concentration in the open air*

In residential areas, with industrial enterprises or heating plants the authorities frequently receive complaints from local residents about annoyance or injuries to vegetation caused by the immissions. The authorities forward these complaints to the management of the plants. The complaint is then checked by carrying out a specific and sensitive analysis (in situ, if possible), before changes in the wind direction and weather situation have set in. Here, the use of mobile laboratories for ambulant analysis is convenient (see page 46). Wind direction and observation of the smoke plume indicate, as a rule, the origin of the waste gas.

A responsible manager of a plant will on his own initiative have the effects of waste gases on the environment tested for various wind directions, weather situations and operating conditions before complaints are received. Here, as well, rapid manual (nonautomatic) analyses, whose sensitivity is also adequate for low toxicant levels (about 10% of the maximum permissible values), are suitable. In this case the relevant situations and sites must be selected where the highest toxicant levels are expected.

If a residential area is subject to the immission hazards of a large number of waste-gas sources, which, depending on the wind direction, act in different ways, both a qualitative analysis of the toxicants (the diagnoses based on damages to plants are not always conclusive) and the setting up of continuous automatic monitoring instruments will be necessary. In this case the wind direction and wind force must also be automatically recorded.

2. *Determination of the basic load of a large area*

The "Technische Anleitung" (see page 7) contains the following legal regulations for effecting this task:

In order to select the individual measurement points a grid (Figure 2.3) is laid out; the individual measurement points are 1 km (± 100 m) apart.

The measurement points designated by a, b, c, d form a measurement group. Measurements are carried out in each group 26 times (every fortnight) per year,

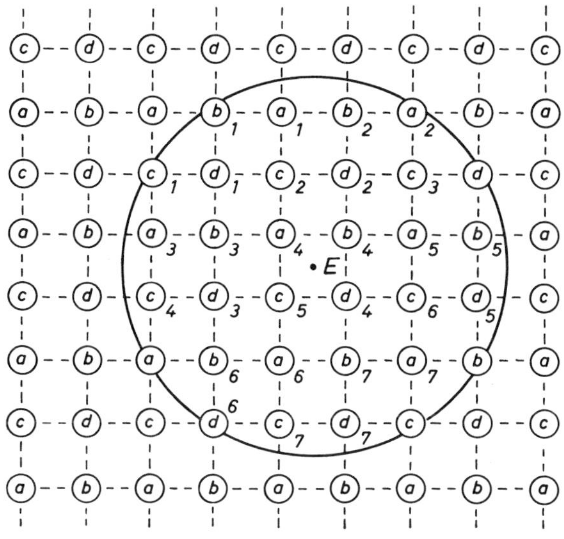

Figure 2.3
Selection of the measurement points

i.e., a total of 104 measurements are obtained for each 4 km^2. On the first day measurements are performed at the sites designated by a, on the second at those by b, on the third at those by c and on the fourth day at the sites designated by d. The measurement days are established independently of the weather conditions, but the results of intense inversion situations lasting a day or longer are neglected. The measurements are performed between 8 and 15 hr.

If in special cases the assessment period has to be shortened to six months (February–July or August–January), the measurements have to be repeated every week in order to obtain the same total number of measurements. For the determination of the basic load in the environment of a plant, it is sufficient to use a circular area with a radius of 3 km (see Figure 2.3). Thus, each fortnight measurements are performed at 28 points in 4 groups 26 times per year.

The "basic load" is calculated from the numerous individual values in accordance with the formula on page 22 and is compared to the maximum permissible ground-level concentrations.

3. *Determination of the sum total of toxicants*

When chronic damages to vegetation are caused only by the pollutants emitted by a single plant, the overall extent of the potential damage is relatively small and no health hazards exist, it is most expedient to select a simple and cheap procedure for determining the sum total of the toxicants (e.g., a rag or bell method). A large number (about 20) of such simple devices is set up at various distances from the plant. These are checked and renewed, for example, each month. The results thus obtained can be supplemented by continuous recording of the wind direction and serve for the assessment of possible compensation claims filed by local residents.

4. *Other air analyses*

When testing the air for purity prior to its industrial processing (e.g., oxygen plants), continuous analysis is employed. This can be supplemented, if necessary, by individual determinations. If the pollutants are especially damaging or hazardous, automatic, continuous or short-interval recording instruments should be used. The best procedure is to connect these to a monitoring device.

The analytical check of manufactured air mixtures requires a very accurate analytical procedure, but its specificity need not be too high. On the other hand, when the function of an automatic recorder is checked by manual analysis on a random sample taken from the instrument, we must select an accurate and specific procedure as the specificity of automatic recorders is not always ensured.

Forensic air analyses require highly specific procedures.

During precipitation (rain or snow) air analysis is of no use whatever.

5. *Investigations at the work site*

In closed work areas, where few but known pollutants may occur at high levels, the analysts are entrusted with the following tasks:

Determination on the basis of numerous individual short-term measurements of whether a health hazard exists and whether respirator devices have to be provided.

Average measurements over long time intervals (e.g., 8-hr samples) in order to determine whether the MAK-value is exceeded.

Assessing possible contamination of products and equipment at the work site.

Checking of monitoring devices at the work site.

Data and analytical results for experiments to improve the air (waste-gas scrubbing or ventilation).

Analyses for detecting leakage in apparatus.

Short-term investigations at the work place to check the MAK-values, whereby an increased accuracy is frequently not required, are facilitated by compact and handy monitoring devices (see page 56).

2.4 Air sampling

As in every analytical investigation, correct sampling is of decisive importance. Here, as well, experience has shown that the most accurate and carefully performed analysis is useless when sampling is not correctly carried out.

Air can be sampled in suitable containers which are later carried to the laboratory, or the air can be immediately introduced into the analytical instrument. The toxicant can be enriched by adsorption on solid surfaces or by absorption, for example, into filter paper or liquids.

In most cases the volume of the air sample must be measured during sampling. In physical procedures, where no enrichment takes place and the air volume used is specified by the apparatus, knowledge of the air pressure and air flow rate is frequently necessary.

The air pressure must also be known when the results are to be referred to 1 m^3 (NTP) because of the higher demands on the accuracy.

In principle, a representative air sample must be taken at a site surrounded by flat terrain without trees or walls and without precipitation.

Isolation and analysis of small enclosed air quantities

When dealing with air bubbles, trapped in a soft material which can be pierced by a needle, the air can be removed by suction via a gastight hypodermic needle of suitable size. Such small gas amounts should be analyzed by gas-chromatographic methods.

For the separation of small enclosed air amounts, which can be liberated by fusion or dissolution (e.g., small gas bubbles in plastic granulates), the following procedure is being applied in the Österreichische Stickstoffwerke.

A gas-generating apparatus for very pure CO_2 is set up (Kipp apparatus for the microdetermination of nitrogen according to the Dumas–Pregel method). The material from which the gas is to be removed is placed into a small flask with an inlet and outlet tube. The outlet tube is connected to a microazotometer containing KOH. The air surrounding the material is expelled by a stream of pure CO_2 until only microbubbles with a negligible volume are present in the azotometer. Then the sample material is fused (provided it can be fused without decomposition) or liquefied by heating with a thoroughly degassed solvent. The gas bubbles released are entrained by the CO_2-stream into the azotometer from where they are transported by a carrier gas into a gas chromatograph and analyzed.

2.4.1 Sampling and analysis as separate procedures

2.4.1.1 *Sampling and storage of the unaltered air sample in glass vessels*

Only in exceptional cases is the air sample bottled at the sampling site for later analysis in the laboratory. This procedure is employed in the very sensitive gas-chromatographic methods, where a sampling volume of 1 to 100 ml is sufficient, and, in general, for toxicant levels above 10 ppm, when the detection methods are sensitive enough for air samples of 1–2 liter (e.g., colorimetric methods).

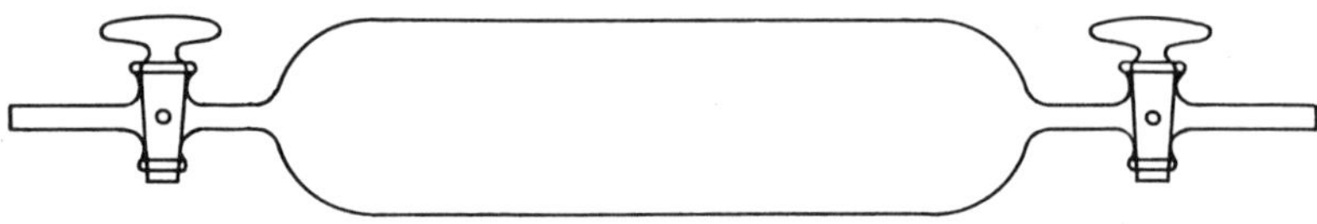

Figure 2.4.1.1a
Gas-collecting tube

Thus, gas-collecting tubes (volume up to 2 liters) with two outlet cocks (Figure 2.4.1.1a) can be used. The air sample is collected in the flow procedure by suction with an aspirator or blowing in with a bellows (see Section 2.4.1.3). Before the sample is taken the gas-collecting tube must be flushed with an air volume at least 10 times that of the sample volume. Gas-collecting tubes with a volume below 2 liters and which are provided with hermetic cocks can be evacuated to a high degree and filled by opening of the cock at the sampling site. It is recommended to cover the vessel with a stocking-like fabric as a protective measure against implosion.

The gas sample in the collecting tube is analyzed by introducing a reagent solution. The tube is then shaken so that the gas constituents are absorbed (or are allowed to react with the absorbing agent), and the (e.g., colorimetric) reagent solution is added. The absorption can be accelerated by adding an inert foaming agent, for example, aryl-alkyl sulfonate solution, in an amount just sufficient to form a fine-bubble foam during shaking (Leithe [151, 155]).

If the required air volume is larger than 2 liters, the air samples can be collected in larger glass vessels (flasks or cylinders). Transport of such cylinders is, however, inconvenient and for this reason the use of plastic bags (see below) is usually preferable. Large glass cylinders are often used for the preparation of air mixtures with known concentrations of toxicants, which are only inert to glass (see page 47).

Air samples can be withdrawn from rigid vessels in different ways. Small volumes (about $\frac{1}{4}$ of the whole volume) can be rapidly displaced by introducing fresh air, provided the point at which the fresh air enters is sufficiently far from the outlet for the air sample, so that dilution is avoided.

Further, part of the sample can be drawn from the cylinder into the analytical instrument without having to compensate the pressure with fresh air; the reduced pressure must, of course, be allowed for when the volume is measured.

Finally, the sample can be removed without introducing fresh air by means of an empty folded plastic bag, hermetically attached to a glass tube (Figure 2.4.1.1b). The bag is rapidly inserted into the cylinder containing the air sample and inflated. The toxicant must not react with the bag material during the short sampling time.

Withdrawal of an air sample from rigid vessels by means of a liquid acting as a seat is permitted only in exceptional cases, for example, carbon monoxide determinations. If, however, there are several flasks filled with the same sample, small volumes

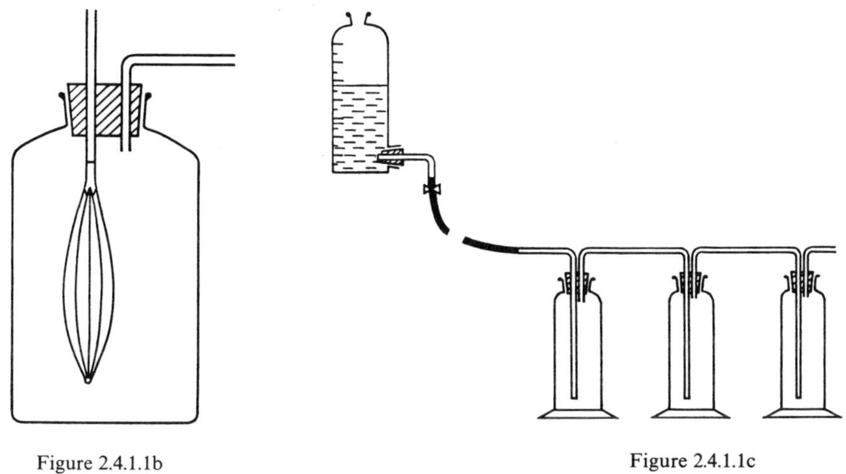

Figure 2.4.1.1b
Sampling with plastic bag

Figure 2.4.1.1c
Sampling procedure using water as a seal

can also be withdrawn by employing a liquid seal in which the toxicant is appreciably soluble. Three flasks are connected in series (Figure 2.4.1.1c), the liquid seal is pressed from a graduated flask with an outlet tube into flask A. The sample is withdrawn from flask C without coming into contact with the liquid seal.

When air samples are stored even for short periods, chemical conversions must be taken into account, especially those occurring at the vessel walls. Oxygen and water vapor may also participate in these reactions. Thus, nitrogen oxides can be converted into nitric acid, SO_2 can be oxidized into sulfuric acid, H_2S into higher sulfur compounds, while organic pollutants can be subjected to various changes.

2.4.1.2 *Storage in plastic bags*

In the USA gas containers made of plastic foils are becoming increasingly popular. These bags vary in size between 1 and 100 liters, are provided with a clamp and a hose connection, and are light and unbreakable. The best procedure is to sample the air with a bellows, lined with inert foil material to prevent the contact of the sample with rubber or corrosive metals.

The selection of a suitable foil material is somewhat complicated. The foil must be inert to the toxicant, so that the composition of the air sample or of the manufactured mixture remains unaltered during storage.

Altshuller, Wartburg, Cohen and Sleva [5] recommend bags made of Mylar (a polyester foil manufactured by Dupont). Air samples containing hydrocarbons, SO_2, ozone and NO_2 can be stored in these bags for at least several hours without appreciable losses. The foil has a high softening point and, thus, cannot be directly heat-sealed; the bags can only be sealed with adhesive tape at high temperatures.

The following heat-sealable plastics can be widely applied according to Clemons [43]: Tedlar 40S manufactured by Dupont and Saran Type 18 fabricated by Dow Chemical. Suitable valve bags made of Saran Type 18 are produced by Vilutis and Co., Frankfort, Illinois.

Polyethylene- and polypropylene-based foils are inadequately resistant against SO_2 and nitrogen oxides.

The water content and temperature of the air sample must be sufficiently far from the dew point in order to avoid condensation on the inner surface of the bags which entails losses of the active substances. It is recommended to condition the bags for some time prior to sampling with the air component (SO_2, NO_2) at a concentration which is about ten times higher than that expected in the air sample. During the storage period a check has to be carried out to ensure that the toxicant concentrations remain constant.

2.4.1.3 *Equipment for withdrawing air samples by suction*

Air is only in exceptional cases sampled by pressure pumps for the direct performance of the analysis or transfer into containers. This procedure involves the risk of air components becoming absorbed by the parts of the pump and the introduction of additional impurities. Sometimes, for example, bellows lined with inert foils may be used for filling plastic bags. Rubber or corrosive metals must not come into contact with the sample.

Air samples are usually withdrawn by suction. For small air amounts (about 2 liters), whose volumes must be simultaneously measured, hand-operated pumps which withdraw 100 ml ($\pm$ 3 ml) per stroke are suitable for operation in conjunction with the Dräger gas monitoring device (see page 56).

In order to withdraw somewhat larger air volumes (about 10 liters) aspirators may be used. These consist of two flasks provided with tubes which are connected by a hose to a clamp (Figure 2.4.1.3a). If the suction effect has to be maintained constant for a longer period independently of the water concentration in the sample, a Mariotte tube is inserted. Hermann [104] designed an aspirator for uniform and low air throughput rates (0.17–0.34 liter/hr) which may, for example, serve for taking daily average samples.

Suction pumps for larger air volumes are selected in accordance with the power connections available at the sampling site.

Thus, for example, the firm Bender and Hobein, Munich, the firm Pfeiffer, Wetzlar and the firm Leybold, Cologne, manufacture electric suction pumps with throughputs of 3 to 30 liters air/min through conventional air scrubbers.

In England the firm Charles Austen Pumps Ltd., Byfleet, Surrey, produces special pumps for air analysis with connections for the power supply or batteries.

For open-air experiments without power-supply connections hose-type compressors (as produced by the firm Bachofen, Basel) with throughputs of up to 10 liters/min

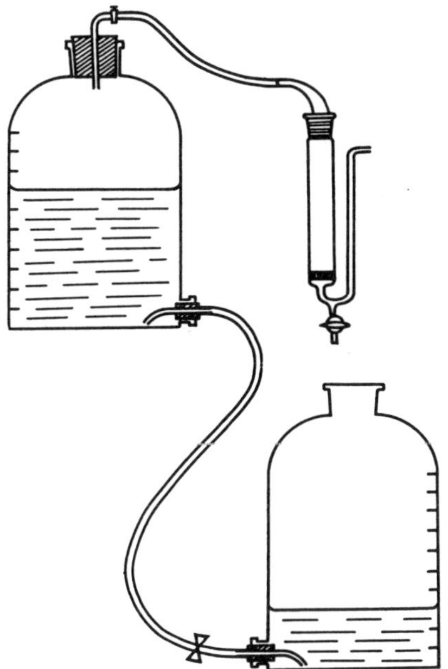

Figure 2.4.1.3a
Aspirator with absorption vessel

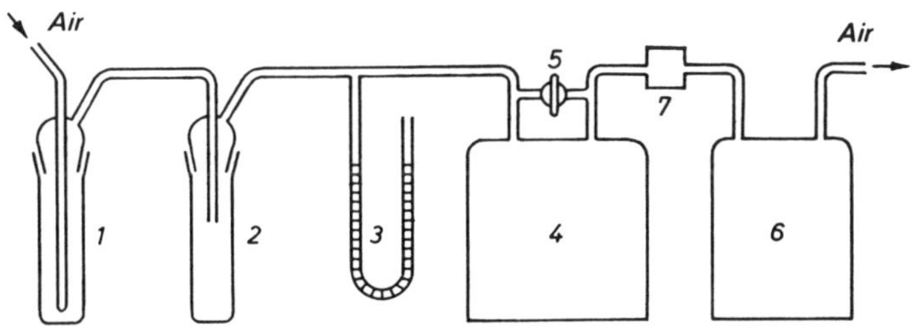

Figure 2.4.1.3b

1 Standard impinger (volume 300 ml)
2 Water separator
3 Open mercury pressure gage
4 High-output pump
5 Regulating stop cock
6 Dry gas meter
7 Oil separator

are used. The compressors are driven by a compressed-air engine. A 10-liter compressed air steel cylinder holding air at 150 atm gage pressure is sufficient for sampling 300 liters of air.

If a suitable water connection is available, conventional water-jet pumps can be used. In many cases an absorption device, whose air resistance is not too high, can be connected to the suction manifold of a motor car during idling.

For measuring the air volume withdrawn by the pump VDI-Vorschrift (Regulation) 2452 recommends the arrangement shown in Figure 2.4.1.3b. Moreover, the air flow rate can also be measured with suitable flowmeters, such as conventional rotameters.

2.4.2 Enrichment of the air component at the sampling site

Instead of transferring the whole air amount sampled into the laboratory, the component to be determined is mostly separated at the sampling site. The component can be determined later in the laboratory or immediately after sampling at the site.

The component is sometimes frozen out but in most cases it is adsorbed by a solid phase or absorbed in a liquid solvent.

2.4.2.1 *Freezing out*

Sometimes high-boiling components or also CO_2 are separated by freezing out at low temperatures. The freezing agents are liquid air or nitrogen at $-196°C$ and dry ice + methanol at $-78°C$. It is, however, inconvenient to bring along the freezing agent when the component is to be trapped in the open air. The fact that the condensate need not be desorbed is an advantage and deep-freezing can be combined with adsorption.

Cold traps are used (see Figure 2.4.2.1), which are immersed in Dewar vessels containing one of the above mentioned freezing agents. The air sample is introduced at a low flow rate (not over 1 liter/min). Mists are often formed during the freezing out of an air stream and must be retained by cotton filters.

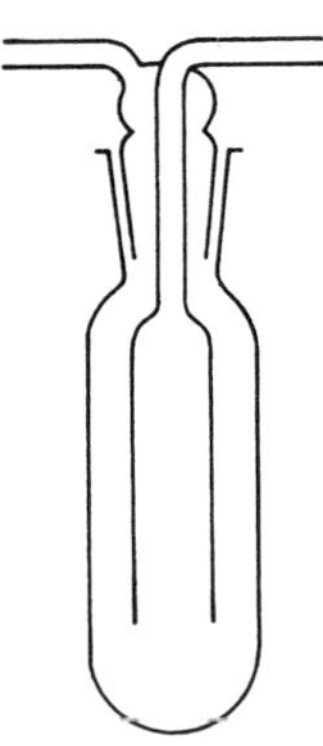

Figure 2.4.2.1
Cold trap

2.4.2.2 *Enrichment by adsorption*

Enrichment of air pollutants by conventional adsorbing agents with the ultimate aim of desorbing them and of determining the pollutants qualitatively or quantitatively is sometimes the preliminary stage of gas-chromatographic analysis. This method will be described in Section 4.5.3. In some cases, however, larger amounts are needed and modified procedures must be used. The adsorbing agent is usually placed in a U-tube which is frequently immersed in a cold bath so as to increase the adsorption efficiency.

Silica gel and activated carbon are the most common adsorption agents. The advantage of the former is its less intense action. The component can thus be desorbed at low temperatures. Activated carbon, on the other hand, has a more intense effect which does not decrease at higher air humidities. It requires higher desorption temperatures and frequently releases the substances adsorbed only after they have been chemically converted. The component adsorbed is released from activated carbon by superheated steam at about 300°C or by pumping the heated (about 200°C) agent in vacuo into several cold traps, cooled in stages to low temperatures (e.g., ice–common salt, CO_2–methanol, liquid nitrogen).

Activated carbon is frequently used for the selective separation of undesirable secondary components. An example is the determination of carbon monoxide in flue gases or in air polluted with automobile exhaust. A tube for preliminary adsorption with activated carbon lets through carbon monoxide but retains the undesirable olefins and higher hydrocarbons. A contact time of 1 sec is usually adequate for complete adsorption.

The adsorption on activated carbon can be increased by impregnating the agent with chemicals, i.e., physical adsorption is enhanced by chemisorption. For alkaline pollutants sulfuric or phosphoric acid, for olefins bromine and for hydrogen sulfide lead acetate serve as chemisorbents.

Frequently, suitable reactions which quantitatively indicate the concentration of the sought substance can be directly carried out with the adsorbate, thus obviating the desorption process. Accordingly, Peters and Straschil [202] proposed the adsorption of NO_2 from waste gases by alkali-treated clay in tubes. The filling is dissolved in water and the nitrite or nitrate formed is determined in solution. These authors also developed a procedure, whereby NO is adsorbed together with NO_2 on silver permanganate or clay impregnated with sodium chlorite. The total nitrogen adsorbed is then determined by means of Devarda alloy.

Compared to liquid absorption, chemisorption on solid surfaces impregnated with reagents has the following advantage: the process rate is very high because of the high efficiency of active surfaces over small areas. Thus, very small amounts of pollutants can be trapped. Consequently, the required air volume is small and the sampling time is shortened or very low toxicant concentrations can be determined in a larger air volume.

Buck and Strattmann [33] obtained an unexpectedly high efficiency of chemisorption of HF traces in manufactured air mixtures on alkaline solid surfaces (see page 208). The authors used silver tubes (100 × 12 mm) filled to a height of 70 mm with silver beads of 3 mm diameter. Sodium carbonate was used as the sorbing agent. 20% solutions were applied to the silver beads which were then dried at 200°C. Air containing HF-concentrations between 2 and 500 μg fluorine/m^3 was drawn through at a rate of 66 liter/min. 95% of the HF introduced was analytically determined. However, according to Leithe and Petschl [159], it is questionable whether fluorine can be determined with such a high accuracy from air containing technical fluorine compounds, for example, gases from superphosphate plants.

The most important practical application of chemisorption on solid adsorbing agents impregnated with dyes is effected by test tubes obtainable on the market. They will be specially dealt with in Section 4.3.

2.4.2.3 *Air analysis on impregnated filter papers*

Sometimes the air sample is drawn through a filter paper impregnated with an absorbing agent. A color reaction occurs either immediately or after applying a reagent, or the filter is extracted and the extract processed.

Among the methods prescribed by the British Factory Inspectorate in Booklets 1–16 (see page 9) for the determination of the concentration range corresponding to the German MAK-values, five are carried out with reagent-impregnated paper strips. The pollutants which can be determined in this way are hydrogen sulfide, sulfur dioxide, phosgene, hydrocyanic acid and mercury. The paper strips are placed into a suction apparatus in such a way that a circular area of 38 mm diameter (in the newer prescriptions 10 mm) is exposed. A hand pump is used whose suction volume is 126 ml per stroke. The equipment can be obtained from the firm Siebe Gormann and Co., Chessington, England. The color shades obtained for different numbers of strokes and pollutants levels can be compared to the color scales provided with the booklet. The booklets can be obtained from H.M. Stationery Office, York House, Kingsway, London WC 2. A similar color scale can be prepared from air mixtures with known pollutant levels and the scale reproduced with color pencils.

A device for inserting appropriate filter sheets with an exposed area of 40 mm diameter is described on page 101 in the section on the use of dust filters of the Membranfiltergesellschaft Göttingen. The bellows pump (manufactured by the Drägerwerke) with a delivery of 100 ml/stroke which is employed for the tube tests can also be used here.

Since paper has no adsorptive capacity and, as a rule, acts merely as a support for the reagents, the latter must satisfy several special requirements. They are either kept in a moistened state or it must be ensured that the humidity of the air does not drop below a certain level. Since the layer is thin, the retention time of air on the reagent is very short and often insufficient for a complete uptake of the pollutants.

A distinct coloration on the side of the paper averted from the air stream indicates that absorption was incomplete.

The color intensity on the paper is also dependent on the air flow rate, which therefore must be equal for each sampling and test series. Otherwise, owing to the increased air flow rate, the pollutant penetrates more deeply into the paper and the coloration is less distinct on the surface.

Direct dyeing by drawing through the air sample until a given color shade is reached is merely a semiquantitative method; the results are less accurate than those obtained with the commercial test tubes. On the other hand, they are much cheaper, especially when large areas have to be covered using numerous single tests.

Continuous dyeing procedures by moving a paper roll impregnated with a reagent along a suction nozzle are described on page 69.

Some time ago the Farbwerke in Hoechst used gas meters (of the size of an alarm clock) in a similar manner. A circular filter paper disk, impregnated with a suitable reagent functioning as a "dial," was covered by a disk which slowly rotated in the clockwise direction. The cover disk contained an open, sector-shaped slit which made one complete revolution every 24 hours like the hour hand of a clock. Thus, at all times a sector of the reagent paper below the disk was exposed. The coloration and the corresponding time indicated the diurnal trend of the pollutant level at the sampling site.

2.4.3 Sampling into solutions

The most important sampling technique with simultaneous absorption of the pollutant consists of passing the sample through a suitable liquid in a wash bottle (bubbling). The toxicant is enriched and can be conveniently processed for the determination, which, when higher accuracy is required, is performed in the liquid phase by titration, colorimetry or electrochemical methods. If a suitable dye is available, the toxicant can be determined during its separation. It is also possible to extract the toxicant completely in order to obtain a stable solution which later can be analyzed in the laboratory. This is more accurate and reliable, since in the laboratory the experimental conditions (temperature, time of development of the color reaction in parallel with calibration, spectrophotometric measurements) can be better controlled. On the other hand, the combined determination of the concentration and the air sample has the advantage that the results are rapidly obtained and that the air volume required can be assessed during sampling.

In chemical analysis the pollutants are scrubbed by the absorption solutions in suitable wash bottles. The latter are selected in accordance with the necessity to obtain the air component (usually present in small amounts) in high concentrations in the solution so that the detection sensitivity can be raised; moreover, the time required to process an adequately large volume should be short.

In most instances the conventional Drechsel bottles (Figure 2.4.3a) do not adequately fulfil the above requirements. They can be used for readily absorbable liquids which rapidly dissolve in the washing liquid (e.g., HCl in dilute hydroxides, NH_3 in dilute acids). The low drag of the air path makes it possible to achieve fairly high air throughput rates (up to 30 liter/min) without there being any need for strong pumps.

At present, impingers, currently used in the USA, are being recommended in Germany for high throughput rates (up to 30 liter/min). The inside diameter of the narrow end of the inlet tube in the Greenburg–Smith impinger (Figure 2.4.3b) is 2.3 mm and the distance from the base 5 mm; the maximum air throughput rate is 1 ft^3/min (28 liter/min). The smaller Midget impinger processes only 3 liter/min (see the book by Jacobs, p. 8).* The impinger recommended in Germany (Figure 2.4.3c) (see VDI-Vorschrift 2452) has a nozzle diameter of 2.5 mm, a nozzle distance of 4 mm and a volume of 275 ml.

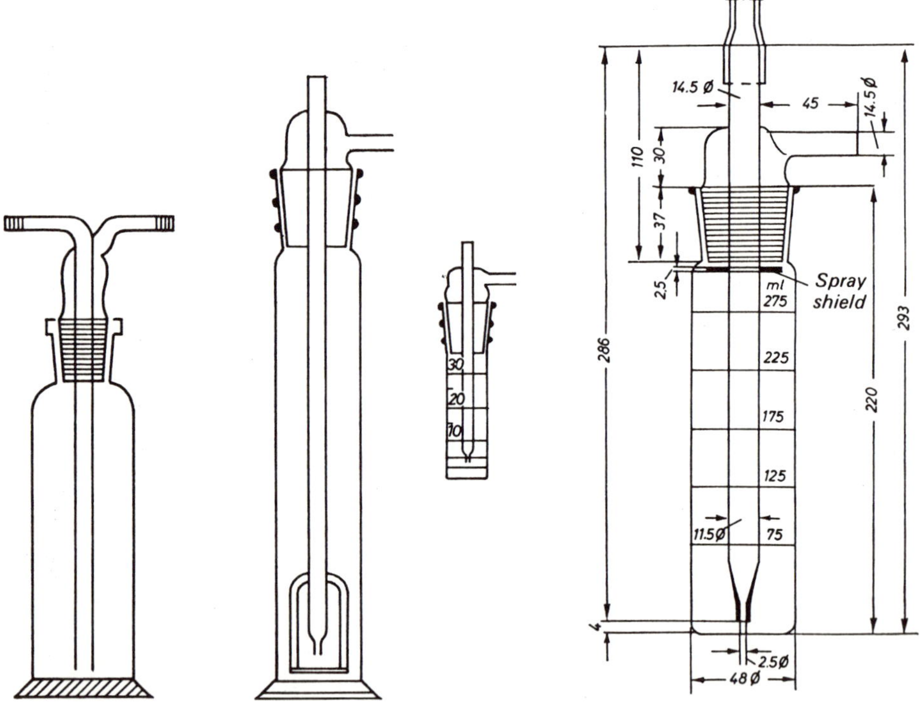

Figure 2.4.3a
Drechsel bottle

Figure 2.4.3b
Greenburg–Smith impinger and the Midget impinger

Figure 2.4.3c
Standard impinger (VDI 2452)

* For a further size reduction of the Midget impinger to the Micro-impinger, containing one third of the washing liquid volume, see Linch and Corn [165a].

Because of the narrow tip of the inlet tube and its small distance from the vessel bottom, the individual gas bubbles impinge on the latter at a high velocity. The instrument was originally intended for scrubbing dust particles from gases (see page 109); later, owing to the high impact rate, a special scrubbing effect for gaseous air pollutants was expected. This effect, however, as was shown by Leithe and Petschl [159], is no greater in practical air analysis than that obtained in Drechsel bottles. The effect of the impinger is completely inadequate (see also Deckert [154]) for the separation of carbon dioxide, which requires a longer time for hydration. Gage [73] also reported that the impinger is less effective for scrubbing organic pollutants than the fritted glass scrubber.

The best wash bottle for the analysis of air is the fritted glass scrubber. Here, the washing effect is enhanced by the increased interface (size reduction of the air bubbles). Frits of various pore sizes are available (G I, G II and G III).* When selecting the pore size, special care must be taken to ensure that some of the pores are not too wide. Scrubbers with glass frits fused around the edge of the vessel require only a small volume of washing liquid. When using these scrubbers, it should be ensured that a

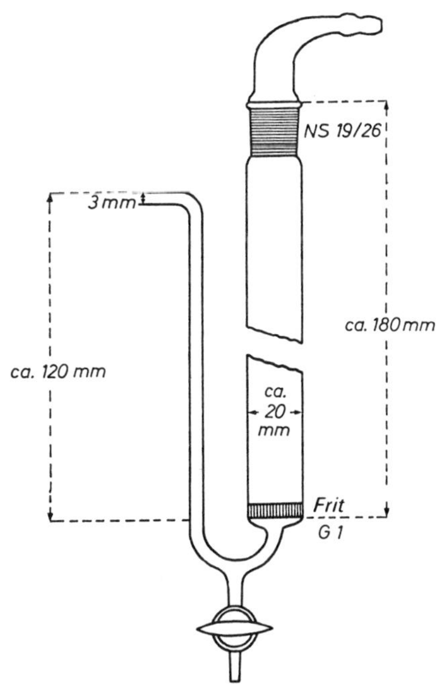

Figure 2.4.3d
Absorption vessel for air analysis

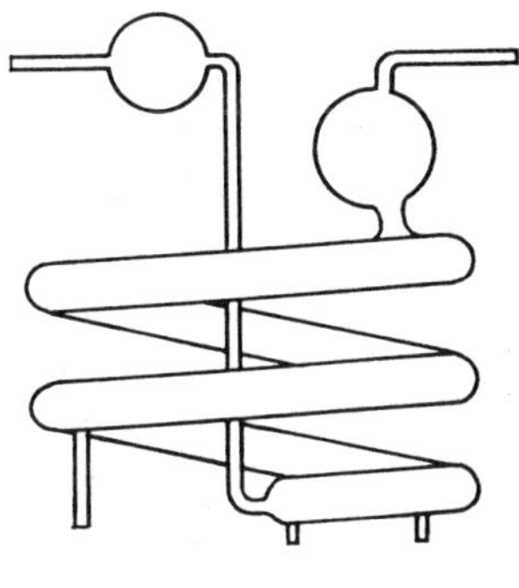

Figure 2.4.3e
Absorption spiral after Winkler

* [These are German brands.]

small overpressure is present in the gas before the washing liquid is introduced, as otherwise the liquid will flow through the frit before the gas is absorbed.

The laboratory of the Österreichische Stickstoffwerke uses a very convenient scrubber with a fused-on frit bottom. The receptacle can be rapidly exchanged and rinsed. Thus, for example, the absorbing agent together with the rinsing liquid can be introduced into a 25-ml measuring flask for subsequent determination (Figure 2.3.4d).

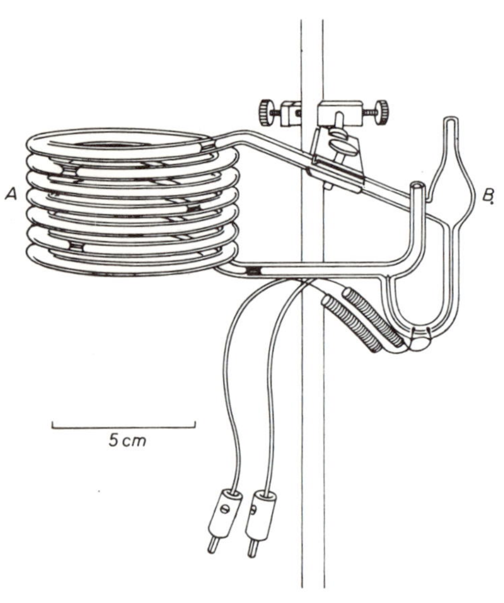

Figure 2.4.3f
Spiral absorber after Holm-Jensen, with electrodes for conductivity measurements

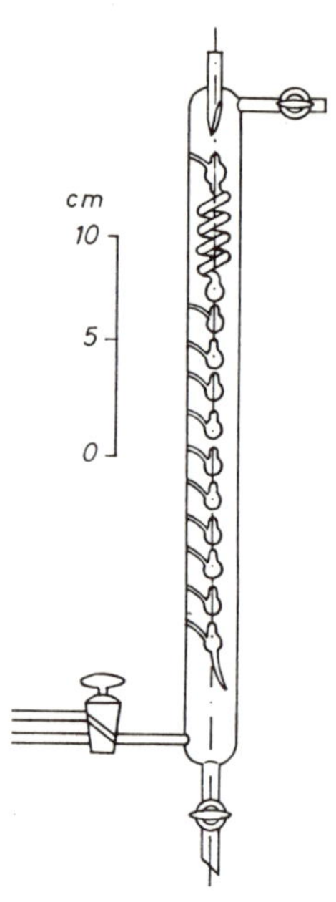

Figure 2.4.3g
Spiral-type dropping tube after Peters

The scrubbing effect can be considerably enhanced by adding a few drops of n-butanol to the washing liquid. When supplying air at an appropriate rate, for example, 3 liter/min, an approximately 7-cm foam layer consisting of air bubbles is formed in a scrubber with a G I frit of 55 mm diameter. The bubble diameter is only 1–2 mm, and their "lifetime" is about 5–10 sec. We thus obtain a large interface and a long residence time at rather high throughput rates (up to 3 liter/min). It is also possible to absorb CO_2 both in concentrations normally occurring in the atmosphere and in the ppm-range present in flowing gases (Leithe [137]). In the analysis of fluorine-containing pollutants from industrial processes foam absorption also proved much more efficient than, for example, absorption in impingers (see page 204).

The residence time of the gas bubbles in the washing liquid can also be prolonged in spiral-shaped wash bottles, but here the gas throughput is low. Holm-Jensen [110] modified the spiral absorber of Winkler (Figure 2.4.3c) and designed a microdevice for the determination of carbon dioxide in air (Figure 2.4.3f). The path of the gas bubbles is considerably lengthened (length of spiral 160 cm), while the required volume of washing liquid is reduced to a minimum (0.5 ml). The maximum air throughput rate, however, is only 15 ml/min.

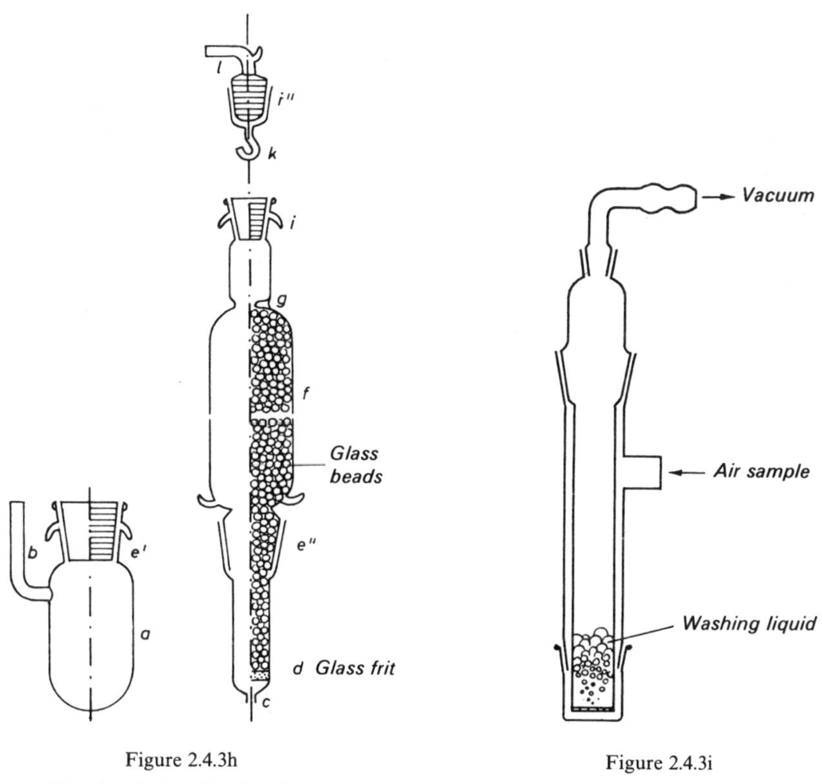

Figure 2.4.3h
Glass bead tube after Pavelka

Figure 2.4.3i
Ehmert scrubber

For other efficient washing devices with a small liquid volume see the papers of Peters [201], Pavelka [199a] and Ehmert [62] (Figures 2.4.3g, 2.4.3h and 2.4.3i).

To test the efficiency of wash bottles several bottles are connected one after the other. An air sample, if possible, of known toxicant level, is passed through the whole series of bottles and each solution is separately analyzed. When the uptake of the toxicant by the first bottle is over 90%, then its scrubbing effect can be considered as adequate. A correction can be used for the residual amount.

2.4.4 Mobile laboratories for ambulant air analysis

For rapid air analysis, for example, after complaints are received or in the event of industrial mishaps, mobile measuring stations are convenient. Thus, among others, the firm Beckmann designed such a vehicle for the various tasks of air analysis. This permits carrying out single determinations or long-term tests with recording instruments. The apparatus is battery-operated via chopper amplifiers or by an external voltage supply. The pump drives in the vehicles are partly provided with current connection (Figure 2.4.4).

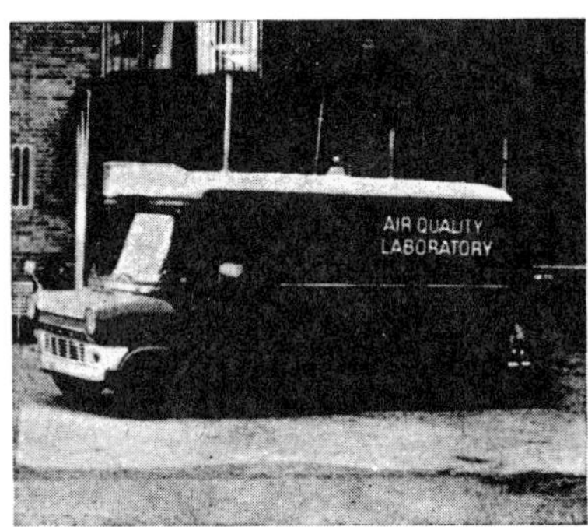

Figure 2.4.4
Mobile measuring station for ambulant air analysis

3

Laboratory Investigations with Prepared Air Mixtures

Air mixtures containing known concentrations of pollutants are prepared and analyzed in the laboratory for various purposes, namely:

1. Methodical experiments for adaptation to and testing of new and known analytical procedures. Frequently, we not only deal with a simple air component whose determination is to be tested but we also add various secondary components in order to check their interference in the analytical procedure.
2. Calibration and testing of automatic measuring instruments. Here, as well, the interference of additional air components is studied.
3. Preparation and analysis of air mixtures for animal or plant experiments and for corrosion tests.
4. Checking the efficiency of technical scrubbing procedures.
5. Carrying out model reactions with waste gases diluted with air in order to establish their future behavior in the atmosphere, for example, due to the effect of light and moisture.
6. Tests to determine the efficiency of respirators.

The requirements for specificity, reproducibility, correctness and sensitivity of the analysis vary from one case to another.

3.1 Preparation of air mixtures with known pollutant concentration

The method selected for preparing the air–pollutant mixtures depends on the amount required, concentration desired, properties of the pure substance and the probable stability of the mixture. The mixture can be prepared in stagnant systems (static method) or in a gas flow (dynamic method). The toxicant can be metered out as such or obtained by its release from known amounts of other starting substances.

The air used for the mixture must be clean; its humidity content should correspond to natural conditions.

3.1.1 Static methods

Principle: The measured amount of pure toxicant is introduced into a suitable container previously filled with pure air under reduced pressure. The container is then filled with air until the required pressure is reached. After a uniform mixture is obtained (in rigid containers by shaking, for example, with balls of inert material, in foil bags by kneading) the dilution can proceed in stages; to do this, a certain aliquot volume of the mixture is withdrawn and transferred to another container which in turn is filled with a known air volume. Foil bags of inert plastics (see page 36) are very convenient; mixtures can, however, be also prepared in pressurized steel cylinders.

Liquid toxicants with an adequate vapor pressure can be fused into ampoules and weighed. These are thrown into an air-filled glass cylinder of suitable size and crushed.

Withdrawal of small gas volumes for dilution

Here we shall consider two devices: the branched gas pipette and gas-impermeable injection needles.

Branched gas pipettes (Figure 3.1.1a) can be used for volumes of down to about 0.1 ml, whereby reproducible values can be obtained. This device has the disadvantage that the whole volume and not portions thereof is used in the dilution process. To fill this gas pipette with the pure gas a moderate overpressure is necessary, which is compensated by opening a stopcock for a short time.

Gas-impermeable hypodermic syringes permit gas dosage, reproducible to some extent, down to a volume of about 50 μl. To introduce the gas into the air stream a

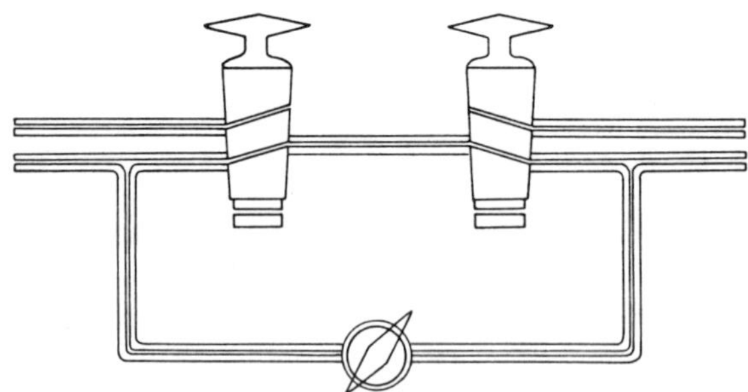

Figure 3.1.1a
Branched gas pipette

circular inlet with a neck is attached to the air line; the rubber cap of an injection bottle (which can be bought in pharmacies) is then drawn over the neck. The syringe is emptied by piercing the rubber cap and slowly injecting the contents. After the needle is withdrawn the cap closes tightly again. The syringe can also be continuously emptied by means of a shaft which is slowly introduced by a motor with a worm drive (Figure 3.1.1b) (manufactured by Sage Instruments).

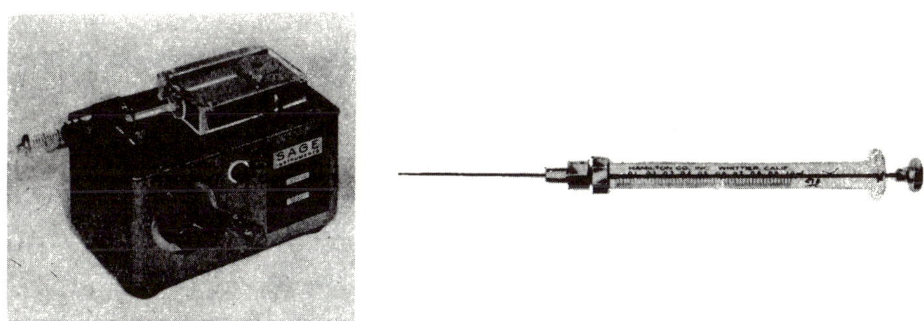

Figure 3.1.1b
Injection motor with syringe

3.1.2 Dynamic methods

If larger air volumes with known pollutant concentration are required, these can also be mixed from flowing gases, whereby a flow meter (Pitot tube or rotameter) is mounted in each gas stream. The mixing process can also proceed in stages. In this case after each mixing stage only one aliquot is further introduced via a suitable Y-tube, while the rest is led to the outside. Dosage pumps (e.g., the gas mixing pump of the firm Woesthoff (see Figure 3.1.2a)), which can produce gas mixtures in arbitrary, accurately adjustable and reproducible ratios, are very suitable for this purpose.

In a uniform gas stream it is also possible to introduce the pollutant continuously from a gas-tight hypodermic syringe with the aid of a motor-driven worm gear (see above).

Preparation of air mixtures by diffusion from liquid components

When larger quantities of air mixtures are to be prepared from foreign components, which at ambient temperatures or after corresponding cooling are fluid, it is also possible to obtain a uniform mixture of any dilution by diffusion of the vapor from a suitable gas container into the gas flow. This method applies especially to high-boiling organic pollutants, whereby the pure liquid is warmed to the appropriate temperature. The method can also be used for gases liquefied by deep-freezing. The diffusion cell of McKelvey and Hoelscher [176] is schematically shown in Figure 3.1.2b, the cell of Altshuller and Cohen [3] in Figure 3.1.2c.

Figure 3.1.2a
Gas mixing pump of the firm Woesthoff

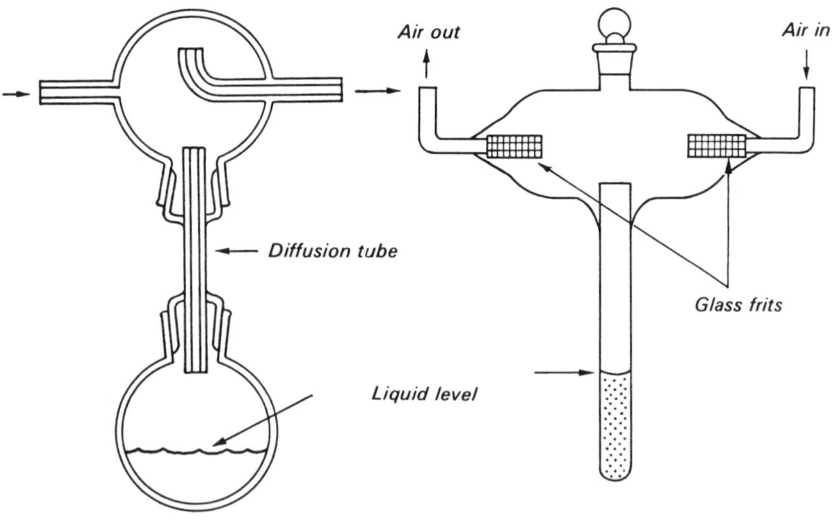

Figure 3.1.2b
Diffusion cell after McKelvey and Hoelscher

Figure 3.1.2c
Diffusion cell after Altshuller and Cohen

The mixing ratio can be adjusted by selecting the temperature and the free surface of the liquid phase. The content of foreign matter taken up by the air is obtained after a certain running time by determining the weight loss or by analysis of an aliquot of the mixture. It can also be calculated by the formula

$$r = \frac{D \cdot P \cdot M \cdot A}{R \cdot T \cdot L} \ln \frac{P}{P-p},$$

where r is the diffusion rate (g/min), D is the diffusion coefficient (see International Critical Tables, Vol. V, 1929), P is the total pressure, M is the molecular weight of the vapor, A is the cross sectional area of the tube, L is the tube length, T is the Kelvin temperature, R is the gas constant and p is the vapor pressure of the liquid.

O'Keeffe and Ortman [194a] produced highly diluted mixtures (see also Scaringelli, Frey and Saltzman [228a]) of air with SO_2, NO_2, hydrocarbons and halogenated hydrocarbons, etc., in the following way: the toxicant is introduced under pressure (or as a liquid) into a teflon tube. The tube is closed at both ends and suspended in the air flow into which the pollutant is to be introduced. The substance loss of the filled tube which is constant over long time intervals when the temperature is maintained constant, is determined by weighing. It can be adjusted to the desired concentration of the air mixture by a suitable selection of the tube dimensions, the temperature and the inflation pressure.

Finally, a liquid or dissolved toxicant component can be uniformly introduced into the air stream by applying it uniformly over a warmed section of the air line. The toxicant is vaporized or introduced as the product of a uniformly and quantitatively proceeding decomposition reaction.

In such cases, when the concentration of the air mixture is not clearly established by the mode of preparation, Leithe and Petschl [159] recommend the withdrawal of a sample from the branch line during preparation. The sample is transferred into a suitable surge vessel and analyzed.

Examples for the preparation of special air mixtures containing toxicants are described in the chapter on special topics (see also Saltzman).

3.2 Labeling of air masses

Saltzman, Coleman and Clemons [219a] recommend sulfur hexafluoride as a means of labeling air to study air movements. SF_6 is an inert and stable marker. It can be detected in concentrations as low as 10^{-5} ppm on a gas-chromatographic column with an electron-capture detector.

4

Special Analytical Methods

4.1 Photometric analytical methods

Photometry constitutes one of the most important methods in air analysis; the toxicants must be converted into colored compounds which are then determined. The reason for the widespread use of this method lies in the large number of specific and very sensitive color reactions available and the short time required to perform a comparatively accurate measurement.

There is no standardized terminology for this method: the terms photometry and colorimetry are mostly used interchangeably; if the measurement is carried out with spectroscopically pure monochromatic light, obtained by resolution of white light with the aid of prisms or gratings, we speak of spectrophotometry.

In the following discussion we assume that the reader is acquainted with the principles of photometry; when necessary, these principles will be briefly reiterated to gain better insight into the special problems which arise in air analysis.

The original colorimetric procedure, whereby the color is compared with model solutions of known concentration either by direct visual comparison with a color scale (Nessler tubes) or adjustment to the same color intensity (Hehner cylinders, wedge or Dubosq colorimeters), is not being used in air analysis, since its sensitivity and accuracy are low and because the method is quite laborious. Exceptional cases, in which comparative colorimetry is used, apply to individual samples of higher pollutant concentrations. Photometers combining optical measurements and electrical recording have proved very successful in air analysis. For the determination of low toxicant levels in the ppb-range we recommend the use of high-quality spectrophotometers with low oscillations of the zero point. For higher concentrations of toxicants (in the MAK-range) cheaper and less sensitive instruments are quite adequate.

The following instrumental parameters are important for attaining a good measurement sensitivity and reproducibility:

1. Adequate stability of the line current.
2. Selection of a favorable ratio of the degree of amplification to the slit width.
3. Measurement at the wavelength corresponding to the absorption maximum.
4. Good mechanical properties, especially low lost motion of the adjustments.

We recommend the MPQ II spectrophotometer of the firm Zeiss, Oberkochen, as an instrument which corresponds best to the requirements of air analysis (Figure 4.1).

The extinction E read off the photometer, the layer thickness d (in cm) and the concentration c of the solution (g toxicant per liter color solution) are connected by the Lambert–Beer law $E = \log(I_0/I_c) kcd$, where k is the extinction coefficient.

The extinction readings E are, as a rule, not converted via the above equation for given values of the extinction coefficient but with the aid of a calibration curve

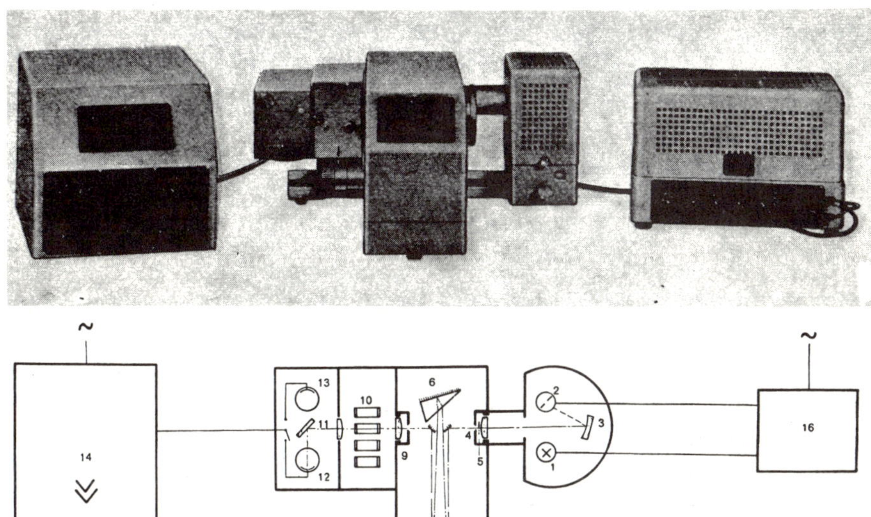

Figure 4.1

MPQ II spectrophotometer of the firm Zeiss, Oberkochen:

1 incandescent lamp;
2 hydrogen lamp;
3 revolving mirror;
4 entrance slit;
5 rotary aperture;
6 prism;
7 mirror
8 wavelength scale;
9 exit slit;
10 cuvettes;
11 tilting mirror;
12 secondary-electron multiplier;
13 photocell;
14 indicator device;
15 indicator scale;
16 voltage stabilizer.

which allows for all the individual parameters (instrument, slit width or light filter, blanks of the reagents). However, it is possible to derive from a known molar extinction coefficient ε (i.e., the fictitious extinction of a 1 *M* solution for a layer of 1 cm thickness) the detection of the limits of a reaction or to calculate the air quantity required to produce a sufficiently intense color for an assessed pollutant concentration. It is most expedient to calculate with a millionth part of ε ($\mu\varepsilon$), i.e., the extinction is referred to 1 μmole of the toxicant.

Example: if for an accurate spectrophotometer and a color reaction with good reproducibility an extinction of E = 0.01 is considered to exceed the zero-point oscillations with a sufficiently high probability (3s-value, see page 27) at, for example, a molar extinction coefficient $\varepsilon = 40{,}000$ ($\mu\varepsilon = 0.04$), layer thickness d = 5 cm, volume of final solution 25 ml (SO_2-determination according to West–Gaeke with para-rosaniline in $Na_2Hg_2Cl_4$ solution, see page 161, a detection limit of 0.08 μg SO_2 is obtained according to the formula

$$C_{(\mu g/\text{liter solution})} = \frac{E \times \text{mol. wt.}}{\mu\varepsilon d}.$$

Thus in 10 liter air 0.003 ppm SO_2 can be detected with certainty.

The calibration curve is obtained by plotting the introduced toxicant amounts in μg as the abscissa and the measured extinction as the ordinate. When the Lambert–Beer law is satisfied, we obtain a straight line, whose slope is given by the fraction E/μg.

Its reciprocal C, "the reciprocal slope" of the calibration curve, μg/E (see VDI-Richtlinie 2451) can serve as a coefficient by which the extinction measured in a sample must be multiplied in order to obtain the toxicant level (μg) in the sample.

The color reaction to be applied must satisfy the following conditions:

1. The color must remain constant after a short waiting time.
2. The color must be stable, if possible for several hours, so that in series analyses no differences arise between the samples analyzed first and last.
3. The Lambert–Beer law (extinction is a linear function of concentration) should be valid.
4. The specificity must be high and the sensitivity towards other pollutants low.

Particularly convenient color reactions are those which take place immediately after the air sample is introduced, so that the experiment can be interrupted after a sufficiently intense color is attained. If this is not possible, i.e., when the color reaction with the absorption solution can be carried out in the laboratory only, it is recommended to let a rapid semiquantitative reaction proceed during sampling in a Dräger tube, provided that the odor or taste does not give sufficient information on the amount of sample required.

The sensitivity of the photometric analytical method, which is especially important in air analysis, can be increased if we succeed in achieving a high extinction with a highly concentrated solution. This aim can be attained in the following manner.

1. Utilization of the lowest possible amount of liquid in the separation of the air component and the lowest possible dilution prior to the color reaction. Usually, a final volume of 25 ml is adequate.

2. Measurement with the maximum possible layer thickness (5-cm colorimetric cuvettes, content about 15 ml).

3. Application of a color reaction which has the highest possible extinction coefficient.

4. Utilization, if necessary, of a larger air sample volume.

The required solution volumes can be considerably reduced by using microcuvettes. Thus, for example, the microcuvettes for the Zeiss spectrophotometer require only 0.2 ml solution per cm layer thickness (1–5 cm), which under certain circumstances can be taken from scrubbers with low liquid volume and corresponding high toxicant enrichment.

4.2 Continuous integration methods with rags or bells

In a region subjected to flue-gas hazards, situations may occur when it is not so important to identify the causes of short-term acute damages, but rather long-term (e.g., 1 month) loads, i.e., possibilities of chronic damage, especially to agricultural plants or forests. The simple, continuous integration methods are then still useful. As mentioned on page 32, these methods are recommended when the extent of damage or the subsequent compensation claims are not large enough to warrant the use of expensive continuous recorders.

Usually, alkali-treated surfaces are employed, such as paper strips or cotton rags impregnated with, for example, barium hydroxide or potassium carbonate solution and glycerol. Fabrics impregnated with PbO_2 are specially used for SO_2. The rags or strips are mounted under conditions very similar to those for the endangered vegetation and must be exposed to the wind from all sides. The rags, impregnated with alkaline reagents, are wound around a solid shell (Liesegang bells). After a certain period has elapsed (100 hr or 1 month) the strips or rags are ashed or extracted. The absorbed pollutant is then quantitatively measured in a suitable fashion.

These procedures yield results which cannot be directly expressed in terms of concentration (ppm); however, rough correlations can be established between these results and those obtained by individual or continuous determinations, so that the average concentrations of the toxicants in question can be estimated and compared with known toxicant effects.

Examples of the application of the rag or bell method will be described in the sections specifically dealing with SO_2 and HF.

4.3 Gas-detection devices with test tubes

The gas-detection devices manufactured by several firms (e.g., Drägerwerk, Lübeck; Auergesellschaft, Berlin; Unico, USA) are based on the adsorption of the tested air pollutant on solid surfaces with the simultaneous occurrence of a color reaction. The Drägerwerk has so far developed test tubes for more than 40 pollutants. They can be used for monitoring the atmosphere at the work place (concentration range of the MAK-values, but sometimes also for the lower concentrations in the MIK-range) as well as for other gas analyses, such as waste gas analysis (stack effluents and automobile exhaust, determination of H_2O, H_2S or CO in refinery gases, synthetic gases or liquefied gases, for detection of leaks near gas lines, for smoke from fire or explosions and for respiration analyses).

The instrument developed by Drägerwerk consists of a handpump (Figure 4.3a) which sucks 100 ml air ($\pm$ 3 ml) per stroke by pressing a bellows equipped with steel springs. The air sample flows through the test tube specifically intended for the relevant pollutant; the color which arises is evaluated to assess the air sample.

The test tubes are made of glass (length 12.5 cm, inside diameter 5 mm). Along the direction of the gas stream (designated by the arrow) the tubes as a rule initially contain a layer of reagents, which serve to retain all the interfering substances. This precleaning layer is sometimes very complex, for example, in the test tubes used for determining CO in automobile exhaust (tube CO 0.5%) it consists of a mixture of

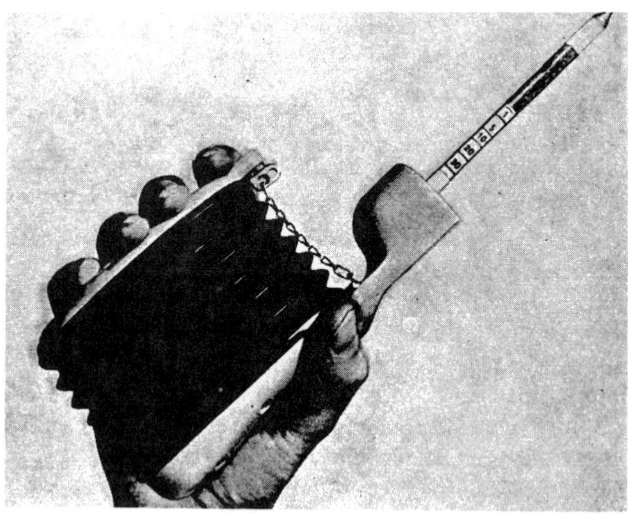

Figure 4.3a
Handpump for the Dräger device

phosphoric acid, 60% oleum and CrO_3, which completely bind the interfering olefins and isoparaffins.

The indicating layer proper follows the precleaning layer. It consists of a carrier (grain size 0.2–0.4 mm), whose solid surface is impregnated with the reagent. For several pollutants more reagents, which are not stable when mixed, are required for the detection. In this case the reagent in question is fused into an ampoule. The ampoule is mounted at the kink of a tube rendered flexible by means of a rubber hose (Figure 4.3b). Immediately prior to the determination the ampoule is broken by bending the tube and its content reacts with the pollutant.

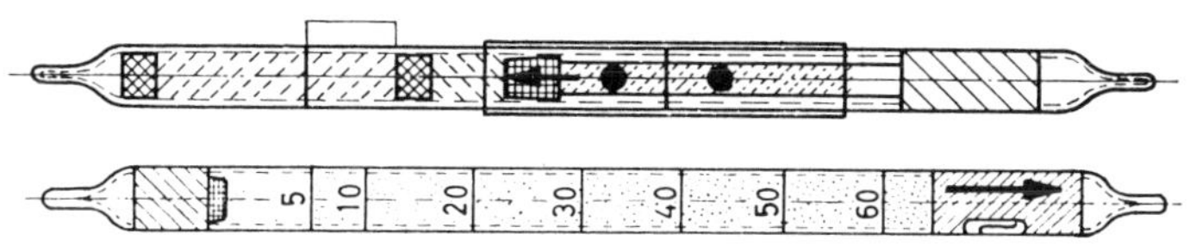

Figure 4.3b
Top: tube with kink; bottom: normal tube

In order to obtain semiquantitative values the reagent is applied in such a concentration over the carrier that with a certain number of strokes (e.g., 5 or 10) the length of the colored layer is a measure of the pollutant concentration in the tested air. Here, concentrations in the range of the MAK-values are usually assessed, though in some cases also lower concentrations (around the MIK-value) can be determined by applying larger air quantities, for example, by increasing the number of strokes or by mounting an automatic suction pump. Further, very low MAK-values of extremely poisonous pollutants can still be detected in 1 liter of air, for example, AsH_3 (MAK-value 0.05 ppm) 0.02 μg, i.e., 0.01 ppm. An amount of ozone (MAK-value 0.1 ppm) as low as 0.05 ppm (0.1 μg) can be detected in 1 liter of air. In such cases the number of strokes is counted until the first colored ring can be seen in the indicating layer, or—in other cases—until a color arises corresponding to a given standard. For the interpretation of calibration curves of test tubes see Leichnitz [150].

Grosskopf [85] classified the Dräger tubes into 6 groups in accordance with the type of detection reaction. Group I comprises reactions proceeding rapidly and completely over a wide concentration range. When the reagent is uniformly applied over the carrier, the layer length is proportional to the amount of pollutant absorbed, independently of the gas flow rate (ml gas tested/min). Thus, when the number of strokes is increased or the air supply is otherwise higher, much lower concentrations can be determined. The following conditions, however, must be fulfilled:

1. The carrier material should also adsorb the pollutant quantitatively from the highest dilutions, but the binding should not be so strong that its reaction with the pollutant is impossible.

2. The reagent must react rapidly and quantitatively with the pollutant to yield an intensely colored reaction product. It should be applied to the carrier material in such a way that at the flow velocities obtained with the handpump the reaction proceeds so fast that the limits of the color reaction are clearly visible.

3. The concentration of the reagent on the carrier must be uniform. It is dependent on the concentration range of the pollutant to be determined.

For different pollutants several test tubes for different concentration ranges have been developed, for example, for CO between 5 and 200 ppm; 10 and 3000 ppm; 0.1 and 1.2‰, 0.3 and 4% and 0.5 and 7.5%.

Group II comprises the tubes mentioned above with ring reactions for very low pollutant levels. Sometimes they are suitable for measuring the lengths of the reaction zones in the sense of group I, when the concentrations or the air volumes used are higher.

Group III comprises organic detection reactions, which, though quantitative, are not strictly stoichiometric; thus, in certain concentration ranges they depend on the gas flow rate. However, here, as well, the concentration range can be extended to detect lower pollutant levels by increasing the air volume.

In group IV the toxicant does not directly participate in the detection reaction but is converted by an auxiliary reaction into the component of the colorimetric reaction (e.g., acrylonitrile via HCN, carbon tetrachloride via phosgene).

In group V the indication is strongly dependent on the gas flow rate, for example, when determining ethanol with chromic acid.

Group VI comprises the so-called equalizing tubes. The sampled air is pumped until a certain color standard (supplied with the tube) is attained. Here, a large reagent excess (e.g., H_2SO_4 for resinification of styrene) is apparently necessary. The pollutant concentration varies inversely with the number of required pump strokes. In order to attain at least somewhat accurate and reproducible results the gas flow rate and the temperature have to correspond as closely as possible to the calibration conditions.

Reproducibility and correctness

The *reproducibility* of the quantitative results is stated by the manufacturer; the standard deviation of the results is about 6–8% in favorable and 10–20% in less favorable cases. This reproducibility is adequate for sanitation evaluations and in many cases for technical purity data. To obtain good reproducibility a sharp detection reaction (if possible, according to group I of Grosskopf), uniform distribution of the reagent over the carrier material and equal grain size of the indicating layer are necessary.

The *correctness* of the results depends on the calibration, specificity of the detection reaction and the elimination of interfering substances. The relevant data are listed in the instructions supplied with the tube.

Table 4.3
TEST TUBES FOR THE DRÄGER GAS-DETECTION DEVICE

Substance	Type of tube	Reaction principle	Measuring range		
			from	to	unit
Acetone	100/b	Dinitrophenylhydrazine	100	12,000	ppm
Acrylonitrile	5/a	Via hydrocyanic acid	5	30	ppm
Ethanol and homologs, tetrahydrofurane	100/a	Chromate/sulfuric acid	100	3,000	ppm
Ammonia	25/a	Mercurous nitrate	25	700	ppm
Aniline	5/a	Furfurol	2	10	ppm
Arsine	0.01/a	Mercury–gold chloride	0.01	0.1	ppm
Benzene	0.05	Formaldehyde + H_2SO_4	0.05	1.4	mg/liter
Hydrocyanic acid	2/a	$HgCl_2$ + acid indicator	2	150	ppm
Chlorine, bromine	0.2/a	o-Tolidine	0.2	30	ppm
Hydrogen fluoride	0.5/a	Zirconium–alizarin lacquer	0.25	15	ppm
Formaldehyde	0.002	Xylene–sulfuric acid	2	40	ppm
Hydrazine (+ NH_3)	0.25/a	Bromophenol blue	0.25	3	ppm
Carbon dioxide	0.1/a	Hydrazine + redox indicator	0.1	6	vol. %
Carbon monoxide	5/b	Iodine pentoxide–oleum	5	200	ppm
Hydrocarbons					
Ethylene oxide					
Vinyl chloride	0.1%	Iodic acid + oleum	0.1	1	vol. %
Mercaptan	2/a	Copper salt	2	100	ppm
Methyl bromide	5/a	Via bromine	5	50	ppm
Nitrous gases	0.5/a	Diphenylbenzidine	0.5	50	ppm
Ozone	0.05/a	Bleaching of indigo	0.05	1.4	ppm
Phenol	5/a	Indophenol	0.5	10	ppm
Phosgene	0.25/b	Dimethylaniline + aminobenzaldehyde	0.25	75	ppm
Mercury vapor	0.1/a	Gold chloride	0.2		μg
Phosphine	0.1/a	Gold salt	0.1	40	ppm
Hydrochloric acid	2/a	$AgNO_3$ + dithizone	2	30	ppm
Oxygen	5%/A	Via CO	5	21	vol. %
Sulfur dioxide	0.01	KI + starch	3.5	150	ppm
Sulfur dioxide	1/a	KI + starch	0.1	20	ppm
Carbon disulfide	0.04	Copper thiocarbamate	10	320	ppm
Hydrogen sulfide	1/b	Lead salt	1	600	ppm
Nitrogen dioxide	0.5/c	Diphenylbenzidine	0.5	50	ppm
Styrene	50/a	H_2SO_4 (resinification)	50	400	ppm
Systox	1/a	Gold chloride + N-chloroamine	1		μg
Carbon tetrachloride	10/b	Via phosgene	10	100	ppm
Toluene	5/a	H_2SO_4–iodine pentoxide	5	400	ppm
Trichloroethylene + others					
Halogenated hydrocarbons	10/a	Via chlorine	10	400	ppm
Steam	0.1	Selenium sulfate	0.1	80	mg/liter

Service life and reuse of the tubes

For fused tubes a two-year service life is guaranteed. After a measurement yielding negative results opened tubes can be used again on the same day.

Tubes with positive results can be frequently kept as a means of verification for a longer period; in some cases sealing with paraffin or plastic caps is prescribed.

4.4 Continuously operating automatic devices for analyzing pollutants

The increasing hazards to the public caused by the sudden presence of pollutants in concentrations injurious to humans and damaging to materials necessitates continuous monitoring at certain sites. At the same time immissions in these areas must be measured over short time intervals and allowed for in order to take the necessary immediate steps. Instead of carrying out continuous manual analyses at such sites, automatic analyzers have recently been developed which continuously measure and record toxicant levels, and emit an alarm signal when the maximum permissible concentrations are exceeded. Many of these instruments are refined versions of apparatus developed earlier for the analysis of waste gases in industrial plants, where the concentrations of the toxicants are higher. Similar problems of the immediate detection and determination of pollutants were already of interest in both world wars because of the use of poison gases. Numerous firms have developed and marketed efficient instruments based on various principles. Because of the ever increasing demands imposed on the analysis of new toxicants and on the sensitivity and selectivity of the readings this field is undergoing rapid development; in and outside Germany new designs are being continuously put on the market.

Automatic analyzers have the following advantages:

1. Rapid determination and recording of the sought toxicant concentration and, if necessary, with an alarm mechanism at the measurement site itself or remote from it.

2. Cut-down in personnel and the possibility of obtaining objective results. On the other hand, a certain amount of work is necessary for calibration, permanent maintenance and control.

3. Possibility of mounting the instruments at sites, which are not readily accessible, and in mobile laboratories.

4. No difficulties in sampling or in transferring the sample from the sampling site to the laboratory.

The following conditions have to be satisfied to ensure proper functioning of automatic instruments in air analysis (see also page 26).

1. The sensitivity required is a function of the harmfulness of the substance analyzed. About 10% of the permissible limit (MAK- or MIK-value) or the expected hazard limits should still be possible to determine. Also the possibility of switching over (possibly automatically) to higher concentration ranges is convenient.

2. A standard deviation of $\pm$ 5–10% is adequate for the reproducibility or correctness. The reproducibility guaranteed by the manufacturers is usually better (mostly about 1 to 2% of the selected measuring range); this data need not necessarily be correct, since systematic errors may occur.

3. High specificity and selectivity, i.e., the sensitivity toward interfering substances must be low.

4. Practically complete uptake of the pollutant; the substance losses in the supply lines and by absorption must be minimum.

5. The zero point must be constant and independent of variations in the temperature, humidity and CO_2-level.

6. Minimum work involved in cleaning (air bubbles, dirt, corrosion products, algae) and filling of reagents.

7. Simple power supply (also at uninhabited sites).

8. Possibility of adjusting the analyzer to the required measurement interval (instantaneous readings or half-hourly mean values).

4.4.1 Classification and survey of methods of automatic air analysis

1. *Physical methods*

These are characterized by the direct determination of a parameter specific for the pollutant without any changes occurring in the chemical composition of the air sample. For the analysis of air those properties are used which are not altered by the normal air constituents. The most important procedures are based on the selective absorption of light in the infrared, ultraviolet or visible ranges. In oxygen determinations very frequently the paramagnetism, which is very specific for this element, is measured. Additive parameters to which the normal air constituents also contribute can only be used when the toxicant levels are high (this mostly applies to the analysis of waste gases); these parameters include the refractive index in the interferometer, the gas density and the thermal conductivity.

The apparatus used in direct physical methods are simple owing to the fact that the volume of the air sample need not be measured, since this volume is solely defined by pressure, and the air sample is not diluted by a reagent.

2. *Chemical methods*

The toxicant to be determined is converted by chemical reactions into a state with characteristic and automatically measurable properties. This group comprises most of the devices used for air analysis:

Colorimetric instruments.

Galvanometric instruments, including arrangements for amperometric titration and polarometry.

Potentiometric instruments whose readings give either directly the concentration (linearizing "single-point titration") or serve as the basis of automatic titration.
Instruments based on measuring the heat of reaction.
Instruments based on the measurement of differences in thermal conductivity.

Automatic devices, whereby first the pollutant has to be separated by gas chromatography and then determined, will be described in Section 4.6.

Chemical methods are more specific than physical ones. On the other hand, the apparatus is more complex, since the ratio of the amounts of air sample and reagent is a characteristic parameter and thus must be maintained. Moreover, these devices are more sensitive, since during the chemical reaction and conversion into a liquid phase enrichment of the pollutant can occur. We distinguish between the following two groups of methods:

a) completely continuous methods in which the gas sample and reagent are continuously introduced in a defined ratio, mixed, allowed to react and measured;

b) semicontinuous methods in which the liquid reagent is periodically pipetted with a time-controlled metering device and supplied to the apparatus, while the air sample is introduced during a certain period. After measurement, the measuring vessel is automatically emptied and filled again. In a special variant of these methods (Imcometer) the air sample is introduced until a certain physical effect takes place; the air quantity consumed constitutes, as in the case of conventional back titration, a measure of the amount of toxicant present. The amount can, however, also be determined by conventional automatic titration from the volume of the reagent solution consumed up to the equivalence point.

3. *Coulometric procedures*

Here, the "reagent" is the electric current consumed during a substance-specific electrochemical process. The current consumed until a certain physical effect occurs is a measure of the pollutant concentration.

In accordance with the *type of indication* we distinguish between direct and compensation measurements. In the former the physical effect which occurs is directly measured and read off; in the latter the original effect is fully compensated by an opposing effect. Its magnitude (e.g., depth of immersion of a wedge attenuating the luminous intensity as the primary effect) is a measure of the concentration.

Recording is usually effected on a moving paper strip. The paper feed is operated by a clockwork which yields the time scale. Depending on the purpose of the analysis, various chart speeds must be selected. The parameter to be recorded is first adjusted on a scale, for example, with an indicator, and then transformed via a mobile stylus which traces out a curve on the moving paper. For this purpose low-energy point recorders (pen recorders) or (after appropriate amplification) direct recorders can be used. Some of the more recent instruments use digital recording.

It is most convenient to obtain results which are directly expressed as concentrations. This is readily done when the measured parameter is a linear function of concentration; when the function is logarithmic (light absorption, potential measurements), recording paper with logarithmic coordinates should be used.

Instead of or in addition to a recorder optical or acoustic signals can also be emitted when a given limit is exceeded. In process engineering it is further possible to utilize the analytical measurement effect for process control after conversion and amplification. This is frequently done when measuring waste gases in plants.

4.4.2 Physical methods based on light absorption

Among the methods based on the direct measurement of light absorption the determinations in the UV and visible ranges are less important in the analysis of air than those in the infrared region. All gases which absorb in the short-wave region are very reactive and can thus be readily determined by chemical methods, especially when present in low concentrations. However, the measurement ranges of apparatus intended for direct color measurement mostly lie in those concentration ranges which correspond to the waste gas itself rather than the dilutions corresponding to the MIK- and MAK-values.

4.4.2.1 *Measuring devices operating in the UV and visible light range*

The *Okometer* developed by Hummel [112a] and manufactured by the firm Withof GmbH, Kassel, is designed to measure the absorption of gases and liquids in visible and UV light. It is primarily intended for the analysis of waste gases. The device consists of an alternating light photometer with two parallel beams (see Figure 4.4.2.1a). The incident light beam is rendered parallel by a convex lens and split by a double-aperture diaphragm into two beams. The latter acquire via a rotating segmental diaphragm a counterface frequency of 6.25 Hz. One traverses the sample cell, the other the reference cell. Both beams are then led to the same photocell. Because of the difference in light exposure due to light absorption in the sample cell light-intensity variations are produced; the corresponding alternating photocurrent is amplified and recorded. The smallest measurement ranges (sensitivity 3% of the range) are 0–700 ppm for Cl_2, 0–200 ppm for NO_2, 0–2 ppm for ozone, 0–0.03 ppm for mercury vapor, 0–100 ppm for benzene and 0–700 ppm for SO_2.

In order to be able to measure in the MIK-range a double-dosing pump for air and color reagent is connected to the sample cell. Thus, using the chemical-colorimetric methods described below, much lower levels (e.g., 0–1 ppm; detection limit 0.05 ppm) can be measured.

As a counterpart to the URAS device which operates in the infrared region (see below) the firm Hartmann and Braun has recently developed a series of LIMAS devices (Licht-Modulations-Absorptions-Schreiber = Light Modulation Absorption

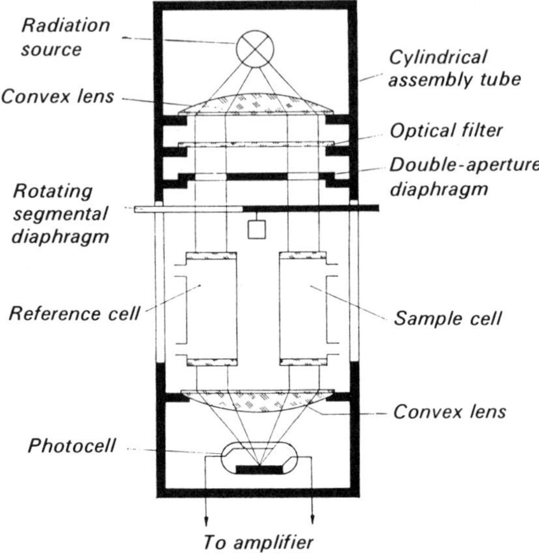

Figure 4.4.2.1a
Okometer

Recorder). These are intended for the determination of the concentration of gases which absorb in the visible or ultraviolet spectral region.

The light path of the LIMAS G (Figure 4.4.2.1b) differs from that in similar instruments in that the incident light is split, selectively filtered and modulated only after having passed the sample cell. Both beams are then successively led to the photocell. Instead of the cell with the gas sample a liquid sample cell (LIMAS K) can be used through which a colored solution resulting from the reaction with the pollutant to be determined is passed.

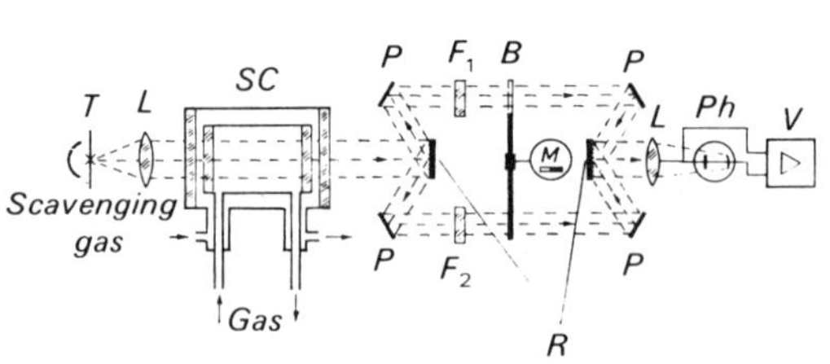

T tungsten lamp;
L convergent lens;
SC sample cell with protective jacket;
P plane mirror;
G grooved mirror;
F_1 measuring filter
F_2 reference filter
D diaphragm disk;
M motor for diaphragm disk;
Ph vacuum photocell;
A amplifier.

Figure 4.4.2.1b
Schematic diagram of the LIMAS G

A similar instrument is the 7004 Double-Beam Flow Colorimeter of the firm Beckman. This device provides a light path of up to 1 m for gases and permits carrying out the determination under pressure (12 atm) and at temperatures of up to 135°C.

4.4.2.2 *Measuring devices operating in the infrared region*

Whereas the light of the UV and visible region is absorbed in a manner suitable for analysis only by comparatively few and rare pollutants, absorption in the infrared region (wavelengths of 2–15μ, corresponding to wave numbers of 5000–660 cm^{-1}) is very expedient, especially when analyzing carbon-containing air components such as CO, CO_2 and CH_4. An additional advantage is that the main air constituents present as elements (N_2, O_2, noble gases) are not absorbed in the infrared region, while the normal air components CO_2 and H_2O, which are absorbed in this spectral region, can be easily removed (see also Karthaus [128]). However, SO_2 and H_2O can also be determined in this way after removal of the interfering substances.

4.4.2.2.1 The URAS device

The principles underlying a comparatively simple automatic measurement technique were established in 1937 by Lehrer and Luft. The instrument designed on the basis of these principles was designated as the URAS (Ultra-Rot-Absorptions-Schreiber = = Infrared Absorption Recorder). This device was first used for recording CO- and CH_4-traces in the synthesis gas in ammonia production. The instrument was later produced by the firm Hartmann and Braun. Since the Second World War, US firms have also employed this measurement principle.

This technique is characterized by the use of nonspecific IR-rays (without spectral resolution) with a double-beam instrument and a substance-specific receiver. The measurement principle is evident from Figure 4.4.2.2.1a: the radiation source consists of two wire coils to which equal currents are supplied. Both beams are periodically interrupted at the same time by a diaphragm disk. The first beam passes through the sample cell, while the second traverses the reference cell filled with a gas which does not react to IR radiation (e.g., N_2). The detector consists of a gas-impermeable double chamber with a metal membrane as partition. The chamber is filled with a mixture of argon and the pure toxicant to be determined (e.g., CO), so that it can only be heated by thermal radiation of wavelength corresponding to that absorbed by CO.

When neither the air sample nor the reference gas contain any IR-absorbing constituents (both beams are simultaneously incident upon the detector), both detection chambers will be simultaneously heated by equal energy impacts corresponding to the unchanged content of CO-active rays. The gases in both halves of the detection chamber expand at the same time and the membrane remains at rest. The same situation prevails when the tested gas contains a component which is also IR-active but absorbs rays of another wavelength (e.g., CH_4). When, however, the

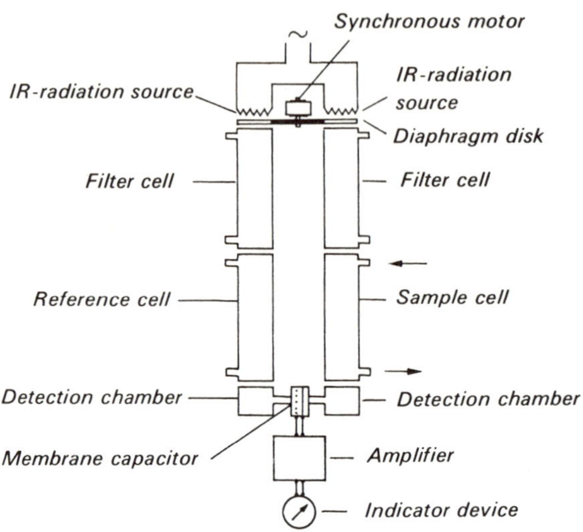

Figure 4.4.2.2.1a
Schematic diagram of the URAS infrared analyzer

air sample contains CO, then it absorbs part of the CO-specific radiation from the beam. In this half of the detection chamber only a residual component remains. Thus, here the gas is heated to a lesser extent and it expands less than the gas in the other half of the detection chamber. The metal membrane, which serves as a partition, is bent; this process is periodically repeated as the diaphragm disk rotates. The metal membrane forms part of a capacitor, whose capacitance periodically changes as distance between this membrane and a second, fixed capacitor plate varies. Thus, when a dc voltage is applied across a resistor, the charging current and the voltage drop across the resistor become periodic. The latter is electronically amplified and indicated or recorded. The scale of the indicating device is directly calibrated in vol. % of the toxicant to be determined.

The reading is specific only for the toxicant to be determined in the detection chamber if the air does not contain any foreign substances exhibiting IR absorption in the same wavelength range (Figure 4.4.2.2.1b). When the absorption of the sought toxicant and the interfering components only partially overlap, the interference can be removed by inserting a filter, which eliminates the interfering component, into both filter cells. This eliminates the common absorption ranges; the reading is now specific, however, weaker.

The sensitivity of the indication depends on the absorbing power of the toxicant (molar extinction coefficient). It can be increased by increasing the length of the sample cell or by applying a higher pressure. The measurement range can also be

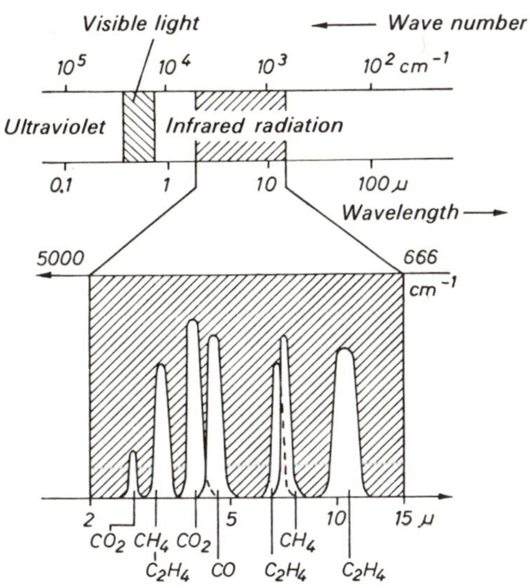

Figure 4.4.2.2.1b
Electromagnetic radiation spectrum with absorption bands for CH_4, CO and CO_2 (schematic)

adjusted to meet specific requirements by varying the electrical amplification. When higher concentrations are to be accurately measured, the temperature and pressure must be allowed for.

4.4.2.2.2 Other IR-devices

The LIRA device developed by Mine Safety Appliances, Pittsburgh, USA, has a detector similar to that in the URAS instrument with the difference that in the LIRA both beams are combined before arriving at the detector which does not contain a partition. Whereas, in the URAS both beams are simultaneously incident upon the detector, a rotating metal diaphragm alternately blocks the radiation to the sample cell and the reference cell, respectively. When the toxicant to be detected (e.g., CO) is not present in the sample gas, the radiant energy incident upon the detector cell does not change during the alternating illumination of the two beams. On the other hand, when the air sample contains CO, the detector is alternately heated to a higher and lesser degree, and the membrane undergoes periodical oscillations.

In the instrument developed by Leeds and Northrup the principle of "negative filtration" is applied. The gas sample traverses the cell A which is sufficiently large to let the beams through in the same manner (Figure 4.4.2.2.2). The two subsequent cells F_1 and F_2 are respectively filled with the pure component to be determined (e.g., CO) and with absorbing argon. T_1 and T_2 are thermopiles. When no CO is present in the sample, the two beams in the thermopiles yield a certain energy dif-

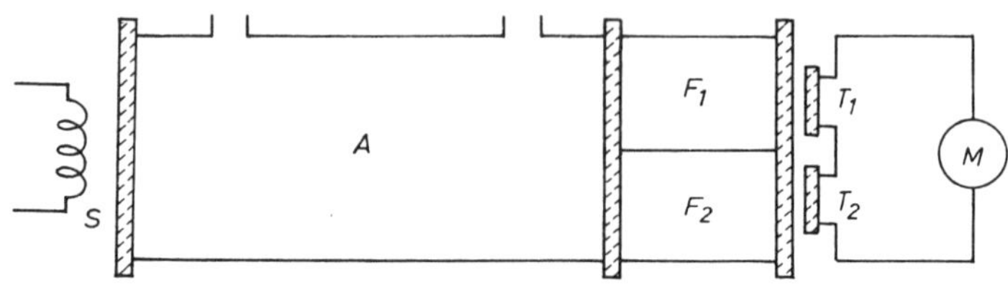

Figure 4.4.2.2.2
IR-device developed by Leeds and Northrup

ference which is taken as the blank. In the presence of CO the blank becomes smaller as the CO-concentration increases.

The principle of the CO-device of the firm Beckman is the same as that of the URAS.

In the Inframeter manufactured by Siemens and Halske the temperature of the gas in the detection chamber is measured without chopping the light and compared with the temperature of the corresponding chamber of the second beam.

The UNOR I measuring device of the firm Maihak, Hamburg, is described by Luft, Kesseler and Zorner [170a]. This single-beam instrument is especially suitable for measuring carbon monoxide traces in the air of coal mines. In contrast to the URAS, where the double chamber of the detector is separated by a membrane into parallel halves and illuminated by a beam passing through the sample cell on the one hand and by a beam traversing the reference cell, on the other, the detector of UNOR I is divided into two chambers of different thickness, arranged one behind the other. Both chambers are filled with the gas component to be determined (e.g., CO) but at different pressures. The pressure and layer thickness can be adjusted, so that in the absence of CO in the sample cells the light beam heats the series-arranged chambers to the same extent; thus the membrane remains at rest and does not transmit a signal. When the sample contains CO, the first chamber, which is primarily excited by the center of the absorption band, is heated less than the second chamber, which mainly corresponds to the edge of this band. Thus, a signal corresponding to the CO-concentration is produced. The sensitivity toward interfering components is comparatively slight.

4.4.3 Automatic colorimetric methods

4.4.3.1 *The Monocolor instrument and the Automatic Smoke Sampler*

We shall first describe two automatic colorimeters in which the air sample is blown onto a paper tape driven by a clockwork at an appropriate speed along the air outlet. The tape is uniformly impregnated with a color reagent, which yields a color characteristic of the pollutant to be determined. The advantage of these analyzers is their simple design, since effecting a uniform tape feed involves less difficulties than metering out a liquid reagent, as done in colorimetry. On the other hand, color measurements by optical means on a paper in incident or transmitted light are much less accurate than in solution. These instruments are mainly used for continuous dust measurements (see page 104) but can also be employed in air analysis; here, however, they are primarily used for higher toxicant concentrations in emission analysis, and only in rare cases for continuous monitoring at work places. The Monocolor MC 1001 of Maihak AG, Hamburg, is schematically shown in Figure 4.4.3.

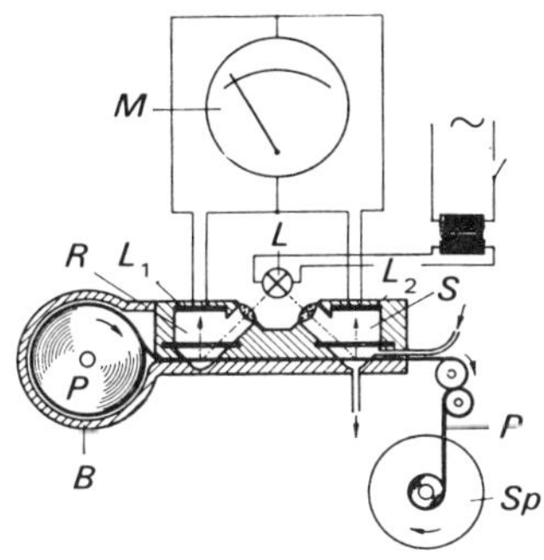

S sample cell;
R reference cell;
L lamp;
L_1 photocell 1;
L_2 photocell 2;
P paper tape impregnated with reagent;
B box for paper tape;
Sp spool head for paper tape;
M indicating device.

Figure 4.4.3
Monocolor (schematic)

The instrument operates according to the following principle: A paper tape impregnated with the reagent (e.g., lead acetate for H_2S) is successively drawn through two cells. One of these, the reference cell, is filled with pure air, while the air sample passing over the paper is blown into the sample cell. Both cells, which are symmetrically arranged are illuminated by a common lamp. The light reflected from both tape surfaces is respectively measured by two photocells. The current difference corresponds to the blackening of the paper caused by the foreign matter in the air sample (e.g., H_2S). The tape feed rate is 1 mm/min; since the diameter of the sample cell is 12 mm, the paper is illuminated 12 min and a mean value over this period is measured. The throughput rate of the sample gas (e.g., 100 ml/min) can be adjusted by means of control valves and measured. The lowest measurement range for H_2S is 0–14 ppm.

A similar instrument, the AISI Automatic Smoke Sampler (Research Appliance Co., Pittsburgh, USA), is described in ASTM D 1704-61 (see page 104) as a dust-measuring instrument and can also be used for determining H_2S when a paper impregnated with lead acetate is inserted.

4.4.3.2 *The Imcometer*

The Imcometer developed by Fuhrmann [71] (Figure 4.4.3.2a; manufactured by Bran and Lübbe, Hamburg) is a special instrument for determining air pollutants

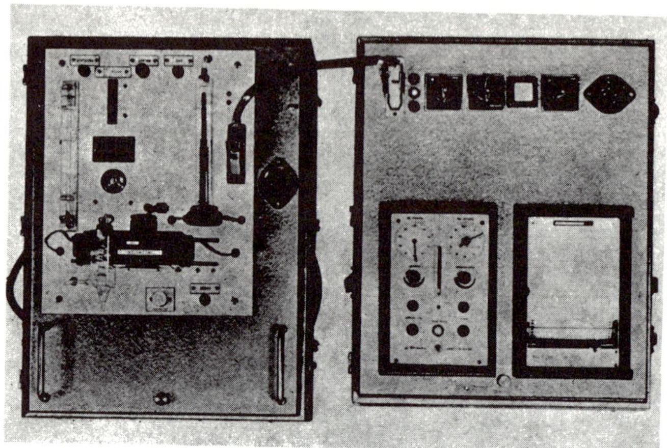

Figure 4.4.3.2a
Imcometer

colorimetrically in the ppm-range. Sampling and concentration range have been specifically allowed for in the design. The device can be used both for measuring half-hourly mean values and intermittent peak values.

The half-hourly mean values are determined in accordance with the flow diagram given in Figure 4.4.3.2b. When the control valve is in position 1, the reagent (6 ml) flows into the sample cell and the blank is compensated. If the blank is too high, a signal lamp lights up. In position 2 the sample gas (20 liter/min) flows 30 min through the absorption system and the colorimetric measurement is effected on two photo-resistors. The reagent that has been used up flows out when the control valve is in position 3.

When short-time measurements are taken, the whole gas-supply period (15 min) is utilized only when the color intensity does not reach an adjustable peak value because the toxicant concentration is too low. If this value is reached earlier owing to higher toxicant concentrations, position 2 is automatically interrupted. Here the measurement value does not correspond to the magnitude of the photoelectric effect, but rather to the time (and thus the sample size) taken until the required effect is attained. This time is inversely proportional to the toxicant concentration. The mode of operation corresponds approximately to a back titration in volumetry in which a certain amount of titer solution is placed into the vessel and the sample solution is introduced and measured from a burette until the equivalence point is reached.

Such short-time measurements for determining the peak concentrations are important for open-air analysis in the vicinity of pollutant sources as well as for determinations at the work place. They can, of course, also serve for the determination of low pollutant levels in industrial gases, whereby it is important to immediately

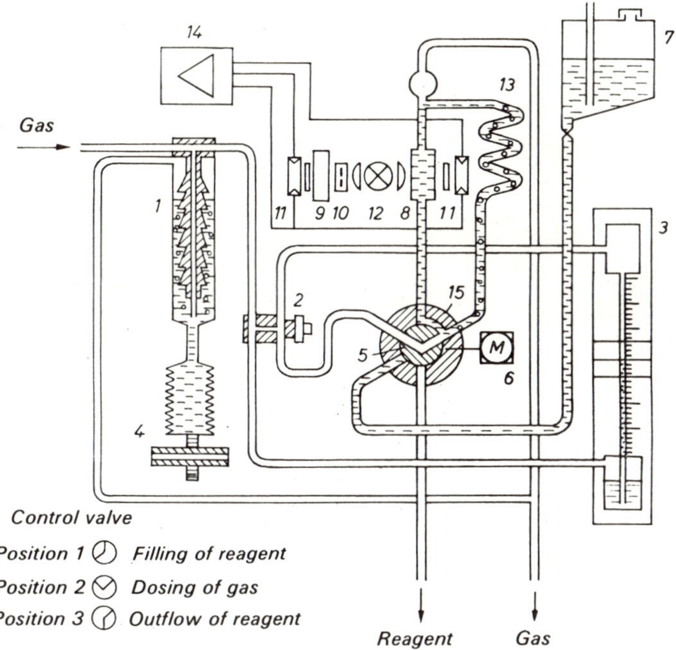

Figure 4.4.3.2b
Flow diagram of the Imcometer

detect sudden penetrations of toxicants, while during normal operation excessive consumption of reagents should be restricted.

Depending on the colorimetric reagent, the Imcometer can be employed for the most widely differing pollutant determinations. Its use for determining SO_2, NO_2 F^- and ozone will be described later on in this book (see the chapter on special topics).

4.4.3.3 *The Technicon Auto-Analyser*

Extensive automation of colorimetric methods of all types has been effected by the Auto-Analyser of the firm Technicon (USA). This technique has proved successful in clinical analysis. Devices of this type are currently available on the market for air analysis.

A typical Technicon arrangement, shown in Figure 4.4.3.3, consists of the following component parts: a sample supply with proportioning pump, a dialyzer instead of the filtering arrangement prescribed for manual analysis, a flow-type heating bath, an automatic colorimeter and a recorder.

The design variant for air analysis consists of spirals filled with glass beads which serve as scrubbers for the uptake of the pollutants. Since the latter are mostly water-

Figure 4.4.3.3
The Technicon Auto-Analyser

soluble, the dialyzer can be dispensed with. The water bath is only necessary when the colorimetric reaction is endothermic. This apparatus is suitable for conventional colorimetric determinations (SO_2, NO/NO_2, oxidants and formaldehyde) in the open air.

4.4.3.4 *The Mini-Adak*

The Mini-Adak (Analytic Systems Associates, Pullman, USA) effects, just as the Imcometer, an enrichment of the pollutant during the colorimetric determination up to a selected measurement range. The device is, for example, designed for the determination of small amounts of fluorine ions (see page 211). For a concentration of 30 μg HF/m^3 the enrichment is attained within 1 minute. The measurement range is adjustable for low concentrations between 0.3 and 500 μg HF/m^3. The apparatus is also suitable for the determination of ozone.

4.4.4 Automatic air analysis based on electrical conductivity

Automatic devices in which the toxicants react with suitable solutions and the change in the electric conductivity characteristic of the absorbed air component is measured, have been used for a long time in air analysis. The main advantage of this method is its high sensitivity; only very small liquid volumes are required in which minute conductivity changes caused by very low toxicant concentrations are still measurable. From the engineering standpoint the measurement bridges involve no difficulties and it is easy to obtain continuous readings. The strong temperature-dependence of the electrical conductivity is eliminated by simultaneously measuring the reagent not in contact with the air sample; thus, we must ensure an accurate equilization of the temperature of the blank and the sample instead of a temperature adjustment.

These instruments are mainly intended for the determination of SO_2, but also H_2S, NH_3, HCl, CO_2, as well as CO and CH_2 after oxidation to CO_2.

A disadvantage of this method is its low specificity when several of the above substances, all influencing the electrical conductivity, are all present at the same time.

The first instrument of this type—the Autometer—was designed by Thomas and described in ASTM-D 1355/60. In Germany the Ultragas 3 of the firm Woesthoff, Bochum, is frequently used. Recently, the Picoflux manufactured by Hartmann and Braun, Frankfurt, has gained widespread application in Germany.

4.4.4.1 *The Ultragas* 3

The mode of operation of the Ultragas 3 of the firm Woesthoff is explained in Figure 4.4.4.1. G_1 is the section formed by two platinum electrodes at which the original reagent solution is measured; G_2 is the corresponding measurement point of the solution after absorption of the pollutant. Both measurement points are adjusted to the same temperature by a tempering bath. A multiple piston pump driven by a synchronous motor supplies air sample and reagent in a constant ratio. Both electrode pairs of platinized platinum form part of a bridge circuit fed by an alternating current. In the case of conductivity changes the pick-up of the potentiometer can be adjusted via a zero adjuster and an amplifier until the bridge circuit is again balanced. The position of the pick-up can be directly transmitted onto a compensation-curve recorder.

The required gas throughput rate is 40–175 ml/min, the liquid flow rate 0.4–1.5 ml/min.

The most sensitive measuring range for SO_2 is 0–1 ppm (sensitivity about $\pm$ 0.02 ppm), 0–5 ppm for NH_3 and 0–50 ppm for CO_2, CO and CH_4.

Special care must be taken to ensure a uniform reagent flow rate, which can be impaired by air bubbles, dirt and algae.

Results obtained in the determination of SO_2 are reported by Thoenes [250] and Hoeschele [108].

4.4.4.2 *The Picoflux*

The Picoflux device of the firm Hartmann and Braun differs from the Ultragas in certain respects: the reagent is fed by capillaries and the rate is controlled by a container in accordance with the Mariotte principle. Since the throughput is independent of the temperature, all the temperature-dependent component parts are placed in a thermostat (40 $\pm$ 0.002°C). The lowest measuring range is 0–0.3 ppm SO_2,

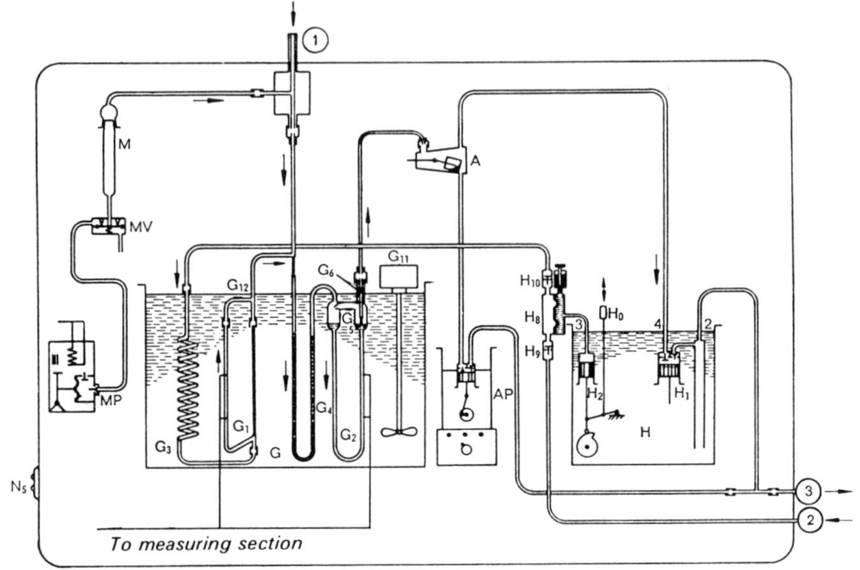

Figure 4.4.4.1
Ultragas 3

A level switch;
P suction pump;
G sample cell;
G_1 measurement section for unconsumed reagent solution;
G_2 measurement section for consumed reagent solution;
G_3 helical glass tube for temperature equalization of the reaction solution;
G_4 reaction tube;
G_5 gas separator for G_2;
G_6 channel;
G_{11} stirrer;
G_{12} connection for G_1;
H pump unit, type NB 175/1;
H_1 pump for the sample gas;
H_2 pump for reagent solution;
H_8 membrane chamber;
H_9 suction valve;
H_{10} pressure valve;
H_0 manual operation of pump H_2;
M trap with activated carbon for absorbing SO_2;
MP diaphragm pump;
MV relief valve;
N_5 mains connection;
1 inlet for gas;
2 inlet for reagent solution;
3 outlet for reagent solution and residual gas.

though measuring ranges 2.5 or 5 times higher can also be selected. If necessary, an integrator can be used to indicate the 10- or 30-min integrated value.

The dead time is 8–10 sec, the half-value time 28–30 sec and the 90%-value 60–100 sec.

The Picoflux can be provided with a very sensitive galvanometric indicator for the determination of H_2S, Cl_2, NO and NO_2 (Engelhardt [64] and Breuer [27a]).

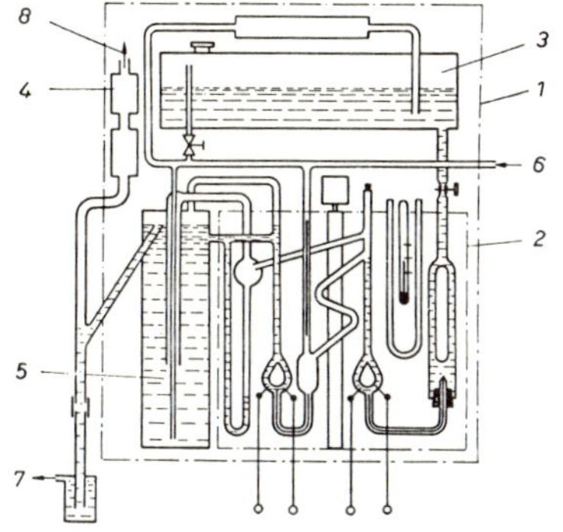

1 analyzer;
2 thermostat;
3 supply tank;
4 pump;
5 flow control;
6 inlet for gas;
7 solution outlet;
8 outlet for reacted gas.

Figure 4.4.4.2
Schematic diagram of the Picoflux gas-trace analyzer

4.4.4.3 *The Acralyser*

The Acralyser (Automatic Chemical Reagent Addition Analyser) manufactured by the firm Beckman, Munich, also effects SO_2-measurement on the basis of differences in the electrical conductivity in dilute H_2O_2 solution. The measuring range is 0–2 ppm; according to the prospectus of the firm, the accuracy of the readings is $\pm$ 5%; the 90%-value is attained after 5–10 min. The reagent flow rate is 1 ml/min and the sample flow rate 2 liter/min. The storage tank provides for a weeklong supply.

4.4.5 Coulometric methods

In these techniques the chemical reaction with the substance foreign to air proceeds in the presence of a reagent, formed directly by electrolysis. The amount of toxicant is determined by measuring the electric current which is supplied until the end point of the reaction is attained.

Coulometric methods possess the following advantages:

1. It is easier to meter and measure electric current than a liquid reagent.
2. It is not necessary to store reagents (which are often subject to decomposition).

The most important prerequisite for applying a coulometric method is an electrochemical reaction proceeding with a current yield of nearly 100%. The reactions taking place on the electrodes should not mutually interfere: the end point of the reaction with the toxicant must be established with sufficient accuracy. The first instrument of this type—the Titrilog—was developed by the Consolidated Electrodynamics Corporation, Pasadena, and is described in ASTM-D 1355. It was originally intended for the analysis of poison gas and has since found special application for the determination of SO_2, H_2S and mercaptans in the ppm-range in the open air. The operation principle is explained in Figure 4.4.5a. The air sample is continuously fed into the anode space, where elemental bromine is generated electrolytically from a KBr solution in dilute H_2SO_4 to the extent that a certain redox potential is produced at the reference electrodes. If this potential is not reached owing to Br-consumption by compounds reacting with bromine (SO_2, H_2S or mercaptans), an electrolytic current is generated which is of such strength that the original Br-concentration in the solution is maintained. The current is recorded and constitutes a measure of the SO_2, H_2S and mercaptan concentrations in the air sample.

The vessel is filled with 2 liter of solution containing 12 g KBr + 353 g H_2SO_4 per liter. Continuous operation for 3 months is possible before renewing the solution. The air throughput rate is 1 liter/min. The zero-point is determined by aspirating air through an alkalized activated-carbon filter.

The measuring range is 0–10 ppm SO_2, 4 ppm H_2S or 8 ppm butyl mercaptan; the reproducibility is 2% of the measured value. In order to obtain selective results, preliminary absorption solutions must be used. Otherwise the presence of elemental chlorine and bromine, ClO_2, NO_2 and ozone causes a decrease in the SO_2-values, since these substances liberate bromine from a KBr solution.

Thus, the Titrilog is neither very selective nor very sensitive. Its advantage lies in the little maintenance required for filling it with the reagent.

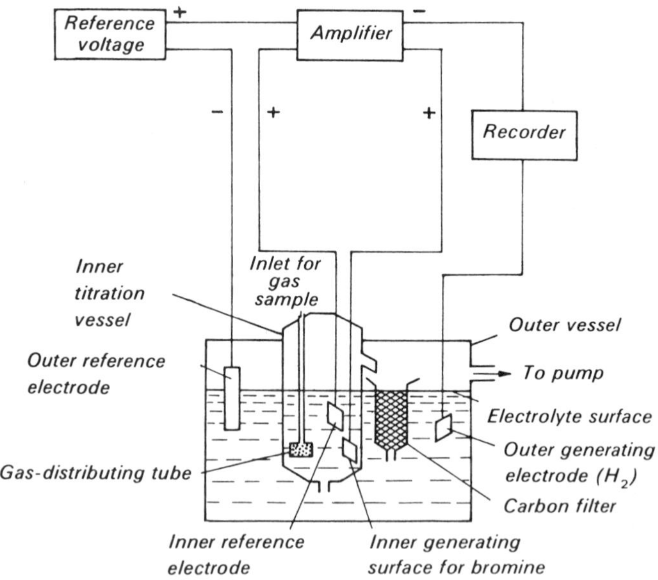

Figure 4.4.5a

Schematic diagram of the Titrilog

Coulometric SO_2*-analyzer of the firm Beckman*

A new SO_2-analyzer based on the coulometric principle is being manufactured by the firm Beckman. The electrolytic cell (Figure 4.4.5b) consists of a platinum anode, a platinum cathode and an activated-carbon bipolar reference electrode, connected in parallel. The electrolyte is a potassium iodide solution, buffered to a pH of 7, through which the air sample is bubbled. A constant current source continuously generates elemental iodine on the anode which is subsequently reduced on the cathode. Thus, an equilibrium concentration for elemental iodine is established and no current is generated at the reference electrode. However, when the air sample contains SO_2, the equilibrium concentration with respect to iodine decreases. Thus, the amount of iodine is not sufficient to transport the charge supplied by the constant current source. Part of the current is compelled to follow the second path, i.e., via the reference electrode. The strength of this branch is proportional to the SO_2-concentration in the air sample.

The measuring range can be adjusted to 0–0.5 ppm or 0 4 ppm, depending on whether we expect SO_2-levels in the MIK- or MAK-range. The limit concentration is 0.01 ppm SO_2 and the reproducibility $\pm 4\%$. The following sensitivities to interfering substances (higher SO_2-indication in ppm in the presence of 1 ppm interfering substance) are given; 0.05 for H_2S and mercaptans; 0.02 for ozone and NO_2; 0.01 for aldehydes, ketones and olefins.

The greatest advantage of the coulometric devices is the use of a stable reagent, which need not be renewed for long periods.

A coulometric instrument developed by the firm Beckman for the determination of

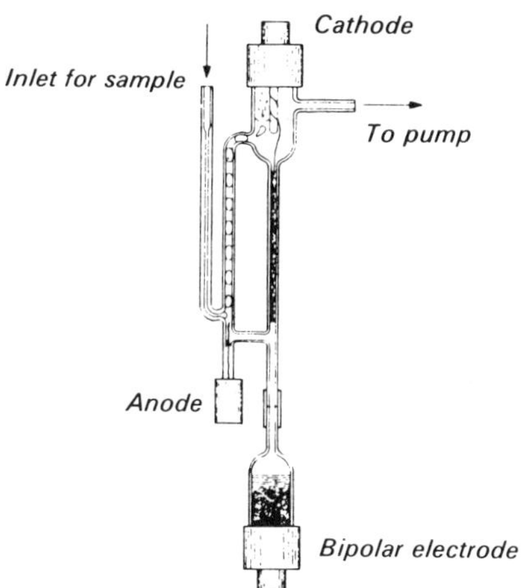

Figure 4.4.5b
The Beckman SO_2-analyzer

ozone is described on page 130. This firm also offers a coulometric CO-measuring device in which the generated substance is iodine.

A coulometric analyzer produced by Consolidated Electrodynamics for determining the water concentration in gas and air samples is treated on page 140.

4.4.6 Potentiometric methods

Among the electrochemical measurement techniques for continuous characterization of solutions potentiometry has distinguished itself by its high simplicity; it is especially used in the analysis of water in order to determine the pH and redox behavior. This method is less popular in air analysis because the potentials measured are not sufficiently specific for the requirements of this type of analysis. Another difficulty, which, however, exists also in colorimetric measurements, is the fact the toxicant concentration to be determined is a logarithmic function of the measured potential difference; this renders the evaluation of the measurement results more complicated.

The last-named difficulty can be circumvented when determining acidic or alkaline constituents by potentiometric "single-point titration" (Leithe [158]). By adding a buffer solution consisting of acids or alkalies with several dissociation constants to the sample the measured pH is converted in such a way that it can be described as a linear function (instead of the usual logarithmic function) of the concentration of the alkali or acid to be determined. An example of "single-point titration" is described in the section on CO and CO_2.

Kündig and Högger [143] proposed the Turicum II recording pH-meter for the semicontinuous determination of SO_2 in the atmosphere. The mode of operation of this device will be described in detail in the section on SO_2.

We further draw the reader's attention to the potentiometric measuring cell employing sulfidized silver electrodes (see the section on H_2S).

4.5 Gas-chromatographic methods in air analysis

4.5.1 Introduction

Gas-chromatographic methods are characterized by the fact that the sample is introduced in the form of a gaseous or vaporized mixture into a carrier gas stream which is then passed in suitable separation columns over solid adsorbing surfaces (gas-solid chromatography (GSC)) or over nonvolatile liquids applied to solid surfaces (gas-liquid chromatography (GLC)). In this process the individual components of the sample move through the column at rates corresponding to their different distribution coefficients between the fixed and mobile (gaseous) phases. The components leave the column in separate fractions and can be determined as individual compounds in the carrier gas. When suitable detectors are used, this technique is very selective (the separation achieved can sometimes be as high as that theoretically possible for several hundred thousand plates) and very sensitive.

Gas chromatography is a separation method, whereby also chemically related substances, such as isomeric or homologous hydrocarbons can be separately determined. This method, however, can also serve for enrichment, since the individual components in the carrier gas occur in higher concentrations at the outlet of the column than in the original air sample. Moreover, it can be directly applied after enrichment processes (freezing out, adsorption or absorption). Difficulties are involved, however, in the identification of the individual substances which are indicated by peaks in the chromatogram. Thus, when analyzing multicomponent mixtures of unknown pollutants, it is frequently necessary to employ another method immediately afterward, which renders their identification possible. The methods mainly used for this purpose are IR-analysis, mass spectrometry or thin-layer chromatography. When the pollutants are known, identification can be effected via the retention time, i.e., the time taken until they leave the column. Other ways of identifying the peaks are described on page 92. The chromatograms are quantitatively evaluated on the basis of the intensity of the detector signal either by measuring (if possible, by mechanical means) the area below the peak (or the peak height, when the experimental conditions are constant) or by means of electronic integrators.

A wide range of chromatographic instruments are available; the manufacturer usually supplies the operation procedures for conventional determinations together with the device.

Other advantages of gas chromatography include the possibility of determining pollutants or components with less distinct chemical detection reactions. Let us give as an example the determination of N_2O, COS and the inert gases, which are identified in gas chromatography by their specific retention times.

Finally, gas chromatography is used when only small sample volumes are available.

The disadvantages of gas chromatography include the complicated apparatus and the necessary experimental skill. An automatic, semicontinuous procedure (successive individual determinations) is possible, but it requires additional expenditure. With the exception of certain special devices, the determination itself has to be carried out in the laboratory. Therefore, in air analysis a simpler and cheaper colorimetric method is preferred when only a specific pollutant is to be determined for which a specific and adequate color reaction exists, with no interference due to accompanying substances.

Since in many cases the pollutant concentrations are very low, the selection of a suitable detector for indicating the individual components leaving the column with the carrier gas is quite important.

Some of the detectors developed for gas chromatography can also be used for the determination of pollutants in the air sample without resorting to chromatographic separation. The highly sensitive flame ionization detector (FID) permits direct and continuous detection of organic compounds with C—H bonds (hydrocarbons, alcohols, ketones, fatty acids, esters, amines, etc.) in very low concentrations (up to 1 ppb), for example, in the air of garages, stores for solvents, etc. (see also Andreatch and Feinland [6]). Very small amounts of chlorine-containing pollutants, for example, phosgene or halogen hydrocarbons, can be found with the aid of an electron-capture detector.

The FID is very suitable because it does not respond to other air components, such as CO_2, H_2O, SO_2, nitrogen oxide, but merely indicates compounds with C—H bonds. Consequently, this device can be also used for the analysis of aqueous extracts or solutions in carbon disulfide.

Gas-chromatographic separations and/or analyses can be effected with small (1–10 ml) air samples which are directly fed into the column. In this manner hydrocarbon amounts as low as 1 ppb can be separated and detected. If the concentrations to be determined are even lower, the pollutants must be enriched. This can be done by freezing out with or without an adsorbing agent, adsorption at normal temperature followed by desorption at higher temperatures or by enrichment on a small preliminary column, filled with a separating liquid on an inert surface, and subsequent introduction into the chromatographic column proper. The air volume and the intensity of the enrichment processes are dependent on the pollutant level to be determined; it can vary between fractions of ppb up to the range of higher MAK-values (several hundred ppm).

For the detection of permanent gases in the ppb-range the Helium-Detector of the firm Varian-Aerograph has recently been recommended. Helium atoms are excited by a beta-source (250 mC). These atoms in turn excite molecules having an ionization potential lower than that of helium.

Ionization potentials in eV			
He	24.5	CH_4	14.5
Ne	21.5	CO_2	14.4
Ar	15.7	CO	14.1
H_2	15.6	SO_2	13.1
N_2	15.5	O_2	12.5
		H_2S	10.4

The ions generated in a 400-V electric field yield measurable currents of about 4×10^8 A if the gases to be determined are in the ppb-range. The Helium Detector has a higher sensitivity than the thermal conductivity cells and even the FID.

This detector is used for air analysis in combination with columns filled with a 5-Å molecular sieve, Porapak or silica gel and Al_2O_3. Highly purified helium is used as the carrier gas.

The following examples will be treated in detail for the analytical application of gas chromatography in the determination of air pollutants:

1. Analytical determination of known organic pollutants in industrial workshops in the MAK-range.
2. Selective determination of acetylene in the presence of other low-boiling hydrocarbons in the intake air of oxygen plants.
3. Investigation of uncombusted hydrocarbons from automobile exhaust in city air.
4. Determination with the electron-capture detector.
5. Determination of normal air constituents.
6. Determination of vapors of solvent mixtures.

Other applications of gas chromatography are described in the chapter on special topics. A survey of the use of gas chromatography in air analysis is given by Altshuller [2].

4.5.2 Determination of organic toxicants in the atmosphere and work sites according to Cropper and Kaminsky

Cropper and Kaminsky developed a gas-chromatographic method which is especially suitable for the determination of toxicants in the MAK-range at work places in the organochemical industry [50]. This technique was included in the new edition of the ICI Manual (see page 9). The authors describe an apparatus and a working procedure which is universally applicable for the most important toxicants. This method will be treated in detail below.

Principle of the method

The air sample is first passed through an absorption tube filled with a suitable liquid phase applied to a carrier material (this corresponds to the fixed phase in the chromatographic column). After the absorption tube is connected to the column the toxicant is expelled with the carrier gas at elevated temperature, introduced into the separation column, separated and detected by means of an FID. The sample volume and detection sensitivity permit recognition of values exceeding the MAK-value.

The FID described by Cropper and Kaminsky has a relatively low sensitivity; the more recent FID's available on the market possess an approximately hundredfold higher sensitivity and are therefore also suitable for concentration ranges considerably lower than the range of the MAK-values. Instead of enrichment in the absorption tube, direct introduction, for example, of a 10-ml air sample into the chromatographic column is possible. However, the authors mention losses in some cases when dosing with injection needles. For this reason they recommend preliminary absorption for higher pollutant concentrations.

The air volume required for preliminary absorption

The air sample volume required for enrichment by means of a preliminary absorption tube with a fixed liquid phase is characterized by the gas-chromatographic behavior of the pollutant. On slow aspiration of the air volume through the tube filled with a suitable separating phase (e.g., silicone grease applied to kieselguhr) the pollutant is first completely adsorbed by the fixed phase and remains in the tube. A detector mounted at the outlet end would first indicate the escape of pure air by a straight reference line. When discharge of the retarded pollutant front concentrated in the fixed phase starts, the detector indication will increase in proportion to the increasing toxicant concentration. In gas chromatography the volume of the carrier gas up to the maximum of this increase is called the retention volume. Experience has shown that for 80% of this volume the pollutant is still completely retained in the tube; this sample volume must not be exceeded if pollutant losses are to be avoided.

Cropper and Kaminsky also report in their paper the retention volumes at room temperature for some organic compounds. These apply to the absorption tube (25 mm silicone–Celite), prepared according to the prescription given below. 80% of this retention volume corresponds to the maximum volume of the air sample used for the analysis.

The retention volume of a substance not contained in the above table in 25 mm silicone–Celite tubes (see below) can be roughly estimated by the formula

$$\log V_{max} = \frac{\text{b.p.}\,(^\circ\text{C}) + 13}{70}.$$

Thus it can be seen that the maximum permissible volume V_{max} is independent of the toxicant concentration in the air sample. Whether this is adequate for accurate

Table 4.5.2
RETENTION VOLUME V_R ACCORDING TO CROPPER AND KAMINSKY

	b.p.°C	Retention vol. V_R, ml
Benzene	80	29
Cyclohexane	81	26
Toluene	110	77
Pyridine	115	110
n-Butanol	118	36
p-Xylene	138	210
Cyclohexanone	155	230
Cyclohexanol	162	230
Aniline	184	580
Nitrobenzene	210	1200

measurements depends on the sensitivity of the apparatus on the one hand, and on the concentration range to be determined on the other. In studies at the work place we shall attempt to place the toxicant indication corresponding to the MAK-value at the lower part of the scale in order to ensure that an appreciable excess concentration will be detected. For high MAK-values a lower sensitivity can be used.

Thus, for example, if 2 μg of a substance gives a full-scale deflection for a toxicant concentration corresponding to the MAK-value of 2 mg/m^3, the toxicant amount isolated from a 500-ml air sample will give a 50% deflection. In accordance with the above formula a sample volume of 500 ml is enriched in the silicone–Celite absorption tube without substance losses when the boiling point exceeds 176°C.

When owing to a low MAK-value or low boiling point the retention behavior of the normal absorption tube is not sufficient to completely absorb the toxicant from the required sample volume, the retention behavior of the fixed phase must be improved. This is done either by freezing the tube or by changing the liquid of the fixed phase. Polar pollutants such as alcohols, pyridine, etc., are more strongly retained by polyethylene glycol, which exhibits a greater polarity. Solid adsorbing agents, such as silica gel, possess even better retention properties. Here, however, it must be established whether the normal desorption conditions suffice to effect a complete removal of the toxicant from the tube.

When several toxicants are present in the air sample, the necessary air volume is determined by the component requiring the largest volume.

Apparatus

The glass absorption tube (Figure 4.5) has a length of 6 cm and an inside diameter of 4.5 mm. The tube exhibits a bulge and an inner constriction about 15 cm from the top of the separation column. The longer portion of the tube, bounded on both sides by a glass wool stopper, contains a 25-mm packing which is prepared as follows:

Celite 545 (Johns–Manville Corporation) of 100–120 mesh BSS (DIN sieve 0.12–0.15 mm) is boiled in concentrated HCl for several hours at 100°C, washed to neutral reaction, dried at 300°C and the original grain size restored. The carrier is mixed in a 7:3 ratio with Silicone Elastomer E 301 (ICI–Nobel Division). All volatile components are removed from the mixture by a nitrogen stream (50–100 ml/min) at 320°C. In some cases (for pyridine and butanol, see Table 4.5.2b, page 86). Polyethylene glycol 400 is used instead of the silicone in the ratio 1:9 parts Celite. When the air pollutants are even more volatile (see above), silica gel of 0.075–0.15 mm (Davison, USA) is used.

Desorption of the toxicant is effected by means of a heating jacket which consists of a copper-sheet cylinder of height 32 mm and inside diameter 7 mm. It is insulated by asbestos tape and wound with a resistance wire which yields an output of 12 W from a low-voltage source (e.g., 12 V). The jacket is connected in series with a variable resistance of 0–10 Ω.

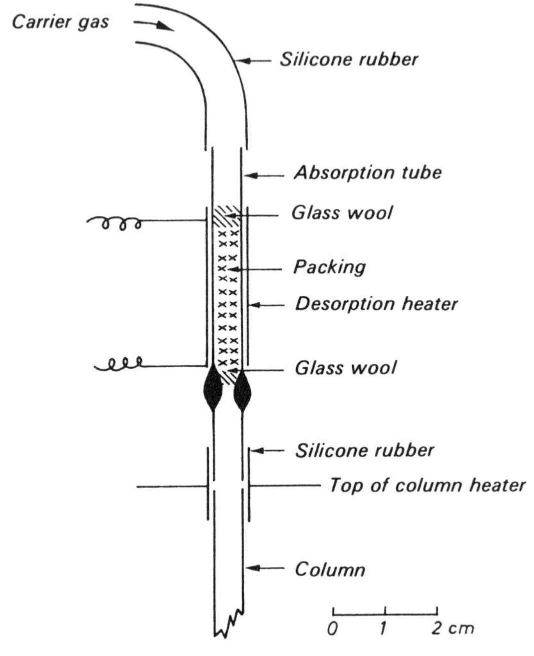

Figure 4.5
Absorption tube with desorption heater according to Cropper and Kaminsky

The chromatographic column has a height of 1 m and an inside diameter of 4–5 mm. The packing is the same as that of the absorption tube but with a coarser grain size (30–60 mesh BSS or DIN sieve 0.5–0.25 mm). Of course, chromatographic columns of the same efficiency supplied ready for use by the manufacturers can also be used for this purpose.

The detector is a flame ionization detector with adjustable sensitivity. The carrier gas is hydrogen–nitrogen 1:1; flow rate 50 ml/min.

Working procedure

The apparatus is calibrated with standard solutions in carbon disulfide; the volume used (0.02 ml) should have a toxicant content corresponding to the MAK-value in the air sample to be withdrawn. This amount should effect an approximately 50% scale deflection when the appropriate sensitivity is selected. Then 0.02 ml CS_2 solution is introduced via an injection needle (as produced by the firm Hamilton) into the heated head of the column. The column temperature is chosen in such a way that the peaks of the toxicant appear within the first 15 minutes. The area under the peaks is determined by the conventional method, i.e., by multiplying the height by the width at half the height.

Collection of air sample

The air volume listed in Tables 4.5.2a and 4.5.2b or calculated by the formula on page 83 is drawn at a slow rate (100–500 ml in 1–2 min) through the prepared absorption tube. It is stored in a test tube with glass stopper and should be analyzed as soon as possible.

Desorption and determination

The tube is connected to the heated chromatographic column. First the desorption heater is switched on for 1 min at full heating capacity (12 W), the carrier gas stream is led through, the detector is ignited and the tube is then heated for another minute. The heating current is then switched off and the carrier gas is further passed through until the expected peaks are attained and a straight reference line appears. The areas under the peaks are measured in the same way as in the calibration procedure.

Calculation of results

If the calibration is carried out with the weight amount of toxicant in 500 ml air corresponding to the MAK-value and the same air volume is also used during sampling, the toxicant concentration x is equal to the MAK-value multiplied by the ratio of the areas under the peaks.

$$x\,(\text{in mg/m}^3) = \frac{\text{MAK-value (mg/m}^3)\,A_s}{A_c},$$

where A_s is the area under the peaks for the sample and A_c is the area under the peaks for the calibration.

Accuracy

The authors report that in a test series most of the values agree with the mean value with an accuracy of $\pm$ 15%.

A survey of the appropriate experimental conditions for some of the more important organic pollutants is given in Table 4.5.2b.

Variant according to Novak, Vasak *and* Janak

Instead of treating the absorption tubes with the air sample as long as no loss occurs (80% of the retention volume), Novak, Vasak and Janak [190] proposed a method whereby the air sample is led through the tube until the absorption equilibrium of the pollutant with the fixed phase is attained. Only then it should be analyzed on the chromatographic column.

Table 4.5.2b

EXPERIMENTAL CONDITIONS ACCORDING TO CROPPER AND KAMINSKY

Compounds under test	MAK-value (ppm)	Vol. of air sample, ml.	Time of sampling, min.	Packing: Absorption tube	Packing: Column	Temp. of column, °C	Sensitivity range used	Standard solution corresponds to MAK-value in specified air volume (except for HCN)
Benzene	25							0.0023 % v./v. benzene, 0.022 % v./v.
Toluene	200	5	0.5	A (see remarks	A	65	Max.	toluene and 0.025 % v./v. p-xylene
Xylene	200			on storage)				in CS_2
Benzaldehyde	—							0.46 % v./v. benzaldehyde, 0.52 %
Benzyl chloride	1							v./v. benzyl chloride, 0.52 % v./v.
Benzoyl chloride }		500	2	A	A	120	Max.	benzoyl chloride, 0.58 % v./v.
Benzal chloride }	None stated,							benzal chloride and 0.64 % v./v.
Benzotrichloride }	taken as 1							benzotrichloride in dry toluene;
								diluted 44-fold with CS_2
Nitrobenzene	1	500	2	A	A	120	Max.	0.010 % v./v. nitrobenzene in CS_2
Aniline	5	200	2	A	A	105	Max.	0.019 % v./v. aniline in CS_2, freshly prepared
Cyclohexane	400	10	0.5	A	A	45	$\frac{1}{10}$	0.09 % v./v. cyclohexane in CS_2
Pyridine bases	None stated, taken as 10	250	1	A	A	90	Max.	0.067 % v./v. pyridine bases in CS_2
Cyclohexanone	50	125	1	A	B	80	$\frac{1}{10}$	0.13 % v./v. each of cyclohexanone
Cyclohexanol	50							and cyclohexanol in CS_2
o-Dichlorobenzene	50	125	1	A	C	95	$\frac{1}{10}$	0.19 % w./v. o-dichlorobenzene and
p-Dichlorobenzene	75							0.28 % w./v. p-dichlorobenzene in CS_2
Pyridine	5	125	1	D	D	75	Max.	0.01 % v./v. pyridine in CS_2
n-Butanol	100	125	1	D	D	60	$\frac{1}{10}$	0.23 % v./v. n-butanol in CS_2
Hydrogen cyanide	10	1000	4	Silica gel	E	20	Max.	0.004 ml of 0.28 % w./v. HCN in methanol

Code for packings: A = Silicone Elastomer E. 301.
B = Poly(ethylene glycol adipate).
C = Octadecylamine-ethylene oxide condensate.
D = Polyethylene glycol 400.
E = Adiponitrile.

The toxicant concentration is calculated from the areas under the peaks, the weight of the fixed liquid phase and the phase equilibrium constant. The phase equilibrium constant is calculated from the specific retention volume in the absorption tube. (See also Selucky, Novak and Janak [228b].)

4.5.3 Gas-chromatographic determination of acetylene traces in air

The determination of the acetylene content in the intake air of air-separation plants (oxygen plants) has become increasingly important for various industrial branches (especially in steel production by the Linz–Donawitz (LD-) process). Among the hydrocarbon traces present in the air acetylene is the most dangerous, since it accumulates in the adsorbers and may lead to explosions. It is regularly analyzed in liquid oxygen in the plant. This analysis was performed by colorimetry before gas chromatography had been developed. Analysis by gas chromatography is much more rapid and sensitive; it can be fully automated by a suitable process chromatograph.

Instead of determining the acetylene concentration in liquefied oxygen, it is advisable to analyze the air already at the intake stage. Here concentrations as low as several ppb must be determined; moreover, we must distinguish acetylene from other hydrocarbons, which are present in much larger amounts but which are harmless. We remind the reader that methane up to a level of 1.2 ppm is a normal component of unpolluted open air. In this method a suitable separation column and a sensitive flame ionization detector (FID) are used. Enrichment is usually not necessary and it is sufficient to directly introduce about 10 ml of air sample into the column.

In order to obtain a straight reference line for the high indication sensitivity which is required we must thoroughly clean the detector and electrodes from dust and oil traces. It is also recommended to pass the carrier gas through water-releasing crystals (e.g., through a cartridge filled with $CuSO_4 \cdot 5\ H_2O$) to achieve uniform moisture. The apparatus must be set up so that it is immune to vibrations and draft.

Larger amounts of higher hydrocarbons, which are gradually collected in the column, can be removed by back flushing.

4.5.3.1 *Various procedures*

Klein [133] applied the Aerograph of the firm Wilken. The 3.4-m chromatographic column (inside diameter 4 mm) contains silica gel with wide pores and a grain size of 0.3–0.5 mm. The separation temperature is 50°C. Nitrogen as the carrier gas is introduced at a rate of 90 liter/min, which is moistened over Glauber's salt. The burner operates on hydrogen (50 ml/min) and oxygen (700 ml/min). The acetylene in a 10-ml air sample appears as a well-developed single peak after 90 sec behind methane, ethane, ethylene and propane, but before propylene and the C_4-hydrocarbons. Acetylene can be reliably detected by this method down to 5 ppb.

Similar good results were obtained with the same apparatus in the central laboratory of the Österreichische Stickstoffwerke.

Kuley [144] reported observations with a FID of the F & M Scientific Co. The chromatographic column (1.35 m × 5 mm) was filled with the following fixed phase: activated alumina (Burrell Co.), 60–100 mesh (i.e., DIN mesh width 0.15–0.25 mm), was treated 15 min at room temperature with NaOH dissolved in dilute methanol, several times washed with methanol and dried at 350°C for 4 hr. Then 7 wt. % of Carbowax 20 M dissolved in methylene chloride was applied and the solvent removed. Nitrogen (60 ml/min) was used as the carrier gas. The separation temperature was 40°C. 10 ml of the air sample were directly introduced into the chromatographic column. The acetylene peak appeared after 6 min before the C_3- and after the C_4-hydrocarbons. The detection limit of acetylene by this method is 12 ppb.

Jones and Green [122] also employed the gas chromatographic determination of acetylene traces in oxygen plants. The apparatus and the FID used were produced by the Perkin Elmer Co. The chromatographic column (1.8 m × 5 mm) contained di-(ethylhexyl)-sebacate as the fixed phase. 3% of this liquid was applied to silica gel (Davison Co.) of 30–60 mesh corresponding to DIN 0.25–50 mm. The authors used a 5-ml air sample; the separation temperature was 50°C with helium (40 ml/min) as the carrier gas. The following sequence appears on the chromatogram: methane–ethane–ethylene–propane–acetylene–propylene–C_4-hydrocarbons.

Decristoforo [55] described an industrial gas chromatograph with the FID of the firm Beckman for the determination of acetylene in oxygen plants. 1 ppm acetylene gives a full-scale deflection.

Theer [248] used an FID of his own design. Since the detection limit of this detector is about 50 ppb, enrichment of the pollutant is necessary to determine lower concentrations. For this purpose the author used small adsorbers containing a 10-mm long and a 1-mm wide layer of silica gel A (grain size 0.125–0.2 mm). The tube was cooled to −78°C by CO_2–methanol coolant and treated with 50–1000 ml air sample. Acetylene is desorbed in an H_2 stream at + 40°C. The chromatographic column was 2 m long and packed with Alusil (Czechoslovak brand) or 0.25–0.135 mm silica gel A. Hydrogen at room temperature served as the carrier gas. The acetylene peak appeared after 2 min between the C_3- and C_4-hydrocarbons. After enrichment the detection limit is 0.1 ppb hydrocarbons; without enrichment this limit is, of course, higher.

4.5.4 Analysis of automobile exhaust in city air

The organic pollutants in city air originating from automobile exhaust are frequently analyzed by gas chromatography. Automobile exhaust consists of a complex mixture of uncombusted fuel residues, cracking products as well as reaction products, such as aldehydes, acids, peroxides originating from partial oxidation processes. Further, in areas with intense insolation photochemically induced reaction products with nitrogen oxides appear as smog components. As an example of such a compound we mention the peroxyacetyl nitrate $CH_3 \cdot CO \cdot ONO_2$ (PAN) and its homologs (page 18) both isolated from the air of smog-polluted Californian cities and produced in simulated experiments under smog conditions. In such intricate separation processes for small amounts of substance gas chromatography is to be preferred.

The isolation and determination of these numerous products is rarely carried out in routine analysis; research workers usually deal with these problems. Our knowledge of the sanitary importance of this complex of pollutants is too limited to warrant the regular determination of the individual component. The isolation and determination of the so-called carcinogenic polycyclic hydrocarbons will be treated at a later stage (page 262).

Research, in which gas chromatography is used to study automobile exhaust, its reaction products in city air and prepared mixtures, is being extensively carried out in the USA. Since both the tasks and the selected and/or available methods differ widely, we cannot give a detailed account of any universally applicable procedure. Therefore, we shall limit ourselves to some of the more important and accessible papers, so that the reader will be able to make his own choice. We shall emphasize papers which not only deal with the analysis of automobile exhaust but also of polluted city air.

In the earlier works (up to about 1960), the somewhat less sensitive thermal conductivity detectors were used. Since the air volume to be led directly into the chromatographic column is limited (in most cases not exceeding 10 ml), the larger amounts required for detection must be enriched from a larger air volume. The flame ionization detectors (FID), whose sensitivity is higher by several orders of magnitude, make it, in many cases, possible to introduce the air sample directly into the column without the need for preliminary enrichment. This applies especially to the lower homologs.

4.5.4.1 *Methods employing preliminary enrichment without an FID*

West, Sen and Gibson [267] developed the following procedure: 20 liter of air are preliminarily adsorbed (0.5 liter/min) in U-tubes at 0°C. The adsorption tube is connected to the chromatographic column via a small tube filled with magnesium perchlorate. The pollutants are desorbed by helium at 200°C. The carrier gas is He at a throughput rate of 50 ml/min. A 2.0 m × 5 mm chromatographic column is used. The fixed phase consists of 20% dinonyl phthalate applied to firebrick. The temperature at which chromatographic separation occurs is 80°C. A thermal conductivity detector with a detection limit of 1 ppm is used. The analytical yield is about 80%. This method can be applied to the analysis of automobile exhaust and for the determination of benzene, CCl_4, n-octane and certain oxygen-containing solvents in air.

The following test conditions are applied in the method of Eggertsen and Nelson [60]. 1–10 liter air is first dehumidified on Ascarite and enriched in a copper U-tube while cooling with liquid oxygen. The adsorbing agent is 30% dimethylsulfolane applied to firebrick. After freezing out the tube is flushed with helium which removes air and methane. The separation column (7.5 m × 5 mm copper tube) is filled with the same fixed phase and operated at 0°C. The thermal conductivity detector has a higher sensitivity than that of conventional devices. The carrier gas is helium at a throughput rate of 60 ml/min. The detection limit is 50 ppb. The acetylene peak appears after 12 minutes behind n-butane and before 1-butene. In areas with heavy traffic the sum total of the C_2–C_5-hydrocarbons was 2 mg/m^3, while in a highway tunnel it was 6 mg/m^3.

The procedure of Farrington, Pecsok, Meeker and Olson [65] is similar. 16-liter air samples are first dried over K_2CO_3 and then passed at a rate of 0.5 liter/min through an adsorption tube filled with 25% di-n-butyl phthalate applied to firebrick. The tube is cooled with liquid oxygen. The chromatographic column has a height of 1.25 m and an inside diameter of 5 mm. Helium at a throughput rate of 80 ml/min was used as the carrier gas. The ionization detector of Ryce and Bryce was used. The separation temperature was 25°C. Hydrocarbons containing 4 or more carbon atoms in their chains are adsorbed and detected. The authors give a detection limit of 1 ppm. The air of Los Angeles during spells of intense smog and test mixtures were analyzed by this method.

4.5.4.2 *Methods employing flame ionization detectors (FID)*

Bellar, Sigby, Clemons and Altshuller [18] employed the gas chromatograph and FID developed by the Perkin Elmer Co. to analyze automobile exhaust diluted with air. The chromatographic column (0.7 m × 3 mm) contains silica gel 40–60 mesh (0.25–0.35 mm) of medium activity which was previously heated at 110°C for 24 hr. The carrier gas here is helium (60 ml/min). The air volume directly led into the column was 1.2 ml. The acetylene peak appears after 6 min behind ethylene but before propylene and the C_4 hydrocarbons. The authors report a detection limit of 1 ppb hydrocarbon. The article contains a description of the analysis of automobile exhaust, polluted city air and prepared C_2–C_4 hydrocarbon–air mixtures in irradiation chambers.

To increase the sensitivity Bellar, Brown and Sigby [17] used the FID combined with enrichment of a 25–300-ml air sample. The pollutants are adsorbed in a 25 cm × 5 mm U-tube filled with 10% Carbowax on firebrick and cooled with liquid nitrogen. After contact with the air sample the adsorbate is flushed for a short time with helium in order to remove air and methane. The pollutants are desorbed in the chromatographic column by immersion into 70°C water. The fixed phase of the column is dibutyl maleate and bis-2(2-methoxyethyl)-adipate applied to a firebrick layer of 0.25–0.30 mm. The carrier gas is helium with a throughput rate of 40 ml/min.

The separation temperature is 36°C. The method was applied for the determination of hydrocarbons (C_2–C_6) in the atmosphere of Cincinatti.

A special method using a FID for the determination of aromatics in city air and automobile exhaust was developed by Altshuller and Clemons [2a]. A 3.1-ml air sample is directly introduced into the chromatographic column (4 m × 3 mm, fixed phase 10% Carbowax 1540 applied to firebrick). Helium at a throughput rate of 50 ml/min is used as the carrier gas. The lower detection limit is 50 ppb. This detection limit is only reached via a 1:2000 dilution of the automobile exhaust in air. The paper reports analyses of aromatics (up to C_{11}) in automobile exhaust, with and without dilution with air, in irradiated mixtures under "smog" condition and city air.

Feinland, Andreatch and Cotrupe [66] employed two chromatographic columns, each measuring 4 m × 5 mm. One of the columns was filled with 20% dimethylsulfolane on Chromosorb, the other with 10% di-isodecyl phthalate on Chromosorb W. Each column was provided with its own FID. The sample volume (undiluted exhaust) was 1 ml. The carrier gas was a 1:1 hydrogen–nitrogen mixture at a throughput rate of 50 ml/min. The determination was carried out at a temperature of 25°C. The purpose of the simultaneous use of two columns is to determine or detect some components which are simultaneously eluted from one column and separately eluted from the other. The analysis reported in this paper gives the level of 27 hydrocarbons (C_1–C_6) in automobile exhaust generated at normal speed (40 mph). The concentrations of the individual components vary from 1 to 450 ppm.

Guerrant [87] designed a portable instrument for determining organic pollutants in the open air with a FID, with and without the application of chromatographic columns. This method enables the continuous detection (without column) of the sum total of organic pollutant of 10 ppb. The same sensitivity is achieved for individual hydrocarbons when 1.6-ml samples are chromatographically separated. The instrument contains three columns which are operated at ordinary temperature. One of the columns is filled with silica gel for methane, ethane and ethylene; the second with 10% hexadecane applied onto firebrick for organic pollutants containing 3 to 5 carbon atoms in their chains, and the third is a capillary column (10 m × 0.5 mm) wetted with hexadecane or squalane, which serves for the adsorption of substances having boiling points between 30 and 130°C. Purified air (25 ml/min) is used as the carrier gas.

To identify the peaks in air-pollutant chromatograms Williams [272] recommended substractive techniques using trapping columns which selectively bind each substance group. After passing the trapping columns the corresponding peaks in the chromatograms will be missing when compared with the nonsubstracted chromatograms. Molecular sieve A for retaining n-paraffins and other unbranched hydrocarbons, and concentrated sulfuric acid for some olefins, disulfide and oxygen-containing compounds are used in such subtraction columns. A dry cartridge with magnesium perchlorate extracts water and most of the oxygen-containing compounds. Further differentiations can be carried out with the electron-capture detector which effects a highly sensitive indication of hydrocarbons with several halogen atoms in their molecule (especially CCl_4, trichloro- and tetrachloroethylene (see below)). For the identification of some hydrocarbons the retention times in columns with tritolyl phosphate and didecyl phosphate can be compared. Williams used 2.5-liter samples which were polluted comparatively slightly. They were enriched in a dry-ice cooled tube with dibutyl phthalate on Chromosorb P. The detection limit thus obtained was 0.1 ppb.

4.5.4.3 *Improved working procedure of* McEven

Considerable progress in the separation and analysis of automobile exhaust and polluted city air was achieved by more complex apparatus (modified Perkin Elmer instruments). McEven [175] gave a comprehensive survey of results on the type and amount of individual hydrocarbons contained in automobile exhaust which were also found in the open air in the ppm-range.

The instrument consists of three chromatographic columns. One of these is an adsorption column for enrichment of C_2 and higher hydrocarbons. The second column is used for subtraction, i.e., for separating aromatics and olefins. The third column is a highly efficient capillary column which effects the actual separation. The separation is facilitated by automatic temperature control (between -25 and $+125$°C). Thus low- and high-boiling hydrocarbons can be characterized in a single chromatogram by peaks of the same form. The FID is used. The carrier gas is nitrogen; the sample volume is 10.6 ml and the entire separation process lasts 22 min.

The adsorption column (10 m × 1 mm) is half filled with alkalized alumina and then further filled with Apiezon applied to Chromosorb. The subtractor column (25 cm × 3 mm) contains $Hg(ClO_4)_2$ on the carrier material "Coast Engineering GC 22". The capillary column (45 m × 0.25 mm) is wetted with Silicone DC 200.

Procedure: The air sample (10.6 ml) first passes the adsorption column cooled with dry ice and trichloroethylene in a Dewar flask. In this column all the hydrocarbons are retained with the exception of methane which first enters the capillary column. After the C_2 hydrocarbons have reached the capillary column, the supply of carrier gas to the adsorption column is reversed. Then the hydrocarbons retained first are expelled by heating and then separated in the capillary column in accordance with the controlled temperature increase. In a second run the content of the adsorption column is first led into the subtractor column, where the olefins and aromatics are removed. The mixture is then introduced into the capillary column. In the second chromatogram the peaks of the olefins and aromatics of the first run are missing so that these compounds can be identified.

The gas sample consisted of automobile exhaust gases from four driving stages (idling, acceleration, cruising, deceleration) and in two engine states (normal exhaust and improved exhaust), and exhaust diluted to the level of foreign substances to be expected in polluted air (2.3 ppm hydrocarbons). Altogether 63 peaks were obtained which, on the basis of the known values, led to the identification and quantitative determination of 22 paraffins, 31 olefins and acetylenes and 5 aromatics. In the samples with pollutant levels corresponding to those in the open air individual components in concentrations as low as 1 ppb could be identified.

4.5.5 Gas-chromatographic analyses with the electron-capture detector

Just as the flame ionization detector responds specifically to organic substances with C—H bonds, while not indicating other compounds such as H_2O, CO_2, CCl_4 or CS_2, the electron-capture detector is especially sensitive to hydrocarbons with several halogen atoms in their molecules (CCl_4, $CHCl_3$, trichloroethylene) as well as nitrogen compounds, nitric and nitrous acid esters in very low traces. This detector was first described by Lovelock [170] and is supplied by several firms as an auxiliary instrument for the gas chromatograph. As an example, we mention here the determination of phosgene traces in the presence of CCl_4 and the detection of the irritant peroxyacetyl nitrate in the smog-laden atmosphere of the cities of California.

To determine phosgene in the presence of CCl_4, Priestly, Critchfield, Ketcham and Cavender [205] employed the Aerograph A 350 B with the ECD of the firm Wilkens. A 0.5–2.0 ml air sample is directly fed into the chromatographic column (2 m × 5 mm, 30% didecyl phthalate on GC 22-Super Support). Nitrogen is used with a throughput rate of 50 ml/min as the carrier gas. The separation temperature is 50°C; the voltage supplied to the detector is 90 V. The detection limit is 1 ppb of phosgene. HCl does not interfere. The retention time of CCl_4 is 8 times longer than for phosgene. It must therefore be blown out of the column after 5 determinations, i.e., before it arrives at the end of the column (see also Clemons [53] and Altshuller [44]).

Darley, Kettner and Stephens [53] isolated peroxyacetyl nitrate CH_3—CO—O—O · · NO_2 (PAN) and peroxypropionyl nitrate C_2H_5—CO—O—O · NO_2 (PPN) from the smog-laden atmosphere of Riverside, California, by gas chromatography, using the ECD of Wilkens. The suitability of the ECD is borne out by the fact that PAN (and PPN) gives only a weak indication in the FID due to the small number of C—H bonds; this compound is readily detected by the ECD because of the strong activity of the nitro group. Accordingly, PNA in the amount of 10 ppb in a 5-ml sample can be indicated without preliminary enrichment. The authors found in these air samples 50 ppb and PAN and 6 ppb PPN. The chromatographic column (90 cm × 1.5 mm) contained 5% Carbowax 400 applied to Chromosorb W. The carrier gas was nitrogen at a throughput rate of 25 ml/min. The separation temperature was 35°C. The PAN-peak appeared after 2 min, the PPN-peak after 3 min.

For the determination of nitrogen dioxide in the air with the aid of the ECD see page 182.

4.5.6 Determination of the normal air components O_2, N_2 and CO_2

Analysis by gas chromatography is of interest when only small air volumes are available. A suitable procedure for determining these components was developed by Lysyi and Newton [173] as well as by other authors. The separation of these gases requires that two columns be connected in parallel; one of the columns (1.5 m × 6 mm) is packed with Molecular sieve 5 A, 60–80 mesh (0.18–0.25 mm). It is used for the separation of oxygen from nitrogen; carbon dioxide, however, is irreversibly bound in this column. The determination of CO_2 is effected by connecting a second column (60 cm × 6 mm) in parallel; this column is filled with silica gel of 20–200 mesh. Upon leaving the column oxygen and nitrogen form a common peak, while the CO_2-peak is distinctly recognizable. The separation temperature is 26°C and the carrier gas helium. A Gow–Mac pretzel-type thermal conductivity cell is used. The sample volume for the determination of O_2 and N_2 is 1–2 ml.

When the air is simultaneously introduced via a two-way gas-sampling valve into both columns, the common $O_2 + N_2$ peak appears on the silica-gel column after 2 minutes and after 6 minutes the O_2- and N_2-peaks are separately eluted from the molecular sieve. After 26 minutes the CO_2-peak from the silica-gel column can be recognized. For an accurate evaluation it is recommended to use an integrator.

If the CO_2-concentration is to be determined with an accuracy $> \pm 200$ ppm, a larger sample volume (10–100 ml) is required. Thus, a tube cooled with liquid air is mounted in front of the silica-gel column; this serves as a cold trap for freezing out CO_2. First air samples of 1 to 2 ml are carried with a helium stream (165 ml/min) to the molecular sieve, while the silica-gel column remains closed. Then the sample is passed over the cold trap into the silica-gel column; during this process CO_2 condenses. The silica-gel column is then opened, the throughput rate of the helium stream reduced to 77 ml/min, the coolant removed and the tube warmed with hot water. According to the authors, the accuracy with a 100-ml sample is $\pm$ 2 ppm. Carbon dioxide can be determined in a similar way by the method developed by Murray and Doe [186].

Argon appears with oxygen in a common peak. For the separation of both gases see Swinnerton, Linnenbom and Cheek [236], as well as Vizard and Wynne [256].

The determination of argon in the presence of oxygen is treated on page 120.

A clearly defined CO_2-peak can be obtained by the column described on page 176 (propylene carbonate–glutaronitrile on Sterchamol).

For the single-stage separation of argon, oxygen and nitrogen see Obermiller and Freedman [191]. The column (3 m × 2 mm) is filled with Molecular sieve 5 A. Hydrogen (30 ml/min) is used as the carrier gas; the sample volume is 0.25–1 ml. At −40°C (the spiral column is placed in a Dewar flask) the A- and O_2-peaks appear. At 50–60°C the N_2-peak is attained after 8 min.

Bober [22] reported a similar procedure for the determination of CO_2, O_2, N_2

as well as CO and CH_4. The author used the Beckman GCM gas chromatograph (complete instrument), which is designed in such a way that two columns can be separately controlled. The adsorption of CO_2 is effected by means of a column filled with activated carbon (1.2 m × 5 mm), while O_2, N_2 and, if necessary, CO and CH_4 are separated with the aid of a column of the same size but packed with Molecular sieve 5 A. The first column is operated at 90°C, the second at 70°C. The sample volume is 0.2 ml. Hydrogen is used as the carrier gas. The two columns are connected in series. The air sample first passes through the activated-carbon column, where CO_2 is retained, while the remaining gases are separated only upon reaching the molecular-sieve column. O_2 appears after 40 sec, N_2 after 50 sec, CH_4 after 100 sec and CO after 120 sec. When separation is completed, CO_2 is eluted from the first column in order to prevent its penetration into and contamination of the molecular-sieve column.

4.5.7 Analysis of vapors and solvent mixtures by gas chromatography

May [179] used the Perkin–Elmer gas chromatograph 116 with a flame ionization detector for the gas-chromatographic analysis of a low-boiling solvent mixtures of esters, ketones, alcohols, aromatic and aliphatic hydrocarbons in air samples. The air sample withdrawn at the sampling site with the aid of large evacuated glass vessels amounted to only 1 ml.

The separation column (2 m × 5 mm) contains 15% polypropylene glycol applied to Celite as the fixed phase. Helium (85 ml/min) serves as the carrier gas, while for the flame ionization detector hydrogen (20 ml/min) is used. The separation temperature is 100°C; the paper tape is fed at a rate of 1 cm/min.

The apparatus is calibrated for qualitative and quantitative analysis with the corresponding pure substances; these are injected via 10–50 μl Hamilton syringes into evacuated 5-liter flasks, which are then filled with air. When the conditions during analysis and calibration are the same, the peak height can be used for the evaluation. Toluene can be used as a standard.

When all the conditions are satisfied, the retention times and peak heights given in Table 4.5.7 are obtained; these values can be used to evaluate the chromatograms. The values designated by an asterisk are obtained at a separation temperature of 50°C, since at this temperature the separation of the components is more efficient.

When the highest sensitivity is used, traces of 0.01 ppm can be detected.

Enrichment of organic pollutants at the work place (this is of special importance when the sampling takes several hours) was effected by Fraust and Hermann [69a] by using an 8 cm × 5 mm tube filled with 0.7 g activated carbon (0.6–2 mm) prior to chromatographic separation. The pollutants are desorbed from the tube by carbon disulfide. The solution can be directly introduced into the chromatographic column. (See also Grupinsky [86a].)

Table 4.5.7
RETENTION TIMES AND PEAK HEIGHTS OF SOME SOLVENTS
(after May)

Solvent	Retention time (min)	1-cm peak height = ppm
Methyl formate	0.9	61
Acetone	1.6	35
Methyl acetate	1.9 (7.8)*	56
Methanol	2.0 (9.8)*	288
Ethanol	2.5	97
Isopropanol	2.8	57
Methyl ethyl ketone	3.1 (12.4)*	41
Ethyl acetate	3.2 (13.3)*	46
Tetrahydrofurane	3.5	50
1,1,1-Trichloroethane	4.0	84
Benzene	5.1	27
n-Propanol	5.5	95
Diethyl ketone	6.0	41
Dioxane	7.7	126
Isobutanol	8.3	64
Isobutyl acetate	9.2	45
Toluene	10.1	39
n-Butanol	11.4	77
Butyl acetate	12.8	58
Ethylglycol	17.4	250
m-Xylene, p-Xylene, o-Xylene	19.3–25.7	67
Amyl alcohol	26.5	
Turpentine	20.5 and 30.2	
Methylglycol acetate	20.7	

* Separation temperature = 50°C.

4.6 Application of mass spectroscopy in air analysis

The principle of this method lies in the ionization of the gaseous sample by electron impact. The ions originating from the atoms, molecules and their fragments are subjected to a magnetic field. The ions are deflected in accordance with their mass and charge, and are classified. The current intensity indicated by the detector is proportional to the number of particles in the relevant class and is recorded.

The characteristic features of this method are the very low amounts of samples required (e.g., 1 μl gas) and the high specificity, since the particles are individually detected. Even small differences in the mass number (e.g., between N_2, CO and C_2H_4) can be unambiguously recognized by means of instruments with adequate resolution.

Mass spectrometers are becoming more readily available at feasible prices and simpler working procedures are continuously being developed. Nevertheless, this technique is applied only in special cases for air analysis, for example, when only very small sample volumes are available. Thus, Neerman and Bryan [187a] studied tiny air bubbles in a glass melt by crushing the glass sample under high vacuum and subsequently analyzing the air in the mass spectrometer. The reproducibility in 1-μl air samples was $\pm$ 1–2% and in 0.05 μl samples $\pm$ 10%. The detection limit was 0.01 μl for N_2 and 0.001 μl for SO_2, CO_2 and argon.

Respiration mass spectrometers (e.g., those produced by the firm Krupp, Bremen) have recently been developed. These are intended to continuously analyze the respiratory air of living beings (e.g., patients under anesthesia).

Highly simplified mass spectrometers are frequently employed to indicate the presence of helium in air when detecting leaks. The "Ultratest" leakage detector, manufactured by Leybold, Cologne, can detect helium amounts down to 0.05 ppm.

An interesting application is the use of compact (5–10 kg) mass spectrometers in rockets for studying the atmosphere at altitudes of 100–200 km.

As recently as 15 years ago hydrocarbons and similar pollutants, for example, from automobile exhaust, were also studied by mass spectrometers. At present, gas chromatographs are mostly used for this purpose. Mass spectrometry can, however, be combined with gas chromatography for the identification of chromatographic fractions or peaks. (See, for example, Quiram and Biller [207] and Quiram, Metro and Lewis [208].)

4.7 Analysis of air by odor

The simplest way to obtain information on the content of foreign matter in air is via the olfactory organ.

Only substances in the form of gases, vapors or fine mists can be smelled. Besides water, glycol, dimethyl formamide and some gases (H_2, CO, NO, N_2O), the lowest saturated aliphatic hydrocarbons and CO_2, there are very few vaporizable substances which cannot be perceived by their odor. The odors occurring in nature originate usually from mixtures of various substances, for example, the aroma of flowers and fruit and food odors from fats and due to roasting.

The ability to effect a qualitative discrimination and memory for various odorant mixtures can be highly developed by training persons having an extraordinary sense of smell. For example, in the perfume industry connoisseurs show surprising ability. The sensitivity of the human olfactory organ differs widely for different odorants.

Table 4.7
ODOR THRESHOLDS
OF SOME AIR POLLUTANTS

	mg/m³
Acetaldehyde	0.12
Acetone	1100
Acrolein	4.1
Ethyl mercaptan	0.0007
Allyl sulfide	0.0007
Ammonia	37
Amyl acetate	5
Benzene	960
Butyric acid	0.0002
Chlorine	14
Hydrogen cyanide	1
Formaldehyde	24
Synthetic musk (trinitrobutylxylene)	0.00008
Ozone	0.1
Pyridine	0.7
Sulfur dioxide	9
Carbon disulfide	2.6
Hydrogen sulfide	0.3
Vanilline	0.0005

Among humans this may vary by several orders of magnitude. The data on various odorants (odor thresholds, i.e., the concentrations at which the odors are just perceived) thus exhibit large variations. The threshold is either expressed in ppm or in weight units.

Taking this into account, let us consider some odor thresholds (mg/m^3) taken from the tables recently compiled by Kaiser [126a] and Storp [244a] (Table 4.7).

As known, the sense of smell in many animals considerably exceeds that in humans. For example, the silk-moth responds to a few molecules of its species-specific substance secreted from the scent-producing glands. In the dog the perception threshold for butyric acid is 6 orders of magnitude lower than in humans. The qualitative power of discrimination of a dog instructed to follow a scent is incomprehensible.

The human olfactory organ is poorly suited for quantitative determinations. We can distinguish between faint, readily noticeable, strong and very strong odors. It is, however, doubtful whether man can learn to determine quantitative (numerical) odor intensities, more so for the reason that the olfactory organ is subject to fatigue.

Therefore, experimental quantitative experiments are based on determining the lowest threshold value, i.e., the odor concentrations which are just perceptible. In quantitative determinations the tested sample is diluted with odorless air until the odor threshold is reached. The various instruments, called odorimeters or olfactimeters, are used only for air dilutions which can be most efficiently prepared. These are tested after introduction into a test chamber or directly into an inhalator. At least two persons must be present; one prepares the gas mixtures and the other tests it without knowning the mixing ratio. When several persons participate in the test and the average is calculated, very reliable values can be obtained.

An odorimeter of the Gaz de France for the odorization of city gas was designed by Angleraud and Borelli [8]; for other instruments see also the section "Odor Analysis" in Jacobs' book, "The Chemical Analysis of Air Pollutants" (see page 9).

May [180] used a conventional gas mask for "sniffing" air samples contained in a large glass flask; a wide, flexible and odorless plastic tube of 10 cm length was drawn over the filter connection. This tube is connected at its other end to an odorless plastic hose reaching to the bottom of the flask.

A quantitative odor test (determination of the odor threshold) is described in ASTM D 1391-57. The equipment comprises graduated 100-ml syringes, washed with an odorless detergent and a mixing syringe with a connecting piece to the sample syringe. The air sample is introduced into the sample syringe and the presumed sample amount is blown into the empty mixing syringe. The content is diluted with odorless air to 100 ml and allowed to stand at least 15 sec. The syringe is then placed near the nostrils and the content injected for 2 to 3 sec. Then the odor is examined. First the order of magnitude is established and then a finer assessment is carried out. The reproducibility is $\pm$ 50% in the most favorable case.

An odor unit is defined as 1 cu. ft (28 liter) of air at the odor threshold. The odor concentration $C = 100/V_s$ is the number of cubic feet the sample will occupy when diluted to the odor threshold; V_s is the volume (ml) of the original sample present in 100-ml of the diluted sample in which the odor is perceptible.

5

Special Topics

5.1 Dust

5.1.1 General

Air pollutants occurring in the form of dust or smoke constituted the first object of study in air pollution control, since in all cases they cause nuisances which are perceived by our sensory organs. Air pollution of this type is frequently a consequence of human activity. Apart from exceptional cases, natural dust, i.e., soil dusts raised by wind or dust originating from volcanoes are far less significant.

The dust level in normal residential areas is about 0.1 to 0.2 mg/m^3, while in residential districts located in the vicinity of industrial areas this level rarely exceeds 0.5 mg/m^3. At work places dust levels up to 100 mg/m^3 are encountered. The MAK-value for inert dust, i.e., dust without any special toxic effects, is 15 mg/m^3.

When assessing damages or injuries due to dust, the various effects of certain substances are taken into account. Thus, the soot content is the decisive factor in the soiling effect, the quartz level in silicosis, the levels of fluorine, lead, mercury, beryllium and other substances in certain poisonings. We must also take into account substances which are considered as potentially carcinogenic (see page 255) or are carriers of radioactivity.

The most important parameter for nonspecific dust effects and the general assessment of injuries due to dust is the diameter of the dust particle. VDI-Richtlinie 2119 distinguishes between three types of dust: Coarse dust (rapid sedimentation rate), particle diameter $> 10\,\mu$; fine dust (low sedimentation rate), particle diameter 0.5–10 μ; very fine dust (sedimentation rate practically zero), particle diameter $< 0.5\,\mu$.

The injurious effect of coarse dust and the coarser fractions of fine dust leads to material damage of linen, clothing, buildings, contamination of residential areas and eye irritation or injury. Moreover, these dusts can damage agricultural and ornamental plants or lower their value (e.g., vegetables). Further fine dust settles on surfaces owing to thermal diffusion.

5.1.2 General methodology

The determination of the dust level without allowance for the chemical composition of the air is essentially the task of physics. Research in different branches of physics has led to the development of numerous methods of dust analysis.

The wide variety of analytical procedures employed is due to differences in dust quantity and quality, local conditions and demands imposed on the rapidity and informational content of the analysis. Consequently, it is impossible to establish a universal method. Thus, for example, the determination of dust in order to obtain information on silicosis hazards will, in principle, differ from dust determinations in the emissions of a power plant with regard to concentration range, particle size and effective substances.

The methods can be classified as follows.

I. Determination of the total dust quantities.
   1. After separation from air:
      a) weighing (gravimetric methods);
      b) particle counting (microscopic methods);
      c) measurement of other parameters characteristic of dust quantity.
   2. Measurement of a physical parameter directly characterizing the dust level in the air sample. This approach is especially suitable for continuous recording.

II. Determination of the dust quantity which precipitates after a certain time from the atmosphere (dustfall measurement).

5.1.3 Gravimetric methods

The most accurate and reliable determinations of the dust quantity are effected by gravimetric methods. The dust-laden air is aspirated, the dust collected on a weighed filter or other separation device and then weighed. The dust can then be used for chemical analysis or other studies (particle size analysis). When large concentrations of coarse dust in room air are to be determined, weighed paper filters (e.g., Schleicher-Schüll or Whatman) can be used. If this determination is to be followed by an assessment of the combustible substances, a weighed, ignited and subsequently cooled filter can be employed. For small dust amounts, for example, in the free atmosphere, light- and moisture-resistant membrane filters made of cellulose ester fibers and having a soft, velvet-like surface, whereby the pore widths and layer thicknesses can be selected (e.g., membrane filter of the Membranfiltergesellschaft, Göttingen), are suitable. These filters completely retain fine dusts with particle diameters below 1 μ; the air resistance of these filters is low. In gravimetric studies a special filter (diameter 5 cm) with suitable holders can be used for a dust deposition of 6–200 mg. After wetting with immersion oil, the filters become transparent and are thus also suitable for microscopic analyses (determination of particle number, particle size distribution and optical properties) (see Bauer [13]).

The polystyrene microsorbane filters manufactured by the firm Delbag are also suitable (see Winkel [276]). The filters are relatively large and air-permeable; the air throughput rate is 20–50 m^3 air/hr. Small dust amounts can also be determined after dissolving the filter in trichloroethylene.

Frequently, the coarse dust fraction (diameter $> 5\,\mu$) which does not penetrate into the lung is of little interest and only the fine dust particles have to be determined. For this purpose small settling chambers, wet scrubbers or small cyclones can be placed in front of the filtration equipment.

The separate determination of the particle size distribution in separated dust can be effected by the pipette method of Andreasen as well as via air classification with the Gonnel classifier (Wolf [280a]). See also Simecek [231a].

Electric or pneumatic pumps of suitable capacity (see page 37) can be used as aspiration devices. When no power source is available, as, for example, in open-air investigations, the air-intake manifold of an automobile can be employed (see page 38).

In the following we shall describe some of the instruments currently available on the market. These are mainly intended for the determination of higher dust levels at work places in industry and mining.

1. The SFI dust-measuring device, developed by the Silikoseforschungs-Institut (Silicosis Research Institute), Bochum. Manufacturer: Dräger, Lübeck; air throughput rate: 3 m^3/hr. The coarse-dust fraction $> 5\,\mu$ is preliminarily separated. The dust is collected on a membrane filter (diameter 47 mm) and weighed. For more details on this instrument see Landwehr [148].

2. The compact dust-measuring device, developed by the firm Lurgi and manufactured by the firm Ströhlein, Düsseldorf. Air throughput rate up to 10 m^3/hr. Smaller dust quantities (below 1 g/m^3) are collected on a filter paper disk or a membrane filter; large amounts (up to 50 g/m^3) are collected in a dust bag and weighed. The lowest measurable dust amounts are 1–3 mg/m^3.

3. The BAT-I collector of the firm Abas, Krefeld, is provided with a cyclone which serves as the precleaner. The fine dust $< 5\,\mu$ is collected on a membrane filter. The air throughput rate is 15 m^3/hr and the device can be operated electrically or by compressed air. This instrument is described in detail by Breuer [29, 30].

4. The "Porticon C 12", developed by Winkel and Coenen [279], and manufactured by the firm Wertebach, has an air throughput rate of 12 m^3/hr and collects the dust on a microsorbane filter of 480 cm^2 filter area which is inserted in the filter box of a commercial respirator (see also Coenen [46a]).

5. The "new dust collector" of Schmidt [239] with a stack of plates serving as a

preliminary precipitator for particles $> 5\,\mu$. Microsorbane filter (11 cm diameter); the air throughput rate is 12 m^3/hr.

6. The Babcock collector of the Technische Überwachungsverein (Technical Surveillance Association) with an air throughput rate of up to 100 m^3/hr.

7. The Gravikon continuous dust-measuring device of the firm Sartorius; air throughput rate 10–50 m^3/hr.

Among the products manufactured outside Germany we mention the "Hexhlet" instrument produced by Casella, London. It has a preliminary precipitator for particles $> 7\,\mu$. The finer particles are separated either on a Soxhlet cartridge or a filter disk and then weighed. The air throughput rate is 3 m^3/hr.

A comparison of the results obtained with some of the above instruments is given by Landwehr [148].

Gravimetric instruments with electrostatic dust precipitation

In the dust separator 9301 of the firm Sartorius, Göttingen, the dust is electrostatically precipitated from an air stream (3.6 m^3/hr) onto an aluminium plate which can take up dust in amounts up to 60 mg. The plate is weighed prior to and after dust deposition on an analytical balance. Because of the constant weight of the instrument, smaller dust amounts can also be weighed.

In the dust balance designed by Gast [75, 76] and manufactured by the same firm the electrostatic precipitator is directly mounted in a microbalance on which dust concentrations as low as 10 $\mu g/m^3$ can be determined. The instrument is sensitive to shocks.

Wet dust precipitation in the impinger

Wet dust precipitation with alcohol in the impinger (see page 43), which is mainly employed in the USA, can also be gravimetrically evaluated by evaporating the alcohol and weighing the residue. However, the separation of the dust particles $< 2\,\mu$ is not complete.

5.1.4 Methods of dust separation and indirect quantitative determinations

When the amount of dust separated is to be determined by methods other than weighing, photoelectric methods can be used for measuring the degree of blackening caused by dust on a filter. The measurements can be continuously performed by blowing the air sample onto a filter tape passing across the opening of a nozzle. On the other hand, instead of direct results only readings which can be converted into dust weights via a calibration curve, plotted on the basis of gravimetric data, are obtained. When the color intensity or particle size of the dust changes, the calibration curve is no longer valid. This uncertainty is not so serious, when we are more interested in the soiling effect than in the dust amount, for example, in the open air. Continuous recording in the open air together with continuous determination of the wind direction permits localizing smoke and dust immissions.

The TM II device of the Hauptstelle für Staub- und Silikosebekämpfung des Steinkohlenbergbauvereines, Essen (Main Office for the Prevention of Dust and Silicosis of the Coal Mining Association, Essen) is based on this principle. The unit is manufactured by Mollidor and Müller, Rodenkirchen near Cologne (see Müller and Thaer [185], and Friedrichs [70a]). The dust-laden air is continuously aspirated through a nozzle via a membrane filter tape (air throughput rate 5–20 liter/hr, depending on the filter thickness). After 24 hr a 120-mm long and 2-mm wide dust strip is formed. The strip is evaluated in the laboratory by photoelectric scanning in reflected light. The instrument also contains a preliminary separator for dust particles $> 5\ \mu$. The speed at which the filter tape is fed can be adjusted to the dust concentration expected. Moreover, means are provided whereby the gaseous pollutants (e.g., SO_2) can be simultaneously measured by colorimetric determinations.

An instrument based on the same principle is described in ASTM D 1704-61; it is designated as the AISI Automatic Smoke Sampler (Figure 5.1.4a). The rolled-up filter tape is passed at certain prechosen time intervals across the opening through which the sample is introduced. Particles of diameter $> 40\ \mu$ were previously removed. The dust collected in the desired time interval is either weighed or the transmitting power or reflectance of the spot is measured in transmitted or incident

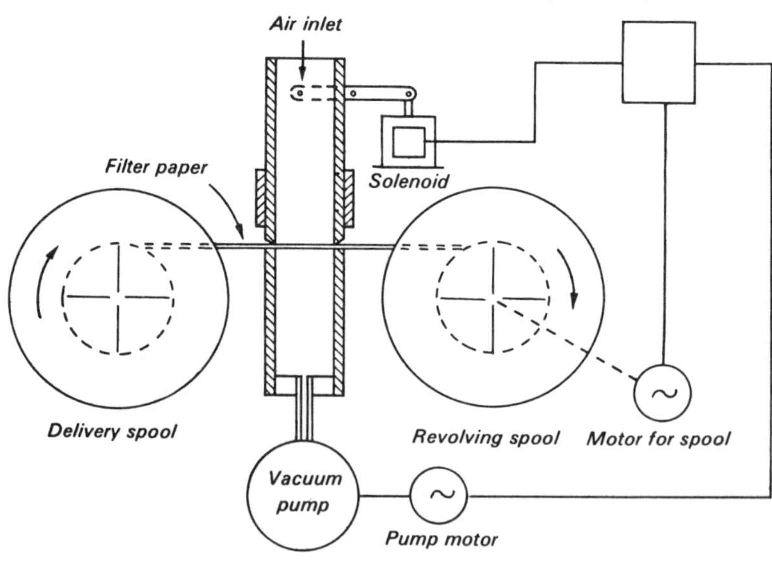

Figure 5.1.4a
AISI Automatic Smoke Sampler

light. Unused paper is used as a reference. By simultaneous weighing and determination of the optical properties the blackening power of the dust or smoke can be determined.

Another instrument, the Kapnograph, collects dust from the continuously aspirated air sample on a filter tape continuously passing the inlet nozzle. The blackening intensity is visually estimated. The lower detection limit is 500 mg/m^3; the air throughput rate is 0.6–1.2 m^3/hr.

The continuous membrane-filter measuring device of the firm Leitz, Wetzlar, operates in analogous fashion; it can be adjusted to running times of 40 min, 8 or 24 hr.

The Bacharach smoke sampler (Bacharach Industrial Instrument Co., Pittsburg) aspirates the soot-laden air through a filter. The blackening obtained is compared with a scale consisting of 10 degrees of blackening (Figure 5.1.4b). This method is officially mentioned in the "Technische Anleitung" (see page 7) and in VDI-Richtlinie 2116.

Beta-ray absorption can also be used for the quantitative assessment of dust which is continuously precipitated on a filter. The "Beta Staubmeter" (Beta dust meter)

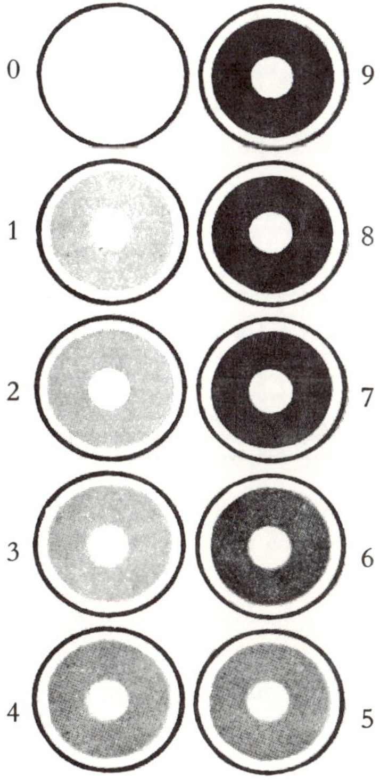

Figure 5.1.4b
Blackening scale after Bacharach

of the firm Verewa, Mühlheim on the Ruhr, and the instrument designed by Aurand and Bosch [10a] and manufactured by Frieseke and Hoepfner, Erlangen, employ this measurement principle. These devices enable the continuous determination of dust immissions in comparatively pure outside air (0.1 mg/m^3) as well as higher dust concentrations in workshops.

A continuous recording device developed by the firm Jouan, Paris, collects the dust on filter paper of uniform permeability. The amount of dust collected is estimated from the reduction in the permeability.

5.1.5 Dust measurements based on physical parameters of the air sample

There are numerous dust-measuring procedures, whereby instead of dust separation, a physical property influenced by the dust level, is directly measured. This mainly applies to optical and electrical measurements.

The soot content of a smoke plume is approximately assessed via its blackening on the basis of the so-called Ringelmann scale. This consists of line gratings of wire mesh with graduated wire thickness, whereby in accordance with the numbers 1–4 on the Ringelmann scale, 20, 40, 60 and 80% of the area (or light) are darkened, respectively (Figure 5.1.5). The wire mesh is placed at such a distance from the observer (about 15 m) that a uniformly gray area is observed and the individual lines of the grating are no longer discernable. The gray shade is compared with the shade of the smoke plume against the white sky (background).

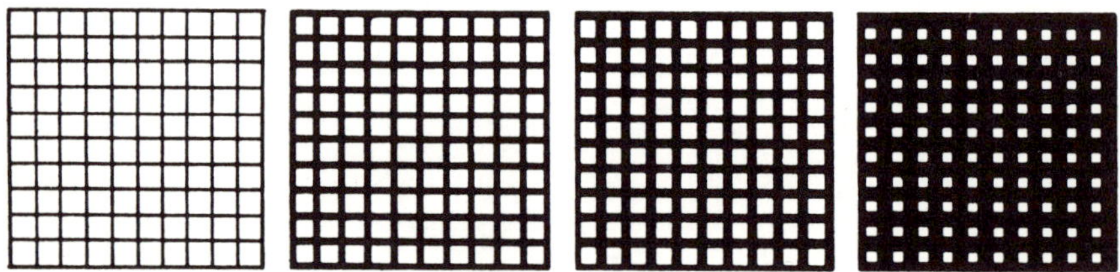

Figure 5.1.5
Ringelmann scale

The Ringelmann scale at one time had widespread application and even today it is occasionally employed (see, e.g., the "Technische Anleitung" TAL). An optical instrument "Telesmoke", based on the same principle, is directed toward the smoke plume. The blackness of the plume is compared with gray areas, corresponding to the Ringelmann scale.

Blackening inside a stack is measured by smoke density instruments manufactured, for example, by AEG, Siemens, Visomat, Sick and other firms.

5.1.5.1 *Apparatus for measuring the Tyndall effect*

The tyndalloscope developed by Leitz, Wetzlar, is based on measuring the light scattered by the dust. The scattered light is received from the illuminated dust chamber at an angle of 30° to the primary light beam and compared as a semicircle in the field of view with a reference semicircle which is split off after the primary light beam passes through a light-attenuation device. An improved version of the tyndalloscope is the tyndallograph of the same firm in which the light scatter is measured photometrically and continuously recorded by means of a double-beam arrangement (see also Stuke and Rzeznik [247]).

A similar device is described in ASTM D 1899-61 T. Here, as well, dust particles whose diameters exceed 40 μ are first separated. The sample is then passed through a smoke chamber in dark-field illumination and a weighed molecular filter. The light scattered by the dust particles falls onto a photocell (secondary electron multiplier), whose photocurrent is amplified and recorded. The instrument can be calibrated with the aid of standard aerosols (di-2-ethylhexyl phthalate in dichlorodifluoromethane). The relationship between the intensity of the scattered light and the weight concentration of the dust in the air sample is found from the dust amount periodically collected on the filter and weighed. Dust concentrations between 1 $\mu g/m^3$ and 100 mg/m^3 can be measured.

5.1.5.2 *Other indirect dust-measuring instruments*

In the "Konitest" designed by Feifel and Prochazka and manufactured by the firm Eckhardt the air sample is led in a turbulent stream through a measuring tube. Due to friction of the dust particles on the inner wall of the tube the latter becomes electrically charged. The current thus produced is amplified and continuously measured. The air throughput rate is 18 to 50 m^3/hr. Amounts between 0.02 and 3 g dust/m^3 can be determined. The instrument is calibrated with dust-laden air of known concentration, particle size and chemical composition (see also Prochazka [206]). A similar instrument is described by Schütz [242].

An instrument of the firm Leeds and Northrup measures via a thermopile the absorption of thermal radiation from an incandescent lamp by the dust particles in the air.

The apparatus of Hasenclever and Siegmann [99], manufactured by the Forschungsanstalt für Strahlenmesstechnik (Research Institute for Radiation Measurement Techniques), Schann (Liechtenstein), consists of an ionization chamber in which a current is generated by weak gamma radiation. The current attenuation due to the dust particles in the air is measured. The lower detection limit is 200 particles/cm^3 (see also Coenen [46]).

In all the indirect methods the conversion of the value obtained to the dust concentration sought is possible only under certain conditions and depends to a considerable degree on the particle size distribution and the substance-specific properties of the dust (color, electrical properties). The correlation between the values obtained and the dust concentration can be determined on the basis of a simultaneous gravimetric analysis. However, these curves apply only insofar as particle size and chemical compositions remain constant. On the other hand, continuous recording is possible by these methods, so that they are especially suitable as relative determinations in monitoring instruments for continuous air pollution control in plants. When certain extreme values are exceeded automatic warning devices are actuated or the control devices put into operation.

For a survey on automatic dust recording techniques see Schütz [242].

5.1.6 Dust determination by particle counting

The principle of this method is the precipitation of the dust from a known (mostly small) air volume on a transparent surface followed by counting of the visible particles under a microscope. Depending on the dust deposition method, we distinguish among four variants.

1. The air sample is aspirated through a filter (membrane filter) on which the dust is collected.

2. The air sample is blown onto an adhesive, transparent surface, to which the dust particles adhere.

3. The particles settle due to thermal diffusion on the cooled parts of the instrument.

4. The particles are separated in a liquid upon aspiration through special impingers, in which the air bubbles impact the glass surface.

The particles are counted by rear or direct viewing in bright or dark fields. Counting by direct viewing can be facilitated by a colored filter. Particles below a certain size cannot be observed under a microscope. For particles $< 1\ \mu$ difficulties arise which can be partially eliminated by applying dark-field illumination. Microscopic examination also gives insight into the particle size distribution, but it must be taken into account that some particles or conglomerates can be mechanically separated during measuring; on the other hand, agglomeration may take place.

The quantity of dust precipitated is usually small, so that the qualitative detection or quantitative determination of certain substances is impossible. However, in some cases, microchemical reactions under the microscope can occur.

The results are expressed in particles per cm^3. To convert these particle densities into concentrations (mg/m^3) decisive for assessing injuries, certain parameters must be known (particle size distribution and densities of the individual substances) for which data is usually sparse. Therefore, as a rule, we must be satisfied with utilizing the particle numbers as relative magnitudes for local comparison of the dust conditions; however, also here, constant particle size distribution and chemical composition are prerequisites.

The following rough approximate data may be used for the conversion:

$$500 \text{ particles/cm}^3 = 2 \text{ mg/m}^3$$
$$2000 \text{ particles/cm}^3 = 10 \text{ mg/m}^3$$
$$20{,}000 \text{ particles/cm}^3 = 100 \text{ mg/m}^3$$

The advantages of the above method are compact and handy apparatus (owing to the small air volumes required), and low time expenditure. On the other hand, in such microdeterminations the problem always arises whether the small air volume is a representative sample. For this reason we must investigate a large number of samples which thus renders the method more laborious. The methods of particle counting are losing in importance and are being replaced by other procedures.

5.1.6.1 *Description of some commercial particle counters*

1. The small Dräger unit. This dust collector consists of a disk-shaped attachment to the handpump for the gas-detection device described on page 56. The attachment contains a membrane filter over which the dust of the air sample (1 stroke of the pump = 100 cm^3 air) is uniformly distributed. The filter is made transparent by wetting in immersion oil. The conventional observations (particle number, particle sizes, crystal forms) can be performed under the microscope.

2. Konimeter. The Konimeter of the firm Sartorius consists of a microscope, in which 2.5–5 cm^3 air is blown via a small piston pump onto the object slide. The revolving object disk, which can hold 36 dust samples, is provided with an adhesive layer to which the dust particles adhere. The microscope (magnification 200 ×) contains a crossline micrometer divided into two sectors of 18° (10% of the total area). Using a sedimentation bell, the nonlung-penetrating dust particles (diameter $> 5\ \mu$) can be separated.

Schedling designed an automatic sample exchanger for the Konimeter [237]. It can be adjusted to the desired time interval and can handle 36 samples.

3. Thermal precipitator (produced, e.g., by Casella, London). The air sample (50–300 cm^3, throughput rate 7 cm^3/min) is drawn through a narrow channel along a centrally mounted heated resistance wire and two laterally cooled thin glass plates (object slides). Due to the thermal diffusion effect the dust particles pass over the central hot zone and settle on the cold glass surfaces. 6 samples can be collected on a single microscope slide. Particles $> 5\ \mu$ may fail to adhere to the collecting slides; on the other hand, fine dust particles are completely precipitated. The particles are not deformed by precipitation and can be examined under the microscope.

4. Impinger. The impingers described on page 42 are mainly employed in the USA. Isopropanol is used as the scrubbing liquid. After drawing through the air sample, the particles contained in an aliquot of the liquid are counted under the microscope. The impinger collects all the larger particles ($> 5\ \mu$) but a considerable percentage of particles $< 2\ \mu$ is not absorbed by the liquid (see Hasenclever [96]).

A survey of results obtained with membrane filters, midget impingers, Konimeters and the Hexhlet instrument is given by Ödelyke [192].

The data obtained by gravimetric measurements and particle counting were compared by Landwehr [148]. The author failed to find any correlation between the results.

5.1.7 Dustfall measurements

The term dustfall is defined in VDI-Richtlinie 2119 as the amount of atmospheric dust precipitated per unit area in a certain time interval ($g/m^2 \cdot$ time).

Though the ability of a dust particle to settle spontaneously on a horizontal surface is mainly determined by its weight and size (surface area), many other phenomena play a part (wind, precipitation, type of terrain, electrical factors). It is thus difficult to draw a sharp distinction between spontaneously and nonspontaneously settling particles.

Current measurement methods do not enable us to distinguish between the atmospheric pollutants which precipitate together with rain and snow and remain in the residue after complete evaporation of the volatile parts, and dustfall.

It is understandable that due to the lack of a clear definition of that part of the dust emission which is to be considered as dustfall, it is impossible to obtain unambiguous dustfall data independent of the external conditions. The values obtained under different conditions and by different methods cannot be strictly compared. Nevertheless, these measurements are still frequently being performed, since they involve comparatively little work and constitute a direct measure of dust damages caused by surface pollution.

5.1.7.1 *Implementation of dustfall measurements*

A collector with known effective surface area is placed in the open at a suitable site and allowed to stand until a dust quantity has accumulated which is sufficient for weighing or other analyses. The instruments are either hollow vessels with a specified cross section in which the dust together with rain or snow is collected and weighed after filtration or evaporation of the water, or, as in the more recent methods, metal foils, which were rendered adhesive by applying a thin layer of vaseline.

No international or national standards exist with regard to the shape of the hollow vessels. Within the German Federal Republic various institutes carrying out such dust measurements have developed their own types; also in the USA and the British Commonwealth various forms are used. Owing to the different aerodynamic factors the values obtained with these instruments, even when referred to the same areas, are not identical, but merely permit a rough comparison in certain cases.

VDI-Richtlinie 2119 describes four such collecting vessels, including the British Standard Gauge, and gives a survey of the advantages and disadvantages of the respective collectors. The ASTM-Method D 1739/62 does not prescribe a specific size for the cylindrical vessel, but the diameter must not be smaller than 6 inches and the height must be between twice and three times the diameter. Moreover, in the American method (in contrast to the German method) the vessel must contain water which is resupplied when the vessel dries up. This creates a uniform effective surface and prevents losses caused by blowing the residue out of the dry dish.

Instruments according to VDI-Richtlinie 2119

The British Standard Gauge (Figure 5.1.7.1a) consists of a trapping funnel made of glass (upper diameter 31 cm, trapping area 760 cm^2) and a 10-liter collecting bottle fitted to the funnel outlet. A wide-mesh wire screen is mounted on top of the device as a protection against birds. The instrument is placed in a 150-cm high tripod.

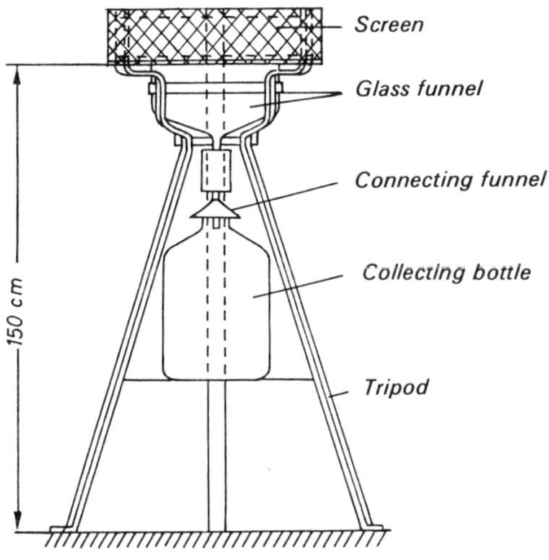

Figure 5.1.7.1a
British Standard Gauge

The instrument (also known as the WABOLU device) developed by Löbner and Liesegang at the Institüt fur Wasser-, Boden- und Lufthygiene (Institute for Water, Soil and Air Hygiene), Berlin, is schematically shown in Figure 5.1.7.1b. The trapping funnel is made of stoneware (diameter 30.5 cm) and has a wide tapered edge. Its trapping area is 730 cm^2. The jacket of the funnel and the 10-liter collecting bottle are installed in a sheet-metal housing. A wire ring serves for protection against birds.

The Hibernia instrument developed by Ost and Mirisch [194b] (Figure 5.1.7.1c) is used by the Technische Überwachungsverein (Technical Surveillance Association). The funnel consists of varnished sheet metal and tapers off somewhat toward the top.

Its diameter at the upper edge is 25.2 cm and the trapping area is 500 cm^2. The funnel and a 10-liter collecting bottle are placed in a 150-cm high tripod. A wire ring serves for protection against birds.

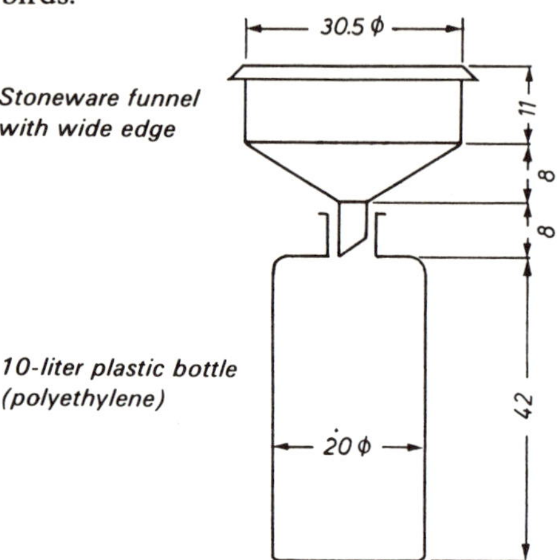

Figure 5.1.7.1b

Dustfall measuring instrument after Löbner and Liesegang (WABOLU-device)

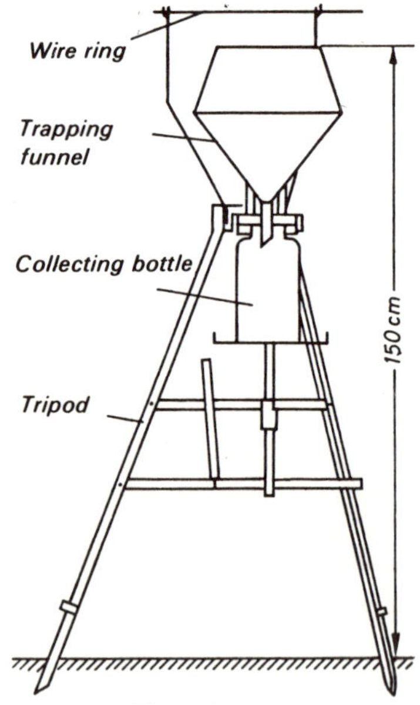

Figure 5.1.7.1c

The Hibernia instrument

In the Bergerhoff instrument of the Landesanstalt für Bodennutzungsschutz, Bochum (Figure 5.1.7.1d), an ordinary household jar for preserves (diameter 89 mm, content 1.5 liter and trapping area 62 cm^2) serves both as the trapping and collecting vessel. A wire screen serves for protection against birds.

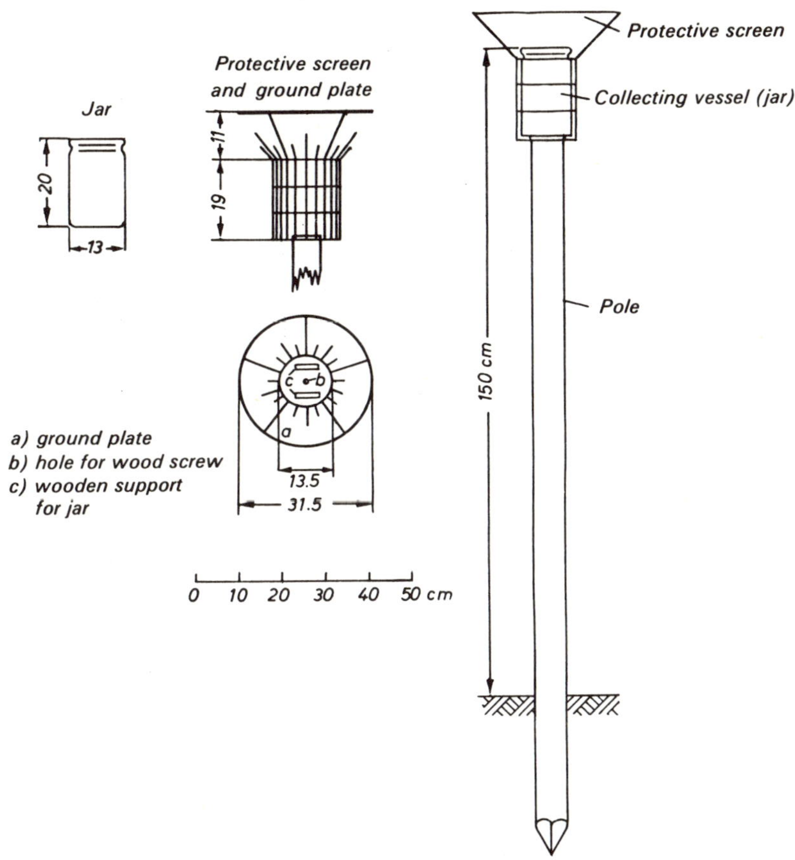

Figure 5.1.1.7d
Bergerhoff measuring vessel (of the Landes-Anstalt)

Among the German-made devices the Bergerhoff instrument is the cheapest and the easiest to handle. Because of its small opening it is especially recommended for higher dust concentrations. In the "Technische Anleitung" it forms the basis for comparison of dust concentrations.

According to Diem [56a] (see also Kirste [131b] and Köhler [135a]) the ratio of the values (referred to equal areas) obtained by the Hibernia, WABOLU and Bergerhoff instruments are about 100:80:110. See also Kettner [131], Schneider and Nagel [241], as well as Köhler and Fleck [136].

Processing

After the sampling period (usually 30 days) the funnels are carefully flushed with some water. The precipitated dust and the water are evaporated to dryness in a weighed dish and the total residue weighed. The results are expressed in $g/m^3 \cdot 30$ days.

The undissolved fractions can be separated by filtration and separately weighed. Moreover, the residue obtained upon ignition and the concentrations of carbon, tar, metals and important anions can be chemically determined.

5.1.7.2 *Diem's adhesive foil instrument*

This device [56a] consists of a 0.07-mm aluminum foil (trapping area 4.0×8.25 cm = = 33 cm^2) and a frame. Pure vaseline (50 mg) is uniformly applied over the foil. The foil is warmed until the adhesive layer melts and then dried at 40°C, cooled in a desiccator at a certain water vapor pressure and weighed. After exposure (the maximum uptake during this period should not exceed 30 mg, i.e., the exposure time is 1–2 weeks) the foils are collected, dried in the desiccator and weighed. If required, the settled dust particles can be morphologically examined under a direct-light microscope.

The method of Diem has the considerable disadvantage that during heavy rain and intense insolation appreciable losses occur, so that the values obtained are frequently much lower than those obtained by methods employing collecting vessels. Even when a roof is mounted over the foil, comparable conditions are not achieved.

While handling of the adhesive foils is easier, the method has another disadvantage, namely, that it must be frequently repeated for long-term experiments. This method is therefore not recommended for determining a reliable and comparable dust level, but rather for testing intermittent local dust conditions at several measuring stations in order to detect dust emittents; here the possibility of a qualitative morphological examination of the dust particles is highly advantageous (see Baum, Hermann and Steinbach [13a], and Baum [13b]).

Kampf and Schmidt [127a] described a method employing transparent, weighed plastic Petri dishes instead of adhesive foils. The precipitated dust can be weighed or photometrically measured.

Installation and evaluation

Dustfall measuring devices should be mounted at a height of 150 cm above the ground. Obstacles (trees, walls of houses, etc.) should be a distance of at least 10 times the difference between their respective heights and the height of the instrument.

Since the results are affected by many sources of random errors which cannot be eliminated, statistical methods (analysis of variance) are used.

Reasonable immission limits for nontoxic dust precipitations are specified in the "Technische Anleitung" on the basis of the following amounts, measured with the Bergerhoff instrument:

Annual mean value

a) general: 0.42 g/m^2 · day (average of the 12 monthly means);

b) in heavily industrialized areas: 0.85 g/m^2 · day (average of the 12 monthly means).

Monthly means

a) general: 0.65 g/m^2 · day;

b) in heavily industrialized areas: 1.3 g/m^2 · day.

In health resorts the dustfall should not exceed 2.5 g/m^2 · 30 day.

5.2 Radioactive substances in the atmosphere

The study of radioactivity and its carriers in the air has gained worldwide importance over the past decade. The measurement of radioactivity is a special branch of physics requiring intricate instrumentation which is beyond the scope of this book, especially since this problem directly concerns only a few chemists working in air analysis. On the other hand, the accumulation and concentration of the activity carriers have much in common with other methods of air analysis.

The radioactivity of the air, especially from the hygienic standpoint, is currently of widespread interest. This is mainly due to the fact that owing to nuclear weapons tests the radioactivity in the atmosphere may sometimes exceed by many times the natural radioactivity level and cause serious concern, even at large distances from the site of the blast. However, in addition to the nuclear blasts the ever increasing number of plants for peaceful uses of atomic energy are becoming a source of danger. Consequently, on behalf of the population the state considers itself obliged to ensure the control of radioactive air pollution at numerous monitoring stations.

Tasks and aims of measuring atmospheric radioactivity

Whereas the analytical determination of ordinary air pollutants is restricted to sites in the vicinity of emission sources (e.g., power plants, factories) or to places, where, because of numerous small, private emission sources (automobiles, domestic stoves), an undesirable accumulation of toxicants is to be expected, the situation is quite different in the case of radioactive pollutants. Here, only a limited (at least for the time being) number of emission sources must be taken into account. These sources are nuclear blasts (tests with nuclear bombs) and reactor accidents which may lead to hazardous air pollutions. However, it is known that the radioactive pollutants do not remain in the environment of the emission sources; they frequently reach the stratosphere due to the tremendous force of the explosion and subsequently disperse over the entire earth with the result that after a short time increased radioactivity can be detected at any point on the terrestrial surface. Consequently, due to the inherent dangers of radioactive fallout monitoring stations are set up all over the world.

In addition, there are smaller radiation sources whose emissions only partly reach the atmosphere and which are thus of local importance only, for example, in the protection of work places. These sources include uranium mines, production and regeneration of nuclear fuel elements and, to a smaller extent, plants producing or processing phosphors and finally, because of the possibility of accidents, all places where radioactive isotopes are used in medicine and engineering.

The hygienist, who is concerned with the health of the population, is not only interested in the radioactive radiation and its intensity; he must obtain information from the monitoring station on the composition (alpha-, beta- or gamma-radiation), duration (half-life) and especially the chemical nature of the carriers (radionuclides). The individual radionuclides widely differ due to their chemical properties, according to retention time and location in the organism or its environment (nourishment), i.e., those factors which determine the biological activity of the organism. The beta-emitter Sr 90 (half-life 28 years) is very dangerous, since it is deposited in the bones (affinity for calcium), where it is strongly retained. Certain radionuclides, such as I 131, are rapidly taken up by pasture animals and enter after a short time into their milk; these radionuclides accumulate in the human thyroid gland.

As long as the atmospheric radioactivity is low and does not appreciably exceed the natural levels, the hygienist need not resort to detailed data. When the tolerance limits are considerably exceeded, analytical information must be made available in short order, so that the necessary protective measures can be taken. For this reason the monitoring stations must be provided with the necessary equipment and personnel.

First we must consider whether the main problem is the intermediate detection of strong radioactive immissions and the activation of alarm systems, as may be necessary in stations located in the vicinity of nuclear reactors. In this case continuous measuring equipment is used which can produce signals without delay. When the pollution source is located at larger distances (nuclear blasts, remote reactors), intermittent monitoring equipment will suffice, whereby the radiation emitters are accumulated, for example, over intervals of 24 hr. A comprehensive determination of a number of mutually occurring radioactive substances is a different analytical task, which may take several days, even when the most accurate equipment is used.

Radioactive pollutants are not determined directly by measuring the radioactivity of the air, since the radiation is, as a rule, too weak. Therefore, substances acting as carriers of radioactivity must be used. There are only a few gaseous carriers (the inert gases radon and thoron from the decay series of uranium and thorium, respectively).

These carriers are responsible for the natural radioactivity of the air which is very low and which, because of the short half-life of these radionuclides, decays after a few days.

The carriers of atmospheric radioactivity occur mainly in the form of dusts deposited on aerosols of sizes ranging from 0.02 to 1 μ. The dust (and thus the radioactivity) is best measured by aspiration through filters. For continuous operation roll-tape filters can be used. The air is then drawn through a slit pressed onto the filter tape, which is fed at an appropriate rate. For noncontinuous operation circular filters (e.g., 6 cm diameter) are inserted into a frame, exposed to the open air and connected to an efficient suction pump.

The filter material must also retain the finest particles and must be air-permeable in order to achieve the air throughput rate required for a large dust yield. Glass fiber filters (e.g., Schleicher and Schüll No. 8) are recommended. If the filter has to be ashed after measurement, synthetic fibers can be used.

Sometimes adhesive foils of about 1 m^2 surface area which collect the spontaneously settling dust are used instead of filters. The foils are ashed prior to the radiometric determination.

Radiometry

The radioactivity is measured in several time intervals so as to distinguish between the rapidly decaying natural activity and the artificial radioactivity. The dust sample is measured both immediately and after 2 days. For this purpose automatic recorders are equipped with a feed device for the filter tape which brings the dust sample after a prespecified waiting time under the counter.

Widely differing commercial counter tubes are used, depending on the type of radiation to be measured. Proportional counters have universal application and are thus very suitable for air analysis; the pulses differ according to whether alpha or beta rays are emitted, so that the alpha and beta activity can be separately determined. If a characteristic curve is plotted for such an instrument, i.e., a preparation with known alpha and beta activity is used and the counts per minute are determined with increasing counter voltage, we must take into account that the alpha activity is stronger at a much lower voltage *a* than the beta activity. The beta activity is manifested only at a higher voltage *b*. Thus the pulse rate of the working voltage *a* can be attributed to the alpha rays while the pulse rate at a higher voltage *b* is associated with the sum of the alpha and beta rays. Thus, we can readily establish where alpha-emitters (e.g., the very dangerous Pu 239) or beta-emitters (e.g., Sr 90) predominate.

Gamma rays are preferably measured with scintillation counters. These convert the individual radiation pulses into light energy, which in turn is converted in photocells into electrical current pulses; the energy of the pulses (pulse height) and the electric potential are proportional to one another. When the output of the secondary electron multiplier is linearly amplified so that voltages of about 100 V can be measured, the various voltage ranges can be scanned by suitable electronic devices (pulse-rate discriminators), which determine the number of pulses within narrow voltage ranges. In this manner we obtain a curve as a function of the pulse frequency and voltage which represents the gamma spectrum of the sample. The presence of certain gamma-emitters can be established on the basis of their characteristic maxima.

Scanning of the whole voltage range with a single band ("channel") takes some time; instruments with several "channels" can be designed in which all the relevant ranges can be simultaneously scanned (multi-channel spectrometer).

Autoradiography of "hot" particles

The radioactivity is usually distributed nonuniformly over the dust surface. If a light-sensitive photographic material is placed on the dust sample and exposed for a sufficiently long time, the resulting "autoradiogram" displays individual points of higher radioactivity which are attributed to the so-called "hot particles."

Chemical separation of radionuclides

When the sample contains a mixture of a number of radionuclides and the individual fractions of the respective elements have to be separated, a radiometric analysis must be carried out. This method comprises two stages: the chemical separation of the mixture into individual substances and the subsequent radiometry of the individual particles. Since the individual radionuclides are present in very low amounts (about 10^{-12}–10^{-14} g) and the solubility product of their deposits is not reached, a sufficient amount of the corresponding inactive nuclides is introduced into the sample (e.g., a knov n amount of Sr or Ba salts). When the element is isolated after homogeneous mixing as a uniform compound (isotopic mixture), its radiation is measured and compared with the results obtained by measuring an isotopic mixture of known concentration of the corresponding radionuclide. This enables us to calculate the level of radioactivity in the sample from the comparative measurements, the weight of the isolated compound and the weight of the carrier material.

As a rule, not only the nuclide to be determined is introduced in the inactive state but also the corresponding inactive compounds of the other nuclides, since otherwise the radiation spectrum would be distorted.

The intricacy of the chemical–analytical methods used depends on the radioactive elements present or to be determined; here gravimetric methods (which are often outdated) are sometimes used.

Thus it is often sufficient to effect a coarse Sr 90 enrichment by oxalate precipitation and measure the β-activity; however, for a more complete separation strontium is precipitated as nitrate with fuming nitric acid, several times reprecipitated and separated from lead, barium and sometimes radium as chromate in acetic acid solutions.

When several radionuclides have to be determined, very complicated separation procedures must be applied; since frequently mixtures of elements, with which the chemist is not familiar (e.g., rare earths, platinum metals) have to be separated.

Separation procedures were developed with ion exchangers whose separating effect was enhanced by EDTA or other, more recent complexing agents; the reader is here referred to the relevant treatments in the literature.

Processing of rain water

The radioactive aerosols present in the atmosphere act under certain climatic conditions as condensation nuclei for water vapor just as other dust particles and reach the Earth's surface via rain or snow. Thus, instead of a "fallout" we are dealing with a "rainout". In order to obtain more information on the resulting radioactive contamination rain water must also be processed and analyzed.

The water is collected, for example, in large PVC troughs. To avoid losses due to adsorption the respective inactive carrier materials are added (e.g., 20 mg, Sr, Cs and Ce salts, each in solutions). Ion exchangers or coprecipitation can be used for enrichment of the radioactive substances after evaporation.

The enriched carrier can be analyzed by the methods described for determining the dust fraction. The alpha and beta rays are determined in the proportional counter, the gamma rays in the scintillation counter. For the identification of gamma rays the gamma spectrometer is also used. The determination of the half-lives can, in simple cases, indicate certain radioactive carriers.

When necessary, the various methods of gravimetry or ion exchange can be used in radiochemical analysis.

5.3 The inert gases

Natural air contains the inert gases helium, neon, argon, krypton, xenon and radon as the carrier of natural radioactivity. The concentrations are given below.

Inert	He	Ne	A	Kr	Xe
Atomic weight	4.003	20.18	39.95	83.80	131.30
Concentration in ppm	5	16	9330	1	0.008

Radon is formed from radium by alpha decay and itself decays with the emission of alpha rays. Its half-life is 3.8 days.

The inert gases are used for many purposes in industry, especially in light engineering and for electronic circuit elements. Argon serves as protective gas in autogenic welding. Helium is now being used, among other things, for leakage detection (see page 97) and as carrier gas in gas chromatography.

From the standpoint of air hygiene the inert gases (with the exception of radon) are of no importance. The analysis of these gases is of significance with regard to their purity, but beyond the scope of this book. Earlier methods, for example, fractional absorption and emission spectral analysis are described by Kahle and Karlik [126].

Peters [200] describes the isolation and determination of inert gases in the residual nitrogen from a normal gas analysis. Nitrogen is bound to calcium metal at 400°C. The residue consisting of inert gases is used for the determination of the individual components by fractional absorption or desorption.

In recent years gas chromatography has found special application in the determination of inert gases (see, e.g., Greene [81]). A survey can be found in the paper of Jeffery and Kipping [120].

Owing to its relatively high level in the atmosphere argon must sometimes be determined in the presence of nitrogen and oxygen. For this purpose Molecular sieve 5 A is used. The separation of argon from oxygen involves some difficulties. Argon can be separated from oxygen at low temperatures (−72°C); the nitrogen is retained by the column (Lard and Horn [149]). Jones and Halford [123] removed oxygen by combustion with hydrogen in a palladium tube over activated carbon. Both methods were applied in the Österreichische Stickstoffwerke, Linz, with good results for 1-ml samples.

5.4 Oxygen-containing air constituents

5.4.1 Oxygen

General

A determination of the oxygen level is very rarely required since this level is more or less constant in the atmosphere. However, in production premises, where large apparatus operate under a protective gas atmosphere (mostly N_2), there is sometimes the risk of oxygen impoverishment due to gas escaping via leaks. In such cases continuous air control can prevent accidents.

The oxygen concentration in workshops should not drop appreciably below the normal value of 20.9%. An oxygen level of 15% already produces harmful effects, while below 10% it is highly dangerous.

The oxygen level in fruit storehouses is sometimes artificially decreased to 4–10% in order to prevent spoiling.

In certain cases it is necessary to determine the oxygen level in the air, when processes are studied which involve a decrease or increase in the normal oxygen level (20.9 vol. %). The present section will deal only with methods having a sufficient accuracy in this concentration range (± 1–2 rel. %). Methods for determining low oxygen concentrations in gases (e.g., induced by entry of air) or high oxygen levels will be only briefly discussed.

For the determination of oxygen in the concentration range of about 20 vol. %

four groups of methods can be used (see also Fresenius and Jander, Handbuch der Analytischen Chemie (Handbook of Analytical Chemistry) Vol. VI aα).

1. Absorption methods with volumetric difference measurement.
2. Titration methods.
3. Physical-instrumental methods.
4. Gas chromatography.

The classical absorption methods of Hempel or Orsat do not require complicated instruments and readily yield values with an accuracy of ± 2 rel. %.

The titration method of Winkler is slightly more involved and time-consuming but is more accurate (± 0.5 rel. %).

The instrumental methods yield the best results as far as reproducibility rapidity and correctness are concerned. These techniques are particularly suitable for continuous measurement. The initial costs are of course higher, but the consumption of reagents and the costs in personnel lower.

Gas-chromatographic methods are important when only small samples are available.

5.4.1.1 *Absorption methods for oxygen determination*

Here it is assumed that the reader is acquainted with absorptiometric gas analyses, such as, for example, the methods of Hempel and Bunte or in the Orsat apparatus (see, e.g., Bayer and Wagner [14]).

A 20% sodium sulfate solution acidified with sulfuric acid is frequently recommended as the liquid seal. The exact gas analysis after Hempel uses mercury for this purpose.

Absorption pipettes with four bulbs are employed to prevent entry of air.

The following absorbing agents are used for oxygen: Alkaline pyrogallol solution: 1 part (by weight) of pyrogallol is dissolved in 6.5 parts (by weight) of 25% KOH. Before use the solution is stored for 1–2 days. Fresh and aged solutions generate small amounts of carbon dioxide. Oxygen is rather slowly absorbed.

Alkaline sodium dithionite solution: 50 g $Na_2S_2O_4$ are dissolved in 250 ml water and briefly shaken with 40 ml 40% KOH.

Yellow phosphorus rods (3–5 cm thick) are used in the commercial phosphorus pipette. The rods are kept under water in the pipette and in the storage flask. Oxygen is absorbed at room temperature within 2–4 minutes. The results are frequently lower by a few tenths of a percent.

Chromic salt solutions are the most efficient but also the most air-sensitive absorption agents. 100 g $CrCl_3 \cdot 6H_2O$ are dissolved in 260 ml water and 40 ml HCl are added. The solution is then shaken with liquid zinc amalgam (3–4% Zn) until the green color changes to blue. The solution has to be stored airtight over amalgam. Further, it is possible to reduce a chromic sulfate solution to chromous sulfate in the Jones reductor.

5.4.1.2 *Volumetric determination of oxygen*

Oxygen can be determined by iodometry (Winkler) or oxidimetry (Leithe).

For the determination of low oxygen or air quantities, for example, in inert scavenging gases with which the air sample is flushed, the iodometric procedure of Leithe is recommended.

5.4.1.2.1 Iodometric determination after Winkler (standard procedure at the Central Laboratory of the Österreichische Stickstoffwerke)

The air sample is introduced at atmospheric pressure into a dry 100-ml gas burette of known total volume. Using a small, hand-operated rubber blower, the following solutions are successively introduced into the funnel: 1 ml foaming agent, 5 ml Solution II, 3 times 5 ml water and 5 ml Solution I. The burette is then shaken 15 min in the longitudinal direction and allowed to stand for 15 min with intermittent shaking. Then 5 ml sulfuric acid (1 part (by volume) concentrated acid + 1 part (by volume) water) are added, the burette is shaken, cooled in cold water and the content flushed with 150 ml water into a titration flask and titrated with 0.1 *N* $Na_2S_2O_3$ in the usual manner.

To determine the blank (oxygen content in the reagents) the gas burette is filled with oxygen-free nitrogen, the reagents are added and the burette is shaken for a short time. The reaction product is titrated as above.

1 ml 0.1 *N* $Na_2S_2O_3$ (sample minus blank) = 0.5 mg O_2 or 0.56 ml (STP).

Reagents: Solution I: 40 g $MnCl_2 \cdot 4H_2O$ made up with water to 100 ml. Solution II: 50 g NaOH are dissolved in 50 ml water and 30 g KI dissolved in 50 ml water. The solutions are combined.

Foaming agent: 1 g Nekal or dodecylbenzenesulfonate or a similar wetting and foaming agent in 100 ml water.

5.4.1.2.2 Oxidimetric determination

According to the procedure of Leithe [151] 70–80 ml of the air sample are introduced into a 100-ml gas burette with allowance for the temperature and pressure. The liquid seal is withdrawn and 20 ml 0.2 *N* $FeSO_4$ (56 g $FeSO_4 \cdot 7H_2O$ are mixed with a few ml of H_2SO_4 and made up with water to 1000 ml) are pipetted through the funnel. Then 2 ml of a 30% $CaCl_2$ solution are added to the funnel (without entry of air), followed by the introduction of about 2 ml of a 1% aqueous solution of foaming agent, resistant against $KMnO_4$ (Nekal, dodecylbenzenesulfonate) and 3–4 ml 25% KOH. The burette is vigorously shaken for 10 min in the longitudinal direction with intense foam formation. The now acidic and oxidation-resistant solution is rinsed from the burette into a titration flask. Then 5 ml 60% H_3PO_4 and 5 ml 10% $MnSO_4$ are added. The mixture is titrated with 0.1 *N* $KMnO_4$ to a pink color which persists for a few seconds.

Taking into account the oxygen content of the reagents, the solution is standardized by introducing ferrous sulfate, $CaCl_2$ and foaming agent into the titration flask. An oxygen-free scavenging gas is blown over the mixture for a short time, 3–4 ml 25% KOH added, the flask agitated a few seconds, the content is acidified with H_2SO_4, and phosphoric acid and $MnSO_4$ solution are added. The solution is titrated with 0.1 *N* $KMnO_4$. The calculations are carried out on the basis of the formula

$$\text{Vol. \% } O_2 = \frac{\text{ml } 0.1\ N\ KMnO_4 \text{ (blank} - \text{experimental result)}}{\text{ml air sample } (0°, 760 \text{ mm Hg})} \times 56.$$

5.4.1.2.3 Iodometric determination of small quantities of oxygen in gases

Leithe [152] proposed the following method for the determination of small amounts of O_2 (exceeding 10 ppm) in gases or small air quantities in scavenging gases.

The oxygen-containing sample is introduced under slight excess pressure into a dry, approximately 1-m long, 500-ml gas burette of known total volume. The burette is then adjusted to atmospheric pressure by letting out excess air via a funnel containing a little water. 1 ml of 1% foaming solution (see page 127) and 18 ml Solution I are blown through the funnel via a holed rubber stopper with connecting hose. The funnel is rinsed, the washing fluid is removed and 6 ml of Solution II are squeezed in. The burette is shaken vigorously for 5 min in the longitudinal direction (about 2–3 jolts per sec). Then 5 ml 50% H_2SO_4 are introduced into the burette, the mixture is agitated, the liquid is transferred into a small titration flask and rinsed twice with a little water. Then the solution is titrated with 0.01 *N* $Na_2S_2O_3$ to a weak yellow color which, after adding 1 ml 1% soluble starch, vanishes.

The blank is determined according to the procedure mentioned on page 123:

$$\text{Vol. \% } O_2 = \frac{\text{ml } 0.01\ N\ Na_2S_2O_3 \text{ (experiment} - \text{blank)}}{\text{ml gas volume } (0°, 760 \text{ mm Hg})} \times 5.6.$$

Solution I: 35 g NaOH, 115 g Seignette salt and 3.5 g KI are dissolved and made up to 1 liter.

Solution II: 100 g crystalline manganous chloride are dissolved in 1 liter of distilled water and acidified with a few drops of concentrated HCl.

In order to increase the sensitivity of the method for determining a few ppm of O_2 or corresponding air quantities the scavenging gases can be subjected to colorimetry with o-tolidine instead of iodometric titration (Leithe and Hofer, to be published). The determination is similar to that of elemental chlorine (see page 212). In this case the air contained in the reagents in the gas burette must be removed by evacuation prior to applying the gas sample. The calibration curve is plotted for very dilute $KMnO_4$ but can also be obtained with metered-out amounts of air-saturated water.

Working procedure

The 500-ml gas burette must be provided with hermetic stopcocks. In order to displace air the burette is filled with an oxygen-free gas (e.g., nitrogen at a throughput rate of

125 ml/min over the "RCH-O_2-Absorbens" of the firm Ruhrchemie AG in a spiral gas scrubber). The burette is evacuated, intermittently and at the end of this process. Then 15 ml distilled water, 0.1 ml 40% NaOH and 1 ml 1% aqueous Nekal solutions are introduced into the burette and the air contained in the reagents is removed after a 10-min evacuation process. Then 1 ml distilled water and 0.1 ml 40% $MnSO_4$ solution (previously freed to a large extent from oxygen by passing through nitrogen) are introduced. The sample is then finally allowed to flow in. The burette is vigorously shaken for 15 min in the longitudinal direction and 0.3 ml of the acid o-tolidine is added. The yellow, oxidation-resistant solution is filtered into a 100-ml volumetric flask made up to the mark; the extinction is measured at 430 mμ in a 10-mm cell.

Reagents: 40% $MnSO_4$ and 40% NaOH in water; saturated o-tolidine solution in 80% H_2SO_4; 1% Nekal solution or another oxidation-resistant wetting and foaming agent (e.g., dodecylbenzenesulfonate) in water; $KMnO_4$ standard solution: 17.80 ml of 0.1 *N* $KMnO_4$ are diluted to 1000 ml in a 1-liter volumetric flask. 1 ml of this solution is equivalent to 10 μl O_2.

In order to plot the calibration curve for the range of 0–10 μl O_2 (corresponding to 0–20 ppm O_2 in 500 ml gas sample), 75 ml water and 0.3 ml o-tolidine solution are introduced into 100-ml volumetric flasks. Then increasing amounts (0–1.00 ml) of $KMnO_4$ standard solution are added and the volume made up to the mark. The extinction is measured at 430 mμ in 10-ml cells with water as the reference solution. The blank is determined just as in the analysis of the gas sample but in vacuum after repeated rinsing with oxygen-free nitrogen.

The "reciprocal slope" of the calibration curve (μl O_2 in 10-mm cells, 100 ml solution, for E = 1) is 28. This value is dependent on the age of the tolidine solutions.

5.4.1.4 *Instrumental methods for oxygen determination in air*

Because of its physical and chemical properties oxygen in the air can be readily determined by automatic instruments.

5.4.1.4.1 Methods based on paramagnetism

The most conspicuous physical property distinguishing oxygen from all other gases is its paramagnetism. This property is an ideal basis for a specific and very accurate determination of oxygen. Owing to its paramagnetism the oxygen molecules are attracted into the magnetic field. As an example, we mention "Magnos 2" of the firm Hartmann and Braun. The mode of operation of this device is shown in Figure 5.4.1.4.1.

The air sample is introduced into an electrically heated cross tube of an annular chamber, half of which is exposed to the field of a strong magnet. The oxygen molecules are attracted due to their paramagnetism to the region of higher field strength. This leads to an air flow (magnetic wind) which in turn partially cools the heating coil. The difference between the electric resistances of both parts of the heating coil constitutes a measure of the oxygen concentration.

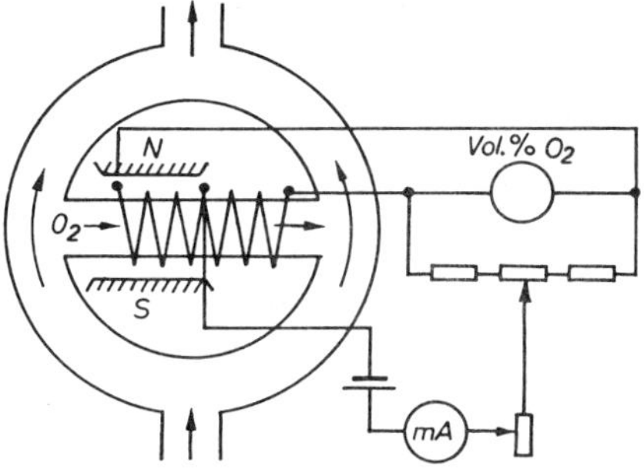

Figure 5.4.1.4.1
Magnetic oxygen analyzer

In the "Magnos 2" a "suppressed" measuring range between 20 and 21 vol. % is available over the whole scale for the determination of the oxygen content under normal conditions. This makes it possible to very accurately detect small deviations from the normal value (about ± 2% full-scale deflection).

The "Magnos 5" manufactured by the same firm is based on the same principle. Similar instruments based on this principle are produced by other firms (e.g., Beckman).

5.4.1.4.2 Determination of oxygen on the basis of depolarization phenomena (polarometry)

The depolarization of galvanic cells by oxygen was originally utilized for the determination of oxygen dissolved in water; it can, however, be also applied for assessing the oxygen content in air and other gases. The air is passed through the electrolyte of a galvanic cell. The method is especially suitable for determining very low oxygen concentrations, but can also be used for higher O_2-levels, for example, in air. These instruments are less accurate but cheaper and more rugged than those operating on the principle of paramagnetism.

Measurement principle

A galvanic cell, consisting, for example, of a zinc electrode, carbon electrode and suitable electrolyte, will yield a constant current in an oxygen-free atmosphere for a very short time only. The current is very rapidly attenuated by polarization, especially at the carbon electrode. However, when an oxygen-containing gas stream is introduced in the immediate vicinity of the carbon electrode, the latter is depolarized and the current is amplified in proportion to the O_2-level. This amplification can be very accurately measured. An instrument of this type is the Auer-TSM 60 for O_2-levels between 0 and 25 vol. %. The measurement accuracy is ± 0.2–0.5 vol. % O_2.

According to a similar polarometric principle (e.g., in the Oxygen-Sensor of the firm Beckman) a polarization voltage of −0.8 V is applied across the electrode system (Au as cathode, Ag as anode). The oxygen diffusing into the electrolyte through a thin teflon membrane produces the cathode reaction

$$O_2 + 4H_2O + 4e^- \rightarrow 4OH^-,$$

while the reaction proceeding at the anode is

$$4Ag + 4Cl^- \rightarrow 4AgCl + 4e^-.$$

The strength of the current generated by this reaction is proportional to the amount of oxygen introduced. In the measuring range from 0 to 25% the accuracy is also 0.2–0.5 vol. % O_2.

5.4.2 Ozone

5.4.2.1 *General survey*

The presence of ozone in the atmosphere is interesting for various reasons. The compound is formed in the air layers at altitudes of 20–40 km. Because of its absorbing capacity for the short-wave ultraviolet radiation, ozone screens the Earth's surface from these highly active rays. The ozone concentration at 20 km altitude is as high as 0.2 ppm, while on the Earth's surface it occurs in the natural atmosphere in concentrations between 1 and 30 ppb, depending on the weather conditions (insolation) and height above sea level. In city air the ozone content is even lower due to the presence of oxidizable pollutants.

On the other hand, in polluted atmospheres the peroxides and ozone levels can be as high as 0.6 ppm. These compounds are formed by the reaction of olefins with oxygen and nitrogen oxides during intense insolation; oxides and peroxides are the most important and active constituents of the smogs occurring in Los Angeles and San Francisco. In Europe, where reduction conditions usually prevail during smog, ozone does not occur in such amounts.

However, ozone can be frequently found at work places. It is generated in various technical processes: irradiation with UV-lamps and X-ray tubes, electrical discharges, in Cottrel dust separation, during electrolytical processes at the anode and upon charging storage batteries. Further, certain pure chemical processes yield ozone as a secondary product; it is formed during the slow oxidation of wet, white phosphorus, upon decomposition of inorganic or organic peroxides and when elemental fluorine reacts with water.

In addition, because of the widespread utilization of ozonized air as an oxidizing and bleaching agent, especially for the purification of drinking water, in refrigerating rooms, for storage of foodstuffs and fruit (about 1 to 3 ppm in the air of the storehouse) and occasionally for disinfection and deodorization of rooms, hazard conditions at the work places may occur. The risk of large-scale plant accidents is reduced, because ozone—in contrast to chlorine—cannot be transported in pure form. It is therefore produced on the spot and strongly diluted with air in the amount required at the place where it is to be used.

Toxicology

In contrast to the favorable effects which are attributed to "ozone-rich forest air" in prospectuses, ozone is actually a toxic and dangerous irritant even when strongly diluted. Levels of 1–2 ppm irritate the mucous membranes and disturb the central nervous system which can lead to bronchitis and headache. The odor threshold is 0.02 ppm. The MAK-value is 0.1 ppm; the MIK-value has not yet been specified.

Damage to vegetation by ozone has been known to occur in California.

A characteristic damaging effect due to ozone is embrittlement and cracking of rubber. This is sometimes utilized for analytical purposes.

Physical properties

Boiling point −112°C; solubility in water (vol./vol. at a partial pressure of 760 mm Hg) at 20°C:0.32.

Chemical properties

Ozone is one of the strongest oxidants known. The standard potential is 2.07 V (in acid solution). The formation of aldehydes when ozone reacts with olefins is specific; this is both utilized in preparative and analytical chemistry.

In higher concentrations ozone oxidizes SO_2 to SO_3 or H_2SO_4 and NO_2 to N_2O_5 or HNO_3. However, in the low concentrations occurring in the atmosphere these reactions proceed slowly.

5.4.2.2 *General analytical guidelines*

The analysis of ozone involves several difficulties. First, due to its harmful effect (MAK-value 0.1 ppm) and the low concentration at which it can be perceived (0.02 ppm) very sensitive methods are required which permit detecting levels as low as 0.01 ppm. When separating ozone from a large sample, it may react with the simultaneously concentrated oxidizable air components (SO_2, NO_2), with which it could coexist in the atmosphere. Moreover, we must take into account that the ozone level in the air depends on the time of the day and weather (insolation), so that the measurement intervals should not exceed half an hour.

Another difficulty arises due to the demands imposed upon the specificity of the determination. There are numerous methods for detecting and determining ozone, but most of these are oxidation reactions which are also characteristic of other oxidants present in polluted air (especially peroxides and NO_2). In air analysis ozone and peroxide are often collectively referred to as oxidants. However, this complicates damage assessment when the proportions of the constituents are not known.

Further, dosing of ozone is complicated both for testing new methods or plotting calibration curves. Ozone–air mixtures of known concentration can neither be prepared from pure substances nor on the basis of quantitatively proceeding conversion reactions. The composition of an empirical air mixture, obtained, for example, by irradiation of the air by ultraviolet rays, can be determined only via a parallel determination with the aid of known methods (e.g., iodometry). Such ozone–air mixtures should not be stored in glass vessels for long periods; Mylar-foil bags are reportedly better for this purpose (see page 36). It is best to work with a flow system, whereby air is sampled both for the standard determination and the calibration curve (see page 51).

Survey of the analytical methods

The great importance attributed to ozone as an air pollutant, especially on the west coast of the United States, has recently led to vigorous research in the development of its analytical chemistry; in the USA special emphasis is being laid upon the design of continuous automatic ozone analyzers.

Among the physical methods spectrophotometric techniques in the characteristic ultraviolet absorption range (250–280 mμ) are used for low concentrations. With the sun as the radiation source the ozone level in the higher air layers can be determined. Such measurements are also possible on the Earth's surface, when the optical paths are defined and monochromatic light sources are used. A special instrument has been developed for this purpose. For higher ozone levels only measurements in the IR range can be used.

Wet-chemical methods are based on the determination of free halogens released from their salts by ozone. The KI reaction is most frequently used in various pH ranges, whereby techniques have been developed which eliminate the interference of NO_2 and SO_2. The iodine released can be either titrated or determined by colorimetry or electrochemically (potentiometry, dead-stop titration, polarometry, coulometry). Likewise, the bromine released by KBr after its reaction with ozone can be electrometrically (amperometry, galvanometry) assessed.

There are also numerous colorimetric and spectrophotometric methods. The blue color formed by diphenylaminosulfonate is common for all strong oxidants but it is not disturbed by the presence of NO_2. The continuous automatic determination in the Imcometer (see page 70) employs the color reaction with N-tetramethyl-p-phenylenediamine.

Certain ozone-specific colorimetric methods are based on the formation of aldehydes after splitting of the olefins via their ozonides. Two methods have been recently proposed: one yields anisic aldehyde as the end product, the other pyridyl aldehyde. Sensitive color reactions exist for the above two aldehydes.

For data on the earlier methods for the determination of ozone the reader is referred to Volume VI a α of the Handbuch der analytischen Chemie (Handbook of Analytical Chemistry), edited by Fresenius–Jander.

5.4.2.3 *Direct spectrophotometric methods*

The ozone concentration in the higher air layers was determined by V. H. Regener [211] and E. Regener and V. H. Regener [212] by means of sounding balloons equipped with an automatically recording quartz spectrograph for measuring the solar spectrum in the absorption range of ozone.

According to Renzetti [213], the ozone level in the near-ground air layers can be determined by UV-absorption measurements. The experimental arrangement comprises a spectrophotometer and a mercury UV lamp with selectively filtered wavelengths (265 mμ corresponding to the absorption maximum for ozone at 254 mμ and 313 mμ as the zero value, where oxygen does not exhibit absorption) placed at a distance of 100 m from the spectrophotometer. The extinction is 0.0125 per 100 m optical path and 10 ppb ozone content.

A commercial UV-ozone photometer is produced by the firm Harold Kruger Instrument Co., San Gabriel, California, USA.

According to Hanst, Stephens, Scott and Doerr [95] ozone can be determined in the IR region at 9.48 μ, i.e., the peak of a specific ozone band with optical path lengths between 72 and 156 m. The extinction coefficient per ppm and meter is 3.74×10^{-4}; it is thus much lower than the corresponding UV absorption value (1.25×10^{-2}).

5.4.2.4 *Ozone determination by iodometry*

The classic iodometric determination after Schönbein which is based on the reaction

$$2KI + O_3 + H_2O = I_2 + O_2 + 2KOH$$

has been subject to numerous modifications (see Fresenius and Jander). Ozone can be absorbed both in neutral (buffered) and alkaline KI solutions. In order to prevent iodine losses from the neutral solution when ozone is present in higher amounts, the iodine released can be bound by adding known amounts of $Na_2S_2O_3$ solutions.

Procedure of Byers *and* Saltzman

Byers and Saltzman [37] (see also Hunold and Pietrulla [113]) absorb air with ozone levels between 0.06 and 0.6 mg/m^3 at flow rate of 1 liter/min in scrubbers containing 1% KI in phosphate buffer (0.1 *M* KH_2PO_4 and 0.1 *M* Na_2HPO_4). After a throughput of 150 liter air the solution is rinsed in a titration flask. H_2SO_4 is added and the released iodine is titrated with 0.005 *N* $Na_2S_2O_3$ against starch. Ozone-free air, prepared by passing the flow over Hopcalite or activated carbon, is used for the determination of the blank.

This method is appreciably disturbed by peroxides as well as NO_2 and SO_2. To eliminate the influence of NO_2 McQuain [176a] used instead of sulfuric acid phosphoric acid saturated with amidosulfonic acid. The mixture is tempered at 22°C and the extinction of iodine is measured at 352 mμ. See also Altshuller, Schwab and Bare [4].

Galster [74] used pH-values $\geqq 8$ in the iodometric titrations to eliminate the interference by NO_2. He added 0.1 g $NaHCO_3$ per 20 ml 0.1 *N* KI and carried out a potentiometric titration in the platinum–calomel cell.

Instead of the visual titration methods the dead-stop titration according to Paneth and Glückauf [196] can be used.

The released iodine can be amperometrically determined also (Ehmert [61]). Here, the current, which is generated at a potential difference of 180 mV between two platinum electrodes and which is proportional to the iodine concentration, is measured.

Schulze [242a] designed an instrument, operating on the same amperometric principle, for the simultaneous determination of ozone and SO_2, which may coexist in polluted atmospheres.

The "Ozone Monitor" manufactured by the firm Beckman is based on the coulometric determination of the iodine continuously released by ozone in an air stream (Figure 5.4.2.4).

Figure 5.4.2.4
"Ozone Monitor" of the firm Beckman

The microcoulometer developed by the firm Mast (Davenport) is frequently used in the USA. The air sample (170 ml/min) is brought into contact with a liquid film (1.25 ml/hr) consisting of a solution of KI, KBr and a buffer. Two electrodes are immersed in the film. The iodine released by ozone generates a current (measured by a microammeter) proportional to the amount of ozone. The instrument has three measuring ranges (0–0.2 ppm, 0–2 ppm, 0–10 ppm). The sensitivity of the instrument is 0.01 ppm.

Cherniak and Bryan [40a] reported on comparative ozone determinations by the Mast apparatus—a colorimetric instrument which measures the extinction of iodine—and an arrangement based on assessing the cracking of rubber.

Colorimetric procedure with iodide in accordance with ASTM D 1609-60

Principle: The procedure is based on the reaction of ozone with iodide. The interference of SO_2 and NO_2 is eliminated. The substance is absorbed in an alkaline KI solution. The sulfite absorbed is oxidized with a small amount of hydrogen peroxide without decomposing more KI. After evaporating the excess peroxide, acetic acid is added which eliminates the interference of NO_2. The freed iodine is spectrophotometrically determined at 352 mμ.

Working procedure

The air sample is passed at a rate of 5 liter/min through a Midget Impinger containing a solution of 1 g KI and 0.4 g NaOH in 100 ml water. After adding 2 drops of 3% H_2O_2 and complete evaporation of the excess, acetic acid is added (1 vol. glacial acetic acid + 5 vol. water) until a pH of 3.8 is reached. The solution is made up to 25 ml and the extinction at 352 mμ is determined after 2 min. The iodine level can be also determined by conventional titration with $Na_2S_2O_3$.

The calibration curve for the spectrophotometric determination is plotted using the reagents after adding standard KIO_3 solution to the absorption solution and acidifying with acetic acid. 1 μg I_2 corresponds to 0.19 μg ozone.

This procedure is applicable for ozone amounts between 1 and 16 μg.

Automatic colorimetry with iodine

Littman and Benoliel [168] describe an older instrument for the continuous colorimetry of the iodine released. The air sample (5 liter/min) is passed through a 60-cm glass tube, filled with borosilicate single-turn glass helixes, against a counter-flow of KI solution (5 ml/min), buffered to pH 7. The iodine released is continuously measured photometrically at 355 mμ. The consumed solution is continuously regenerated by passing over an activated carbon bed. 1 ppm ozone corresponds to a 90% scale deflection of the recorder.

5.4.2.5 *Ozone determination by bromometry*

The reaction

$$2KBr + O_3 + H_2O = Br_2 + O_2 + 2KOH$$

can also be used for the determination of ozone. This reaction has the advantage that it is less oxygen-sensitive than the iodometric reaction. It is mainly used for continuous electrometric procedures.

Lübke and Damaschke [51] describe a continuous amperometric method, whereby the potential difference between a Hg–Hg_2SO_4 electrode and a platinum electrode (413 mV) yields a current proportional to the bromine content. A reading of 5 μA corresponds to 0.1 ppm O_3. The current can be amplified for recording. The absorption solution consists of 0.25 *N* KBr in 3 *N* H_2SO_4. The gas throughput rate is 20 liter/hr. Ozone concentrations between 10 and 500 ppb can be continuously measured. The instrument is calibrated with a solution containing 0.5 μg $KBrO_3$ per ml.

Hersch and Deuringer [106a] also used the reaction of ozone with KBr yielding free bromine. Bromine enters a galvanic cell consisting of a platinum grid (anode) and activated carbon (cathode). Bromine oxidizes the activated carbon via the reactions

$$O_3 + 2Br^- \rightarrow O_2 + O^{-2} + Br_2$$

$$2e^- + Br_2 \rightarrow 2Br^-$$

$$\ldots\ldots C + O^{-2} \rightarrow \ldots\ldots CO + 2e^-.$$

When both electrodes are connected via a galvanometer, a current is indicated (Faraday's law); a measuring range of 0–200 ppm can be achieved.

The cell is schematically shown in Figure 5.4.2.5. The electrolyte consists of a 3 *M* KBr, 0.001 *M* NaI, 0.1 *M* NaH_2PO_4 and 0.1 *M* Na_2HPO_4 solution.

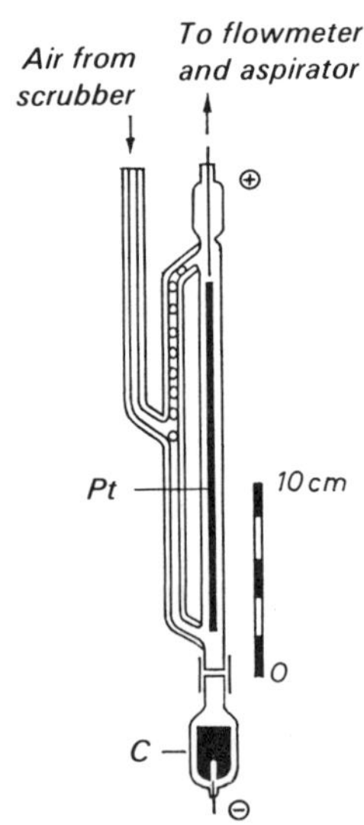

Figure 5.4.2.5
Galvanic cell after Hersch and Deuringer

The air sample (200 ml/min) must be humidified before entering the cell to prevent evaporation of the electrolyte. For this purpose a small wash bottle, containing 0.05 *M* H_2SO_4 and 0.15 *M* $KMnO_4$ (to remove sulfur dioxide), is inserted before the cell. NO_2, 1 ppm of which can simulate 30 ppb O_3, is removed by scrubbing with $KMnO_4$ in a deep-wash bottle with a sintered-glass sparger (to increase the retention time), reaching about 30 cm below the level of the permanganate solution.

When the O_3-level in the air is normal, the activated carbon cathode is renewed only after several months of continuous operation. The electrolyte solution is not used up.

When O_3-levels exceeding 150 ppm have to be analyzed, it is most expedient to use a gas distributor, which introduces only an aliquot of the sample into the cell.

Since the electrolytic reaction obeys Faraday's law, calibration of the cell with air of known O_3-level is no longer necessary.

The elimination of SO_2 and NO_2 by absorption in chromate-sulfuric acid applied to silica gel is described by Saltzman and Wartburg [220].

5.4.2.6 *Colorimetric methods*

Method with diphenylaminesulfonate

The redox indicator diphenylaminesulfonate yields a turquoise color with ozone, a violet color with gaseous chlorine and peroxides and a yellowish-green color with NO_2. This method surpasses the iodometric methods in two respects: the redox potential is better adjusted to ozone (0.85 V instead of 0.59 V for the I_2–KI_2-reaction) and the colors produced by the interfering substances differ.

Bovee and Robinson [26] report the following procedure: the air sample is passed at a rate of 2.8 liter/min (1 cu ft/min) through the Midget Impinger, containing 10 ml of the absorption reagent, until a distinct color appears (10 min). The absorption reagent is a solution of 1% sodium diphenylaminesulfonate in 0.02% $HClO_4$. In the event of losses due to evaporation the volume is made up to 10 ml. The photometric determination is carried out at 593 mμ in a 75-mm cell.

The calibration curve is obtained from ozonized air produced by irradiation by a mercury-vapor UV lamp. The O_3-level of this air is determined with neutral KI solution. In the 75-mm cell 0.3 ppm ozone in 28 liter air (i.e., 18 g O_3) yields an extinction of 0.48. The molar extinction coefficient is thus 1700. The detection limit is about $\pm$ 0.2 μg ozone; 400 μg NO_2 simulate 20 μg O_3.

Method with N-tetramethyldiaminodiphenylmethane

According to Arnold and Mentzel [9], this base reacts in dilute acetic media with several strong oxidants. However, as in the case of phenylaminesulfonate, the colors produced differ; with ozone a violet, with chlorine a blue and with NO_2 a yellow color is produced, while no color appears with H_2O_2. The authors applied the reagent on paper for the qualitative detection of ozone. According to Fuhrmann [71] (see also Galster [74]), however, this reagent is also suitable in citric acid solution for the continuous colorimetric determination at 565 mμ in the Imcometer (see page 70). The molar extinction coefficient is about 40,000.

Colorimetric methods based on the formation of aldehydes from ozonides

The formation of aldehydes via olefin ozonides can be also used for the specific determination of small ozone concentrations.

Method of Bravo *and* Lodge *via anisaldehyde*

The procedure of Bravo and Lodge [27] is based on the cleavage of 4,4′-dimethoxystilbene $CH_3O \cdot C_6H_4 \cdot CH{=}CH \cdot C_6H_4 \cdot O \cdot CH_3$ by ozone to form anisaldehyde, which together with fluoranthene and trifluoroacetic acid produces an intense blue.

Working procedure

In order to effect a microdetermination the authors use 3 ml of a solution of 5 mg 4,4′-dimethoxystilbene in 100 ml sym-tetrachloroethane in Ehmert scrubbers (see page 45). The air sample is passed for 15–60 min at a rate of 0.1–0.15 liter/min. An aliquot of the solution (1 ml) is mixed with 1 ml fluoranthene solution (firm Fluka, 5% chloroform) and 0.8 ml trifluoroacetic anhydride in a 10-ml volumetric flask. The mixture is allowed to stand for 5 min and then made up to the mark with CF_3COOH. The photometric determination is carried out at 610 mμ, using the reagents without the sample as the reference solution.

Trifluoroacetic acid is poisonous and strongly corrodes the metal parts of the spectrophotometer when the cells are not hermetically closed.

The calibration curves are plotted by weighing anisaldehyde and measuring the color intensity in accordance with the above prescription (the authors indicated a 98% yield, assuming that 1 mole of ozone yields only 1 mole of aldehyde in the reaction with stilbene). The molar extinction coefficient is 35,000, the detection limit in 8 liter air is 10 ppb.

The following substances do not interfere, provided they occur in amounts below 1 ppm: NO_2, peroxyacetyl nitrate, methylhydroperoxide; SO_2 does not interfere even at slightly higher concentrations.

Method of Hauser *and* Bradley *via pyridine-4-aldehyde*

Principle: In order to avoid the difficulties involved in the use of CF_3COOH (see above) Hauser and Bradley [100] employed 1,2-di(4-pyridyl)ethylene in glacial acetic acid. The pyridine-4-aldehyde formed is determined with the aldehyde reagent 3-methyl-2-benzothiazolinone hydrazone, described on page 232. In this case, however, it was more appropriate to work without the addition of $FeCl_3$. Instead of the blue color an intense yellow reaction product is formed and the extinction measured at 442 mμ.

Working procedure

Sampling: The air sample is drawn for 30 min at a rate of 0.5 liter/min through a fritted scrubber containing 15 ml absorption solution (an impinger yielded a much poorer absorption). The fritted scrubber is connected to a small wash bottle containing water to protect the gas meter and the pump against acetic acid vapors. The resulting amount of absorption solution is weighed or 15 ml of it are pipetted into a test tube. 1 ml of the color reagent is added, the mixture is heated for 20 min in a boiling water bath, cooled and the extinction is measured at 442 mμ.

Reagents: 0.5 g of the absorption solution 1,2-di(4-pyridyl)ethylene (Fluka) are dissolved in 100 ml glacial acetic acid. The pyridine-4-aldehyde required for calibration can also be obtained from the same firm.

Color reagent: 0.2 g 3-methyl-2-benzothiazolinone hydrazone hydrochloride (British Drug House) is dissolved in 100 ml water.

Calibration curve: Taking into account the yield established by the authors, i.e., 2.75 μg pyridine-4-aldehyde per μg ozone (61% of the theoretical yield under the assumption that 2 moles of pyridine-4-aldehyde are formed per mole ozone and dipyridylethylene), the calibration curve can be obtained with measured amounts of pyridine-4-aldehyde in glacial acetic acid. In a 15-liter sample the limit concentration is about 20 ppb.

Other colorimetric determinations for ozone (oxidants)

The reader is here referred to the relevant literature on other colorimetric procedures. With these methods not only ozone but also other oxidizing pollutants (peroxides) are determined.

Dorta-Schäppi and Treadwell [58] determined low ozone levels by investigating the time-dependence of bleaching of highly dilute indigosulfonic acid solutions.

Egerow [59] measured the fluorescence in fluorescein regenerated by oxidation of fluorescin (fluorescin is the colorless reduction product of fluorescein). Haagen-Smit and Brunelle [91] used the analogous reaction with phenolphthalein (see also Bender and Breidenbach [19]).

The red color produced due to the formation of ferric thiocyanate in the reaction of ferrous thiocyanate with oxidants was discussed by Todd [252].

For a comparison of seven methods for determining ozones + peroxides (KI, phenolphthalein, diphenylaminesulfonate, fluorescin, dimethoxystilbene, NO/NO_2 and rubber cracking) see Hendricks and Larsen [102a].

5.4.3 Air humidity

The water vapor (or humidity) content of the atmosphere is—next to the oxygen level—the most important factor for life as far as air is concerned. It does not merely bear significance with regard to the meteorological conditions for well-being and health but for other vital phenomena as well. In this connection we mention the reduced keeping qualities of foodstuffs and many other materials (metals) at high humidities and the effect of humidity on the properties of textiles, paper, etc. Air, in which the water vapor is removed by drying, is frequently applied in engineering. Accordingly, humidity measurements are very often carried out in air analysis.

The ability of dry air to take up water vapor is a function of temperature. At the dew point the air is saturated with water vapor; undercooling leads to separation of water, while heating makes additional uptake of water possible.

The water vapor content of air is either expressed in terms of the partial pressure of water vapor (in mm Hg) or as mg H_2O in 1 liter humid air (absolute humidity) (see Table 5.4.3). The relative humidity, i.e., the percentage of humidity contained in the air with respect to the maximum water uptake at the given temperature (shown in Table 5.4.3), is also important. Numerous instruments yield results directly in terms of relative humidity.

Table 5.4.3

WATER CONTENT AND WATER VAPOR PRESSURE IN WATER-SATURATED AIR BETWEEN 0 AND 40°C

(relative air humidity = 100%)

Temperature	mg H_2O/liter	mm Hg
0	4.8	4.6
5	6.8	6.5
8	8.3	8.0
10	9.4	9.2
15	12.8	12.8
17	14.5	14.5
20	17.3	17.5
22	19.4	19.8
25	23.0	23.8
27	25.8	26.7
30	30.3	31.8
32	33.7	35.7
36	42.0	44.6
38	46.0	49.7
40	51.0	55.3

5.4.3.1 *General methodology of humidity determinations*

Methods for determining the air humidity were developed in almost all branches of physical and chemical analysis. When selecting a method for a given task, we must make allowance for interfering pollutants (dust, substances creating mists) which are always present. Furthermore, the method selected depends on the humidity range to be expected and the sampling procedure (individual or continuous sampling).

When drawing individual samples into a glass vessel, we must bear in mind that the glass surfaces firmly retain a weighable water film.

The individual samples can be determined by gravimetric methods (absorption in a weighed amount of drying agent). Among the volumetric methods we will describe alkalimetry with magnesium nitride and the Karl–Fischer titration method. Small air samples are analyzed by gas chromatography. Dräger tubes are used for orientative determinations.

Continuous recorders are most frequently used in the analysis. The most widespread instrument is the simple hair hygrometer. Instruments measuring IR absorption, conductivity, heats of reaction and coulometry are being used for more accurate determinations. Among the earlier instruments, temperature-drop psychrometers, lithium chloride sensors and dew-point instruments will be described.

5.4.3.2 *Gravimetric determination*

When passing the air sample through a weighed tube filled with drying agent, the amount of water vapor taken up will depend on the efficiency of the drying agent. The residual water after passing the agent is given (in mg H_2O/liter air) in the table below. These numbers apply only for a sufficient holding time.

Drying agent	NaOH 97%	$CaCl_2$ anhydr.	Silica gel	$Mg(ClO_4)_2$	P_2O_5
mg H_2O/liter air	0.8	0.36	0.006	0.002	0.0002

The air throughput rate should not exceed 50–100 ml/hr per ml drying agent. Air used as a flushing gas must be predried with the drying agent used in the determination.

5.4.3.3 *Volumetric determination with magnesium nitride*

According to Roth and Schulz [215], humid air frees ammonia when passed over magnesium nitride:

$$Mg_3N_2 + 6H_2O = 2NH_3 + 3Mg(OH)_2 .$$

The ammonia is trapped in 0.05 *N* HCl and titrated. 1 ml 0.05 *N* HCl is equivalent to 2.7 mg H_2O.

Magnesium nitride (grain size 0.2 mm) is mixed with glass wool and placed into a U-tube. The tube is first flushed for 30 min in order to remove traces of NH_3. Then the air sample is introduced at a throughput rate of 3–5 liter/hr.

Karl Fischer titration

Principle: The method can also be used for determining the humidity of air. It is based on the reaction of water with iodine and sulfur dioxide in the presence of pyridine and methanol which probably proceeds as follows: $C_5H_5N \cdot I_2 + C_5H_5N \cdot SO_2 + C_5H_5N + CH_3OH + H_2O = 2C_5H_5N \cdot HI + [C_5H_5NH]SO_4 \cdot CH_3$. Hence, the end point of the reaction is reached when the brown color of iodine vanishes.

The titration liquids (iodine solution in methanol, solutions of SO_2 and pyridine in methanol) can be bought. For prescriptions of their preparation see, for example, Medicus–Poethke [181]. The titer of the solutions (water equivalent) is determined daily by introducing known amounts of water into anhydrous methanol or using standard substances containing water of crystallization (e.g., oxalic acid).

The simplest way to perform the titration is to introduce the air sample into an absolutely water-free gas-collecting tube. Then a few milliliter anhydrous glycol or methanol are added, the buret with the iodine solution is connected to the gas-collecting tube and the solution is titrated until a permanent brownish color is reached. A small excess pressure of a dry gas, e.g., through a bellows via a P_2O_5-tube, can be applied to the buret.

Determination in the Dräger tube

The Drägerwerke manufacture a test tube for the determination of small water amounts in air and inert gases. It is filled with a colloidal solution of selenium in concentrated sulfuric acid. The solution, which is originally yellow, turns red as the particle size increases due to uptake of water. With ten strokes of the pump 1–40 mg H_2O/liter can be determined. When a 50 liter sample is used, 0.1–0.5 mg/liter can be determined in half an hour (air throughput rate 100 $\pm$ 10 liter/hr). The calibration curve (see Communication No. 27 concerning the Dräger gas-detection instrument) is given in Figure 5.4.3.3.

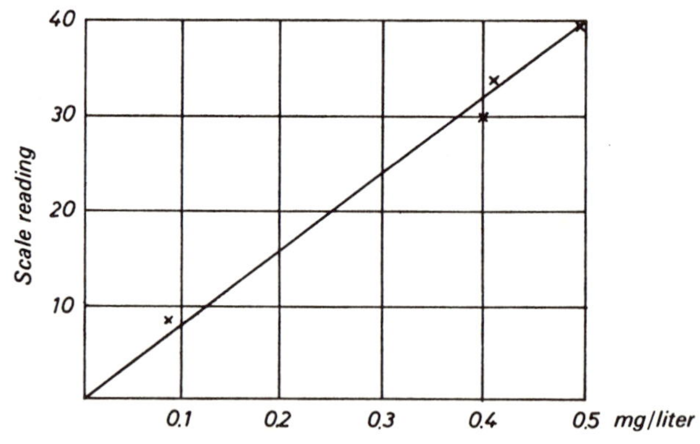

Figure 5.4.3.3

Calibration curve for the determination of water vapor in pressure gas; gas sample 50 liter, gas throughput rate 100 liter/hr.

Ordinate: scale readings on the tube. Abscissa: mg water vapor per liter expanded gas.

5.4.3.4 *Determination of water in air by gas chromatography*

In order to achieve a symmetric water peak without strong tailing, the nature of the fixed phase (in particular, the carrier substance, which must be sufficiently water-repellent) is of special importance.

Aubeau, Champeix and Reiss [10] obtained symmetric peaks by using 1-m columns (inside diameter 3 mm) filled with 20% polyethylene glycol 1,500 on 12 g Fluoropak (teflon powder for chromatography). Helium (100 ml/min) at 90°C was used as the carrier gas. For water contents below 100 ppm the water should be concentrated in a 50-cm absorption column at 0°C and desorbed at 100°C. Burke, Williams and Plank [35a] employed a similar method. See also Kirkland [131a].

Effective gas-chromatographic separation of water vapor from air, CO_2 and gaseous hydrocarbons can be achieved with solid modified polystyrene carriers (Poropak of the firm Varian-Aerograph) (see also Hollis [109]). Castello and Munari [39a] used the electron-capture detector in their analysis.

Methods have also been reported, whereby the water vapor is first converted into acetylene (reaction with calcium carbide), which is then determined by the techniques on page 87. (See also Knight and Weiss [134].)

5.4.3.5 *Other instrumental methods*

The hair hygrometer

The operating principle of the hair hygrometer is based on the changes in length of cleaned animal hair; the length increases by approximately 2% with an increase in the relative humidity from 0–100%. The swelling processes take a considerable time, so that the indication may be retarded by as much as 10 min at average temperatures or much longer at lower temperatures and humidities below 20%. Heating above 60°C leads to irreversible changes in the hair substance.

Nevertheless, owing to their simplicity and low cost, hair hygrometers are still frequently used, especially at small meteorological stations. The maximum obtainable accuracy with proper maintenance is ± 3% of the relative humidity.

Determination by infrared absorption

The URAS device described on page 66 can be also used for humidity determinations, especially in redried air. The detector cell contains a mixture of argon and water vapor. At the most sensitive adjustment the measuring range is 0–0.25 g H_2O/m^3, which gives a reproducibility of the readings of about ± 0.5 mg/m^3.

Humidity measurements based on electrical conductivity

The humidity of the air taken up by hygroscopic substances can affect their electrical conductivity. This effect is used in the Picoflux (see page 73) for the determination of the water vapor content of the air. The measuring liquid consists of strong sulfuric acid (94–99%) H_2SO_4). At this concentration the uptake of water effects the strongest increase in the electrical conductivity, as illustrated by the curve shown in Figure 5.4.3.5. Since the measuring process involves a considerable enrichment of the water vapor from a large air volume, a high sensitivity can be achieved. The most sensitive measuring range is 0–20 mg H_2O/m^3 and the least sensitive 0–20 g H_2O/m^3. The air, of course, should not contain other pollutants which influence its electrical conductivity.

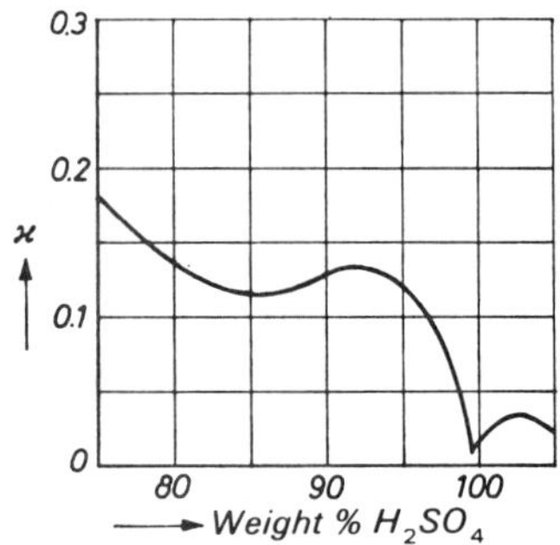

Figure 5.4.3.5

Electrical conductivity as a function of the H_2SO_4 concentration

The electrical conductivity of semiconductors can also be affected by the humidity of the air. Instruments based on this principle are being manufactured by the firm SINA, Zürich, under the name Equi-Hygro-Scope. The temperature-dependence of the conductivity is compensated by a differential connection with a reference cell of constant air humidity. The measuring range covers relative humidity from 2–100% at temperatures between −10 and + 60°C, the reproducibility is ± 0.5%, the correctness (accuracy) of calibration ± 2% and the response time 5–50 sec.

Coulometric instruments

Principle: The air humidity taken up by phosphorus pentoxide or highly concentrated phosphoric acid can be decomposed by dc electrolysis. The requisite quantity of electricity is 192,800 coulomb per mole H_2O (Faraday's law). Thus, the current required for adjusting and maintaining a certain electrical conductivity of the P_2O_5-cell at a constant level constitutes a measure of the water amount in a constant air flow entering the cell.

The design of the P_2O_5-cell is similar to that of the LiCl-sensor (see below) and is described by Keidel [130].

Instruments of this type are being manufactured by Consolidated Electrodynamics, Pasadena, and Beckman. Water vapor concentrations between 0 and 20,000 ppm corresponding to dew point between −76 and + 20°C can be determined. Organic vapors and gases, insoluble in phosphoric acid, do not affect the results. The response time (63% value) is given as 30 sec, while the most sensitive measuring range is from 0–10 ppm, since here, as well, the enrichment factor is high.

Hygrometers based on thermal effects

The heat generated upon introducing wet air into concentrated sulfuric acid can be measured by the Thermoflux instrument of the firm Hartmann and Braun and converted into the water content of the air. The apparatus is especially suitable for the continuous recording of this quantity. Concentrated sulfuric acid continuously flows (12 ml/hr) via a tempering screw into the reaction space, where it is mixed with a tempered air stream (300 ml/min). The flow rates of the acid and air sample must be accurately adjusted. The heat generated (1 g H_2O vapor generates 1.0 kcal in 96% H_2SO_4) is measured by a thermopile and automatically recorded. The most sensitive measuring range is 0–0.25 g H_2O/m^3; other ranges comprise 0–2.5 g/m^3 and 0–20 g/m^3.

An instrument developed by Mine Safety Appliances, USA, measures the heat of adsorption on an hygroscopic surface. The air sample is divided into two equivalent flows. A chamber filled with a drying agent is mounted along the respective flow paths. One of the air flows is completely dried before it enters its chamber. Under such conditions the undried air stream releases its water in the chamber with liberation of heat. After 2 minutes the air paths are switched. The dried air flow now enters the humid warm chamber and cools the latter by removing the air amount and heat absorbed in the previous phase. Thus, in each cycle one chamber is heated, the other cooled. When a powerful battery consisting of a large number of series-connected thermocouples is installed in each chamber, the thermoelectric voltages can serve as a measure for the heat quantities respectively generated and absorbed in each phase or for the humidities causing these effects. The air throughput rate is 9 liter/min; the measuring ranges are 0–50 ppm or 0–5000 ppm H_2O. The reproducibility is 1% of the full-scale deflection.

Lithium chloride humidity sensor

This device consists of a metal tube over which a glass wool fabric impregnated with LiCl solution is drawn. A heating wire, to which an ac voltage is applied, is wound around the tube. The air sample passes along the outer wall. A thermometer is inserted inside the metal tube.

The hygroscopic lithium chloride absorbs water from humid air, thus increasing the electrical conductivity of the salt. A stronger current is now passed through the wire which heats up the reagent. Water evaporates again until a temperature is established at which water uptake and water release are in equilibrium. This temperature is measured and/or recorded and serves as a measure for the air humidity.

Psychrometer

Here, the air sample flows past two thermometers, one of which is freely suspended in the air space, while the bulb of the other is covered by cotton cloth saturated with water. The cooling effect produced by evaporation results in a temperature difference between the two thermometers which increases with decreasing relative humidity.

The sought water vapor partial pressure P_x at the temperature t indicated on the dry thermometer is calculated via the formula $P_x = P_w - 0.5 \times p(t - t_w)/755$, where P_w is the saturated water vapor pressure at the temperature t_w of the wet thermometer and p is the barometer reading. The factor 0.5 specifically applies to the Assmann psychrometer.

Air with a water vapor concentration adjusted via salt solutions (see below) can be used for calibration. With the aid of tables P_x can be converted into absolute and relative humidities. The temperature differences $t - t_w$ can be also measured with thermocouples or resistance thermometers.

Dew point instruments employing mirrors

The dew point of an air sample is determined by passing the air over a mirror surface which is slowly cooled from the other side. At the dew point condensation occurs immediately on the surface; this temperature is measured and compared with the original temperature of the air sample.

In modern instruments, the clouding is measured photoelectrically. The Peltier effect (cooling of a thermoelement caused by passing an electric current through it) can be utilized for cooling the mirror surface.

The necessity to clean the mirror frequently after analyzing dust-laden air is a disadvantage of the method.

5.4.3.6 *Preparation of air with a certain humidity*

Air samples of known water content, which are frequently required for testing, calibration and model experiments, are most readily produced by bringing pure air into equilibrium with solutions of known vapor pressure. For this purpose we can use, for example, dilute sulfuric acid solutions of the following concentrations:

H_2SO_4 content (wt. %)	0	17.9	26.8	33.1	38.4	43.1
rel. air humidity (25°C)	100	90	80	70	60	50%

H_2SO_4 content (wt. %)	47.7	52.5	57.8	64.5	69.4	
rel. air humidity (25°C)	40	30	20	10	5	%

If saturated salt solutions are to be used, the following salts are available at 20°C:

Salt	KNO_3	NaCl	$Mg(NO_3)_2 \cdot 6H_2O$	$MgCl_2 \cdot 6H_2O$	LiCl	LiBr
rel. air humidity (20°C, %)	94	75	53	33	12	6

In addition, the mixing procedures discussed on page 47 can be applied.

5.5 Sulfur-containing air pollutants

5.5.1 Hydrogen sulfide

5.5.1.1 *General*

Hydrogen sulfide occurs in nature as the putrefaction product of sulfur-containing natural substances, especially of animal and vegetable amino acids. It appears in intestinal gases and together with ammonia in liquid manure pits, lavatories, etc. In small concentrations it is present in some hot spas and volcanic gases, while in many natural gases (e.g., about 14% in the natural gas of Lacq, southern France) it occurs in larger concentrations.

In industry hydrogen sulfide is contained in crude coke-oven gas, in water gas and many natural gases. It occurs as a contaminant in the waste gases produced in the production of sulfate pulp, artificial silk (rayon), staple fiber, rubber, as well as carbon disulfide, sulfur dyes or other sulfides. Furthermore, it is a normal waste product in petroleum refining and coal-tar production. Traces of hydrogen sulfide are found in the air of slag-slaking plants, foodstuffs plants and during anaerobic processing of waste waters.

Physical properties: boiling point −60°C, solubility in water (20°C, 760 mm Hg) 6 g/liter.

Chemical properties: Hydrogen sulfide is combustible and can even explode when mixed with air. The gas is soluble in alkalis, thus forming alkali sulfides, and precipitates the heavy-metal sulfides, which are practically insoluble in water, from their salts. The aqueous hydrogen sulfide solution is acidic. The dissociation constants of H_2S are $K_1 = 5.7 \times 10^{-8}$ and $K_2 = 1.2 \times 10^{-15}$.

Hydrogen sulfide is oxidizable. Depending on the experimental conditions and the oxidant, hydrogen sulfide is oxidized to elemental sulfur (e.g., with iodine solution), SO_2 (combustion in air) or H_2SO_4 (with chlorine or bromine in the presence of water). The gas is qualitatively indicated by its typical odor and by blackening of paper wetted with lead acetate.

Physiological and toxicological properties

Hydrogen sulfide is characterized by its rotten-egg odor. Depending on the individual person, the odor threshold varies between 0.025 and 0.1 ppm. After long-term exposure the organism becomes accustomed to it. In cells (in particular, enzymes) H_2S acts as a poison.

The MAK-value has been recently reduced and is now 10 ppm; the continuous MIK-value is 0.15 mg/m^3 (about 0.1 ppm) while the intermittent MIK-value is 0.3 mg/m^3 (about 0.2 ppm). The latter is not permitted to occur more than 3 times in 24 hr. The same values are given as immission limits in the "Technische Anleitung."

The damages caused by hydrogen sulfide to agricultural plants are much slighter than the injuries inflicted upon humans.

5.5.1.2 *Survey of* H_2S*-determination methods*

The known qualitative indication of H_2S with lead acetate paper was developed to determine this gas quantitatively. For individual determinations the air sample can be aspirated through a filter paper disk impregnated with lead acetate. Continuous determination can be effected by blowing the air sample via a nozzle onto a moving filter tape which is impregnated with the reagent (Multicolor of the firm Maihak, see page 69).

Two methods exist for determining H_2S-concentrations in the MIK-range; in both cases a cadmium hydroxide suspension is the absorption agent. The reaction product is subjected to colorimetric analysis, in one case using methylene blue and in the other molybdenum blue.

H_2S in the MAK-range can also be potentiometrically determined with the aid of a silver sulfide electrode.

For an automatic determination of H_2S on the basis of fluorescence quenching of tetrahydroxymercurifluorescein see Andrew and Nichols [7].

5.5.1.3 *Procedure according to* Jacobs, Braverman *and* Hochheiser

Principle: Jacobs, Braverman and Hochheiser [117] determined low H_2S-concentrations (1–20 ppb) by absorption in $Cd(OH)_2$ suspension. After sampling, this suspension is dissolved in acid, simultaneously carrying out the reaction with methylene blue. The losses due to absorption and oxidation of H_2S, respectively, are lower than 10% and are not allowed for.

Apparatus: Greenburg–Smith Impinger (see page 42 with suction equipment for an air throughput rate of 27 liter/min.

Solutions: The absorption mixture consists of 4.3 g cadmium sulfate ($3\,CdSO_4 \cdot 8H_2O$) dissolved in water. 0.3 g NaOH dissolved in water is added and the mixture made up with water to 1 liter.

Amine solution in sulfuric acid: 12 g N,N-dimethyl-p-phenylenediamine are added to and dissolved in a cooled mixture of 30 ml water and 50 ml concentrated sulfuric acid; 25 ml of this stock solution are diluted with 1:1 (vol.) H_2SO_4 to 1 liter.

$FeCl_3$ solution: 100 g $FeCl_3 \cdot 6H_2O$ are dissolved in water and made up to 100 ml.

H_2S test solution: a stock solution (1 ml = 100 μg H_2S) prepared from 0.71 g $Na_2S \cdot 9H_2O$ in 1 liter and standardized by iodometry, is diluted with water to 1:100 (1 ml = 1 μg H_2S).

Working procedure

The air sample is introduced at a rate of 27 liter/min (1 cu. ft/min) for 30 min into the impinger containing 50 ml dilute absorption solution. Then 0.6 ml of the dilute amine solution and 1 drop of the $FeCl_3$ solution are added and the mixture is transferred into a 50-ml volumetric flask and allowed to stand for 30 min.

The blank solution is prepared from 45 ml fresh dilute absorption solution, 0.6 ml dilute amine solution in H_2SO_4 and 1 drop of $FeCl_3$ solution. The volume is made up to 50 ml. This serves as the reference solution in the colorimetric determination.

The extinction of the sample solution against the reference solution is determined at 670 mμ. When the measured extinction does not lie on the calibration curve, the sample solution and blank solution can be diluted.

The calibration curve is plotted using the H_2S test solution corresponding to 0–9 μg with 45 ml absorption solution, 0.6 ml amine solution and 1 drop of $FeCl_3$, as described above.

Concentrations above 20 ppb can be determined in a smaller impinger with a smaller sample volume (e.g., 45 liter).

5.5.1.4 *Determination of lower* H_2S-*concentrations according to* Buck *and* Stratmann

Principle: The authors [31] attempted to develop a method having a detection limit between 10 and 20 μg H_2S/m^3 (i.e., lower than the odor threshold). The relative standard deviation in the MIK-range should not exceed 5% of the measured value. Thus, the sample volume is increased to 1 m^3 and is introduced into the impinger (yield about 93%) over a period of 30 min. The hydrogen sulfide is trapped in an alkaline $Cd(OH)_2$ suspension and subsequently released by $SnCl_2$ in HCl (weight loss of about 25% is practically constant). It is then introduced into ammonium molybdate and colorimetrically determined as molybdenum blue.

Apparatus: Impinger with a narrow bottom part, which fits into a centrifuge (Figure 5.5.1.4a). The nozzle opening of the impinger is 2.50 ± 0.05 mm. When connected to a pump with a capacity of 2–3 m^3/hr and an air throughput rate 1000 ± 50 liter/30 min, a constant underpressure of 140 ± 10 mm Hg must be established. The impinger is filled with 25 ml of a solution containing 2.15 g cadmium sulfate ($3CdSO_4 \cdot 8H_2O$) in 250 ml water and 25 ml of a solution containing 1 g NaOH in 250 ml water.

Working procedure

After introducing 1000 liter air into the lighttight impinger (see Lahmann and Prescher [147a]) over a period of 30 min, the suspension in the bottom portion of the impinger is centrifuged (15 min at 3000 rpm). The time interval between sampling and centrifuging must not exceed 4 hr. The multipurpose vessel with the residue is placed into the apparatus (Figure 5.5.1.4b) for the separation of hydrogen sulfide in a nitrogen stream (7–10 liter/hr). The nitrogen is cleaned by an activated-carbon filter and a glass wool filter impregnated with cadmium sulfate solution. The $SnCl_2$ solution in hydrochloric acid (30 ml of a solution of 100 g $SnCl_2 \cdot 2H_2O$ in 1 liter 12 *N* HCl) is introduced into the reversely connected scrubber W^1(4) and lifted by the nitrogen

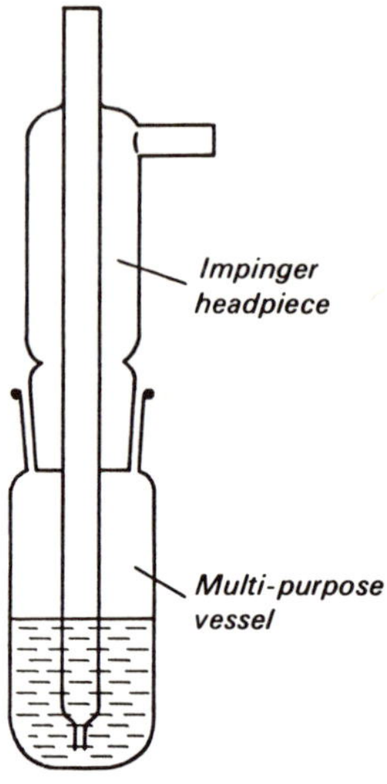

Figure 5.5.1.4a
Special impinger for determining microgram amounts of hydrogen sulfide

stream into the vessel with the cadmium hydroxide suspension. The excess HCl vapors are removed from the decomposition products in scrubbers W^3(6) and W^4(7), each containing 30 ml dilute $SnCl_2$ solution (100 g $SnCl_2 \cdot 2H_2O$ in 600 ml concentrated HCl and made up to 1 liter with water), and then proceed to scrubber W^5(8), containing 50 ml ammonium molybdate solution (3 parts by vol. of a 3.33% aqueous ammonium molybdate solution + 2 parts by vol. of a solution of 0.5 g urea in 1 liter *N* H_2SO_4). The ammonium molybdate solution must be frequently renewed. Decomposition ceases after 20 min continuous passage of nitrogen. The molybdenum solution is rinsed into a volumetric flask and allowed to stand for 20 min until the color reaction is completed; then the extinction is measured at 570 mμ and a layer thickness of 1 cm against water. At extinctions above 1.0 an aliquot of the molybdenum blue solution must be diluted.

In the calculation the extinction obtained from the calibration curve is multiplied by 1.07 because of the incomplete absorption and by 1.32, which is the loss factor associated with oxidation by air.

The calibration curve is plotted by adding a pyrosulfite solution containing about 0.6 g $Na_2S_2O_5$/liter. The solution is standardized by iodometry. Amounts between 0.1 and 0.8 ml are added to the measuring arrangement shown in Figure 5.5.1.4b, and H_2S is converted and measured as molybdenum blue.

The molar extinction coefficient for hydrogen sulfide as for molybdenum blue is

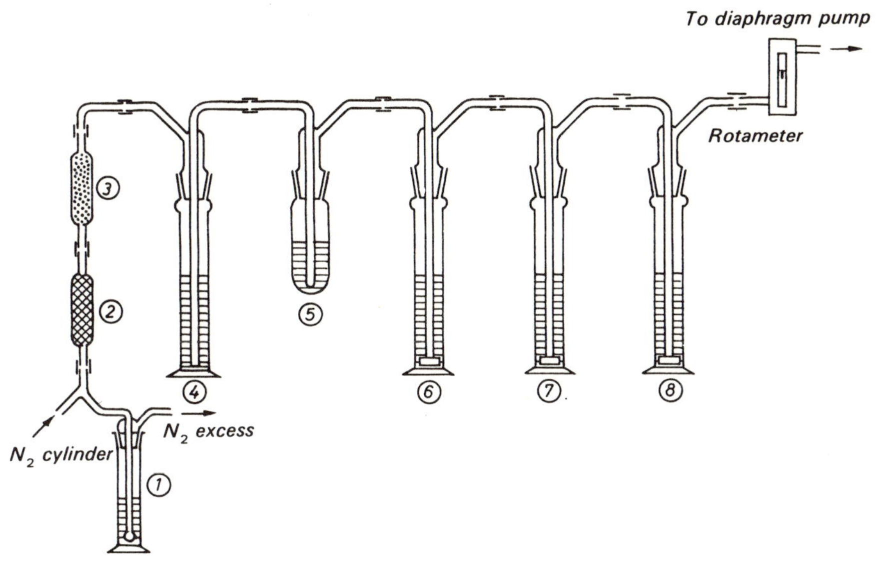

1 Scrubber
2 Activated carbon filter
3 Glass wool impregnated with calcium sulfate
4 Scrubber W_1
5 Multipurpose vessel W_2
6 Scrubber W_3
7 Scrubber W_4
8 Scrubber W_5

Figure 5.5.1.4b
Apparatus for the removal of hydrogen sulfide from sulfides.

10,000 and the "reciprocal slope" (see page 54) is k = 162. Fifty μg H_2S/50 ml correspond to an extinction of 300.

The authors give 20 μg H_2S/ml air as the detection limit and $\pm$ 5% as the relative standard deviation in the MIK-range.

NO_2, SO_2, CS_2 and mercaptans in the concentrations expected in the air do not interfere.

Prescher and Lahmann [205a] combined the isolation of hydrogen sulfide according to Buck and Stratmann with the determination as methylene blue according to Jacobs, Braverman and Hochheiser (page 144).

5.5.1.5 *Other methods*

Method employing lead acetate paper in accordance with the British Factory Inspectorate, Booklet 1.

Apparatus: The paper strips (5 × 10 cm^2) impregnated with lead acetate are inserted into a suction device (see page 40), which exposes a circular area of 40 mm diameter. The sample is drawn at a rate of 126 ml per stroke through the paper (up to 5 strokes) until a distinct color is formed. The color is compared with the color standards of the British Factory Inspectorate (distributed by the Stationery Office, York House, Kingsway, London WC 2) or with a color scale copied with color pencils from the results of own tests. An air sample containing 6 ppm H_2S produces after 5 strokes (630 ml) a pale yellowish-brown color, while 17 ppm yield the color of milk chocolate and 40 ppm a deep chocolate color.

Preparation of the lead acetate paper strips

10 g lead acetate are dissolved in 90 ml water. 5 ml glacial acetic acid and 10 ml glycerol are added. The paper tape ($5 \times 10\,cm^2$, Whatman No. 1) is immersed in this mixture for 1 min. The excess liquid is allowed to drip off and the papers are dried at room temperature in H_2S-free air. 2.5 cm from both the upper and lower edges are cut off and discarded. The papers are stored in glass-stoppered glass vessels.

Potentiometric methods in the MAK-range

An instrument for the potentiometric determination of low H_2S-concentrations in air (0.5–5 ppm, $\pm$ 0.5 ppm) was developed by Oehme and Wyden [193]. The sulfide-sensitive indicating electrode consists of a silver rod covered with a layer of silver sulfide.

5.5.1.6 *Determination of hydrogen sulfide in the presence of sulfur dioxide*

Sulfur dioxide and hydrogen sulfide are both stable when mutually present in low concentrations in air. Whereas some methods, for example, iodometry determine both components jointly, a separate, mutually noninterfering determination of both pollutants is of interest because of the different nature of their damaging effects.

Buck and Gies [35] developed a method for the separate determination of the two pollutants. Hydrogen sulfide is retained in a sorption tube filled with glass beads which are coated with a thin layer of silver sulfate. SO_2 is not absorbed in the tube and can be determined by one of the conventional methods.

A mixture of equal volumes of saturated Ag_2SO_4 and 5% $KHSO_4$ solutions is poured over the glass beads (diameter 2–3 mm) in an H_2S-free atmosphere. The excess liquid is allowed to drip off on a Nutsch filter and the moistened beads are dried in a drying oven. The beads are packed to a height of 10 cm in an absorption tube with a ground-in joint.

The air sample is introduced at a throughput rate of 3–4 liter/min.

For desorption the sorption tube is connected to the apparatus of Buck and Stratmann, described on page 147. 25 ml stannous chloride (100 g $SnCl_2 \cdot 2H_2O$ in 1 liter concentrated HCl) are drawn into the tube; the hydrogen sulfide is freed from the excess HCl in two successive scrubbers ($SnCl_2$ in dilute HCl) and finally allowed to react in the last scrubber to produce molybdenum blue.

5.5.2 Sulfur dioxide

5.5.2.1 *Sulfur dioxide (general)*

Presence

In nature sulfur dioxide occurs in appreciable concentrations only in volcanic gases. On the other hand, it is without doubt the most widely distributed and important air pollutant originating from industry and domestic emissions. This substance is formed upon combustion of sulfur-containing materials and thus occurs in the flue gases of installations for the generation of heat, vapor or electric power from sulfur-containing fuels. It is also formed upon roasting of sulfide-containing ores and sintering of sulfur-containing fine ores. In the chemical industry this gas is produced in the contact process for the production of sulfuric acid and occurs in the form of the residual SO_2 which is not converted into SO_3.

Table 5.5.2.1

SO_2-EMISSION 1962

IN THE GERMAN FEDERAL REPUBLIC

Bituminous coal heating	2,340,000 tSO_2
Lignite heating	550,000 tSO_2
Oil heating	850,000 tSO_2
Ore sintering plants	120,000 tSO_2
Chemical industry	60,000 tSO_2
	3,920,000 tSO_2
1964	4,600,000 tSO_2
1965	5,400,000 tSO_2

Sulfur dioxide is used in many industrial branches: production of sulfuric acid, sulfite lye in sulfite cellulose, bleaching and preserving agents, coolants in refrigerators and as an extractant in the petroleum industry.

Important physical properties

Boiling point (760 mm Hg) 10.0°C.

Soluble in water.

Partial pressure, mm Hg	Temperature	
	10°C	30°C
760	154 g/kg	75.6 g/kg
10	3.7 g/kg	1.85 g/kg
1	0.63 g/kg	0.31 g/kg
Specific gravity at 0°C and 760 mm Hg	2.926 kg/m^3	
Molecular weight	64.06	
Molar volume at 0°C and 760 mm Hg	21.89 liter	

Chemical properties

Sulfur dioxide dissolves in water to form sulfurous acid (H_2SO_3). The first dissociation constant of this acid is $K_1 = 1.7 \times 10^{-2}$ and its second dissociation constant $K_2 = 6.2 \times 10^{-8}$.

With oxidizing agents it is converted into SO_3 (or H_2SO_4). Depending on their strength, reducing agents convert it into elemental sulfur (e.g., in the Claus furnace) or H_2S (e.g., stannous chloride or phosphorous acids).

5.5.2.2 *Hygiene and toxic effects*

Sulfur dioxide is a strong irritant, which can be perceived by its odor and taste even when highly diluted. The taste and odor thresholds as well as the irritant effects due to higher SO_2 concentrations differ widely among individual persons. Therefore, the data given in the literature can differ by several orders of magnitude.

According to VDI-Richtlinie 2108, sensitive persons can perceive 0.3 ppm SO_2 by its taste and about 0.3–1 ppm by its odor.

The effect of sulfur dioxide on the respiratory organs is enhanced by the presence of water vapor (fog) and smoke. This is apparently due to the fact that gaseous SO_2 is mostly dissolved in the mucous membranes of the mouth and nose, while in aerosol form it can penetrate into the internal respiratory organs, where it is converted into sulfuric acid. However, under the influence of water, soot and flyash this process partially takes place already in the aerosol stage. As was mentioned on page 4, respiratory air, polluted by SO_2, especially during persistent fogs, has an adverse effect on persons suffering from bronchitis and can lead to a significantly higher mortality rate. It was recently shown in London, however, that a reduction in the smoke emission by improving domestic heating without a simultaneous decrease in the SO_2-discharge considerably reduces bronchial diseases.

Effect of sulfur dioxide on vegetation (see also page 19)

The effect of SO_2 on plants is also very important. Concentrations as low as 1–2 ppm may lead in a few hours to acute damage in the form of localized destruction of leaf tissues (necrosis). In sensitive plants chronic damage may occur at 0.3 ppm SO_2 or higher. A concentration of 0.15 ppm is taken as the tolerance limit, even for the most sensitive plants. Evergreens, leguminous plants, and rye are particularly susceptible to these hazards.

Other damages caused by sulfur dioxide

Although iron in netural, even humid atmospheres, displays little tendency toward corrosion, moisture, which is acidified due to the uptake of SO_2 from air, strongly enhances corrosion and can lead to very costly material damages. Acid-sensitive structural materials, such as limestone and concrete, are also strongly attacked in an SO_2-containing atmosphere.

Limit values

A MAK-value of 5 ppm is specified for SO_2. The MIK_D-value is 0.5 mg SO_2/m^3 (0.2 ppm), while the MIK_K-value (which must not occur more frequently than once in 2 hours) is 0.75 mg/m^3 (0.3 ppm). The corresponding immission limits of the "Technische Anleitung" (see page 7) are 0.4 and 0.75 mg SO_2/m^3, respectively.

5.5.2.3 *Technical possibilities for reducing the* SO_2*-discharge*

Waste gas from combustion processes

In principle, the sulfur dioxide in waste gases can be reduced by two methods: removal of sulfur in the fuel prior to combustion or removal of SO_2 from the waste gases.

The first method is applied in the coking of bituminous coal. A considerable portion of the sulfur contained in the coal is concentrated and utilized in ammonia liquor, while another portion is used in the crude coke-oven gas, from which it can also be obtained as a concentrated secondary product (via the gas-cleaning substance) and processed, for example, into sulfuric acid. This contaminant can also be removed from natural gas, when the latter contains large amounts of SO_2.

For low-boiling coal tar and petroleum products (motor gasoline, diesel oil) high-degree desulfurization is possible and, in the case of engine fuels, even necessary. On the other hand, the high-boiling petroleum fractions (fuel and tar oils) can be desulfurized only by destructive pressure hydrogenation, which makes the fuel much more expensive.

Removal of sulfur dioxide from waste gases of combustion processes

The removal of sulfur dioxide from waste gases generated upon combustion is of extreme importance because of the hazards involved in the discharge of large amounts of SO_2, for example, by thermal power plants or large long-distance heating systems. The problem is being intensively studied in Great Britain, Germany and the USA, but so far no technically and economically feasible solution has been found. Two methods are mainly employed: wet scrubbing and adsorption.

Wet scrubbing

Sulfur dioxide can be washed out of flue gases with the aid of aqueous alkaline solutions or suspensions by various techniques. The gases can be scrubbed with milk of lime or $CaCO_3$ suspensions followed by removal of the product ($CaSO_3$ and $CaSO_4$), or by means of ammoniacal solution with subsequent processing of the ammonium salt solutions.

The first process is preferred in Great Britain. Sulfur dioxide is converted into $CaSO_4$ via suspended limestone in the presence of oxidation catalysts; the end product is discharged into the Thames (Battersea–B.E.A.-process) or allowed to settle in a sedimentation tank and the sludge pumped into the sea (Howden–ICI-process). In both cases a 90% scrubbing effect is achieved; however, since the flue gases acquire inadequate buoyancy owing to cooling at the stack mouth and are thus insufficiently diluted upon reaching the ground in the vicinity of the stack, the number of complaints about odorous annoyance did not decrease much, the more so because of the fact that the humid smoke plume is visible when no provisions for fume removal are made.

The use of ammoniacal scrubbing liquids leads to products which can be utilized. Processes based on this principle have been developed in the USA (Cominco Process) and Russia, and are being successfully employed, for example, by the Österreichische Stickstoffwerke, Linz, for lowering the SO_2-emissions from a sulfuric acid plant. The gas is scrubbed with an ammonium sulfite lye, the pH of which is adjusted in such a way that the sulfur dioxide is largely retained, while free ammonia is not yet appreciably discharged.

Dry scrubbing

In order to avoid the above difficulties (cooling of the flue gases), attempts were made, especially in Germany and the USA, to develop dry scrubbing processes; sulfur dioxide is adsorbed at higher temperatures and then desorbed.

Alkaline dusts (e.g., from dolomite lime hydrate) can be blown in but they must be separated again in high-efficiency dust separation plants. Furthermore, flue gases can be desulfurized by means of the filter ash of lignite containing 25–30% CaO (Wickert and Still Processes).

Adsorption methods with activated carbon are more advanced. We shall here describe the "clean-air" process and the sulfacid process. (See also Spengler [235], Jüntgen [124], Johswich [121] and Bienstock, Field, Katell and Plants [20].)

The clean-air method

A cheap semicoke, which can also be prepared from peat, is used as the absorbing agent. It removes sulfur dioxide from the flue gas at 80–200°C in the adsorption zone. In this process sulfur dioxide is converted into sulfuric acid. The adsorbing agent containing the acid is recovered in the desorption zone at 350–450°C. Sulfuric acid is reduced again to SO_2 which is, however, now present as 40% gas. Part of the coke is oxidized in this process but the major portion can be again used for adsorption. The 40% SO_2 must be further processed on location, i.e., a chemical plant must be present near the flue-gas emitter. The main advantage of this method is that dust separation is foregone. At present two large plants with outputs of 30,000 and 50,000 m^3/hr, respectively, are being investigated.

Other drying processes

Two drying processes are mainly being studied in the USA. The process developed by the Pennsylvania Electric Company (Penelec Process) consists of catalytic afteroxidation at 480°C over vanadium contacts. The SO_2-mists formed are electrostatically precipitated as sulfuric acid or, after addition of ammonia, as ammonium sulfate (Kiyoura [132]). In the other process (Bureau of Mines) sulfur dioxide is adsorbed by activated alumina at 330°C and reduced by generator gas to H_2S, which is processed in the Claus furnace into elemental sulfur.

The sulfacid process of Lurgi is a "semidry" process. The cooled raw gas releases its sulfur dioxide at 55°C as 10% sulfuric acid in a wet activated-carbon bed. The acid can be concentrated by hot flue gases. No data are as yet available on the quality and usability of the acid obtained.

Present state of desulfurization of flue gases

In addition to the technical difficulties involved in the absorption or adsorption of SO_2 from flue cases and the costs of the process (the increase in the electricity costs is more than 10%), the main problem is the elimination or utilization of sulfuric acid obtained from sulfur dioxide in the flue gas. Taking into account the production of sulfuric acid, we must bear in mind that the total amount of SO_2 in flue gases considerably exceeds the demands for this reagent in an industrial area and that even the transport of concentrated sulfuric acid is unprofitable. For this reason the isolation of elemental sulfur is much more favorable; provided it is sufficiently pure, sulfur can be processed into sulfuric acid in a central plant. Gypsum or sulfate-containing lime sludge can be piled on slag dumps, but this must not lead to pollution of the groundwater. By no means should permanent or even provisional removal of waste gas pollutants endanger the purity of water.

Thus, the problem of desulfurization of flue gases is not yet solved. For the time being the only practicable way of removing SO_2 from flue gases is by resorting to sufficiently high stacks.

5.5.2.4 *Sulfur from flue gas as a nutrient for plants*

Surprisingly enough, plant physiologists and agricultural biologists express the view that the sulfur emitted from stacks can be utilized as a nutrient for cultured plants. The sulfur is not used in the form of SO_2 for the direct nutrition of the plant (sublethal SO_2-doses, because of the low amounts, cannot be used) but as sulfates penetrating the soil upon precipitation (see Linser [167]).

Sulfur is an indispensable element for plants; it is contained in some amino acids, mustard oils, thiamin and other vital substances. With each harvest 60–100 kg SO_3 per hectar are extracted by the plants from the soils; considerable losses also occur due to leaching. The conventional commercial fertilizers (superphosphate, ammonium sulfate and sulfate-containing potassium salts) can make up for the loss but the increasing use of "ballast-free" mixed fertilizers decreases the sulfate content of the fertilizer. Thus, the sulfur deficiencies to be expected have to be replenished by fertilizing if the air does not provide sufficient amounts of the vital sulfur via precipitation (in Central Europe about 30–300 kg SO_3/ha · year are supplied in this way).

Reduction of the SO_2*-emission in the production of sulfuric acid* (*Bayer double contact process*) (see Werth [263])

Whereas in the normal contact process for sulfuric acid only 97–98% SO_2 are converted into SO_3, the rest being emitted by the plant, the Bayer double contact process converts 99.6% or more, so that 90% of the air pollution can be eliminated. The principle of the process consists in freeing the reaction product after 90% conversion of the SO_3 has been achieved by absorption. The residual gas is again heated, passed through a contact stage, so that of the remaining 10% SO_2 (referred to the initial amount) such a quantity is converted that the total conversion reaches 99.6%.

For the most part, when a new plant is provided with equipment for the double contact process and for processing SO_2-rich roasting gas (9%), excessive initial and production costs need not be expected. However, installation of such equipment into an existing, older plant or into a plant processing SO_2-poor gas mixtures does involve greater difficulties.

Determination of sulfur dioxide

5.5.2.5 *General*

Determination of the SO_2-level is certainly the most important and most frequently performed procedure in the analysis of the free atmosphere. Numerous methods are available for this purpose; the selection of a specific method will depend on the aim of the analysis as well as the time and means at one's disposal.

Whereas in long-term investigations on chronic damages the integrating bell and rag methods are still being applied, continuous recording instruments must be used when acute effects and definite hazards are expected. The recorders mainly operate on the principle of electrical conductivity and can be mounted in vehicles which are immediately driven to the site in question. Other automatic analyzers are based on colorimetric, potentiometric or coulometric methods.

A rapid determination of high concentrations at work places is possible by the less accurate gas-detection instruments (e.g., of the firm Dräger). Tube 1/a (0.1–20 ppm) is suitable for measurements in the MAK-range. In the presence of SO_2 starch is colored by elemental iodine released by iodate due to the reducing effect of this gas.

For more accurate determinations of the SO_2-level in the free atmosphere by individual analyses there exist several established and recommended methods. VDI-Richtlinie 2451 "Messung der Schwefeldioxid-Konzentration" (Determination of Sulfur Dioxide Concentration), states the silica gel method of Stratmann and the somewhat simpler pararosaniline method according to West and Gaeke. In both methods sampling and the determination in the laboratory are carried out separately. However, there also exist colorimetric methods, whereby analytical results can be obtained simultaneously with or immediately after sampling.

For larger SO_2-quantities (e.g., daily averages) titration methods can be used or the SO_2 can be gravimetrically determined as $BaSO_4$.

5.5.2.6 SO_2-*determination on impregnated filter papers*

Principle: Huygen [115] (see also Pate, Lodge and Neary [198]) used alkali-treated filter papers, which are washed out after sampling (see page 160). In the washings SO_2 is determined according to the method of West and Gaeke.

Preparation of the impregnated filter papers: circular filter papers (Whatman No. 1) of 5.5 cm diameter are immersed in the absorption solution described below; the excess solution is removed by dabbing. In this way an uptake of about 13 mg solution per cm^2 filter area is achieved. The filter papers are dried at 110°C.

Absorption solution: aqueous solution containing 20% KOH and 10% triethanolamine or glycerol.

Working procedure

A treated filter is inserted into a funnel-like suction device (see, e.g., page 41) and the air sample is aspirated at a rate not exceeding 16 liter/min (1 liter/$cm^2 \cdot$ min). The air humidity must not be lower than 30%. The air sample should not contain more than 10 mg SO_2/m^3; thus, no more than 10% of the alkali on the filter should be neutralized.

After sampling the filter is placed on a frit, washed with water after adding a small amount of sodium chloromercurate and the SO_2 determined colorimetrically in the filtrate according to the method of West and Gaeke.

The above authors report that more than 95% of the SO_2 present is detected. The procedure can be used both for short-term determinations and for routine daily and weekly analyses.

Filter paper method with zinc nitroprusside for the determination of SO_2 *at the work place*

This method is described by Hands and Bartlett [94] and recommended in Booklet No. 5 of the British Ministry of Labor. Filter paper impregnated with an ammoniacal zinc nitroprusside solution turns red, when SO_2-containing air is aspirated through the filter. In investigations at the work place between 1 and 20 ppm SO_2 can be detected in a 360 ml air sample.

Preparation of the filter papers

50 ml of 6% aqueous solution of crystalline zinc sulfate ($ZnSO_4 \cdot 7H_2O$) are mixed with 50 ml of a freshly prepared 10% aqueous solution of sodium nitroprusside.

Ammonium acetate is added in portions under agitation until the zinc nitroprusside is dissolved. Then 20 ml glycerol are added. Into this solution 2.5-cm wide strips of filter paper are immediately immersed (before zinc nitroprusside is again precipitated). The excess solution is allowed to drip off and the filter papers are left to dry in clean air. The test papers can be stored in stoppered containers for at least one month.

The determination is carried out by inserting a strip into a filtering device with an inside diameter of at least 1 cm; an air sample of 360 ml is aspirated at a rate of 90–180 ml/min. For reference a color scale, obtainable via the British H.M. Stationery Office or prepared with air containing 1–20 ppm SO_2 can be used. The firm Tintometer Ltd., Salisbury, Wiltshire, offers disks of standard tints.

This method is highly specific for sulfur dioxide. Interference due to hydrogen sulfide is only very slight.

5.5.2.7 *Acidimetric determination of* SO_2 *as* H_2SO_4 *(British Standard Method)*

This method is described in the British Standard 1747, III (1963) as a 24-hr average test. For instantaneous tests or smaller air volumes the method is not sensitive enough.

Apparatus

Dust filter (filter paper Whatman No. 1 in holder); 125-ml gas scrubber. The glass must not release alkali amounts into the solution which can interfere with the determination. Prior to use, a new scrubber has to be treated for 24 hr with 1:1 HCl and thoroughly rinsed with water. The absorption solution must not change its pH after standing for 24 hr. The scrubber is connected to a $CaCl_2$-drying tower, a dry gas meter and finally to a suction pump (diaphragm type) with a throughput of 1.4–2.2 m^3 air in 24 hr (75 liter/hr).

Reagents: The absorption solution is prepared by diluting 10 ml 30% H_2O_2 to 1000 ml; 50 ml of this solution are titrated against 0.1 ml mixed indicator (pH = 4.5) to gray and the remaining 950 ml are adjusted to pH 4.5 with a 19-fold amount of alkali or acid. The solution is stored in polyethylene flasks.

The mixed indicator of the British Drug House (pH = 4.5) is recommended (color at pH 3.5 orange-yellow, pH 6 blue, pH 4.5 gray).

Titration solutions: 0.01 *N* Na_2CO_3 and 0.01 *N* HCl.

Working procedure

50.0 ml of the neutralized absorption solution are introduced into the scrubber. In 24 hr about 1.6 m^3 air (accurately measured) is sucked through. The solution is titrated with 0.01 *N* Na_2CO_3 to the gray tint of the blank. 1 ml 0.01 *N* Na_2CO_3 is equivalent to 320 μg SO_2.

A titration error of $\pm$ 1 ml titration solution corresponds to an error of $\pm$ 0.01 ppm SO_2 in 1 m^3 air sample.

Titration of SO_2 *as sulfate with Thoron*

Kündig [142] used 0.001 *N* $Ba(ClO_4)_2$ with Thoron as the indicator for the titration of the sulfate ion formed by the absorption of SO_2 in H_2O_2 (see also Fielder and Morgan [67]). The color change from yellow to pink is much sharper than in acidimetry. The titration can be most accurately carried out in a photometer at 550 mμ.

Apparatus: The sulfuric acid mist is removed by first passing the air through an impinger containing concentrated phosphoric acid. Sulfur dioxide is absorbed in two impingers of height 15 cm and inside diameter 2 cm connected in series.

Solutions: The absorption solution consists of 1% H_2O_2 adjusted to pH 9 with a small amount of NH_3.

Indicator: 0.080 g Thoron (sodium salt of 1-arsenophenylazo-2-naphthol-3,6-disulfonic acid) are dissolved in 200 ml deionized water. In the titration 2 drops are added. The optimum pH in the titration is 3–4, while the solvent ratio of water to isopropanol should be 1:4. Heavy metals interfere; sodium salts in concentrations exceeding 0.1 *N* blur the color change.

0.001 *N* $Ba(ClO_4)_2$ solution: the equivalent amount of barium perchlorate is dissolved in 200 ml CO_2-free water and the volume made up with isopropanol to 1 liter. The titer is determined with 0.001 *N* H_2SO_4.

For the automatic determination with the Auto-Analyzer see Persson [199b].

Working procedure

20 ml absorption solution are introduced into each of both the absorption impingers. The air sample (at least 100 liter) is passed through at a rate of 1–2 liter/min.

The content of each impinger is separately evaporated in quartz dishes along with 20 ml of the original absorption solution for the blank. The residue is dissolved in 1 ml water and 1 ml glacial acetic acid. Then 8 ml isopropanol and 2 drops of Thoron solution are added. The titration is carried out with the aid of a photometer at 550 mμ until the color changes.

An error of 0.1 ml corresponds to $\pm$ 3 μg SO_2 or $\pm$ 0.015 ppm for 100 liter air.

5.5.2.8 *Iodometric-colorimetric* SO_2*-determination according to* Zepf *and* Vetter

The iodometric methods used in the determination of large SO_2-concentrations in the sulfuric acid industry can also be employed in the refined micromodification of Zepf and Vetter [281] for the determination of low SO_2-levels in the open atmosphere. It was previously widely used in the laboratory of the Österreichische Stickstoffwerke but has since been substituted by the method of West and Gaeke. The method has the following advantages: during the introduction of the air sample the instant at which the test procedure is stopped can be directly determined and, furthermore, it is possible to carry out in situ measurements. The main disadvantage is that large NO_2-amounts can interfere.

Principle: A 0.0001 *N* iodine solution containing starch is used. The bleaching caused by the SO_2-content is compared with the color of an iodine-starch solution treated with SO_2-free air, since pure air changes the blue color to a light violet tint.

Working procedure: The air sample is drawn by means of an aspirator (throughput rate 0.5 to 2 liter/min) through a tube with 10 bulbs or through a fritted scrubber, filled with 25 ml 0.0001 *N* iodine and 25 ml starch solution. Then the air freed by the SO_2 is introduced into a second tube with 10 bulbs or a fritted scrubber containing the same absorption solution. The air volume to be used must be measured in such a way that no more than one third of the iodine amount in the scrubber or tube is reduced. Usually, a volume of 20–30 liter air is used.

After absorption of the sulfur dioxide each of the two liquids is introduced into a Hehner cylinder. The volumes in both cylinders must be equal. Liquid is withdrawn from the reference cylinder until the color intensities of both solutions as viewed from above are the same.

The SO_2-amount present is calculated according to

$$\mu g\ SO_2 = \text{ml } 0.0001\ N \text{ I soln.} \left(1 - \frac{\text{ml reference solution}}{\text{ml sample solution}} \times 3.2\right).$$

Bokhoven and Niessen [24] developed an automatic analyzer based on the principle of the decoloration of an iodine solution by SO_2. In order to eliminate interferences by nitrogen oxides and ozone, hydrogen is introduced into the air sample, which is then passed over a platinum catalyst heated to 100°C (see also Bassett and Davies [12a]).

5.5.2.9 *Silica gel reduction method according to* Stratmann

Principle: In this method [245] SO_2 is extracted from the air sample at the sampling site by silica gel in an adsorption tube. The adsorbed SO_2 is desorbed in the laboratory at about 500°C and reduced in a hydrogen stream to H_2S, which is then determined as molybdenum blue.

Sampling apparatus

The following devices are employed in the sampling procedure: a small fritted scrubber with 5–10 ml 80–90% phosphoric acid which removes dust, SO_3 and H_2SO_4 mists from the sample; a 26 cm × 14 mm quartz tube filled with 10 g purified silica gel for adsorption of SO_2; a suction pump connected to the mains or to a storage battery (e.g., of an automobile) and finally a gas meter. Polyethylene hoses are used as connecting pieces.

Apparatus for desorption in the laboratory

The desorption apparatus (Figure 5.5.2.9) consists of hydrogen storage vessels (steel cylinder and laboratory gas meter), a furnace to heat the adsorption tube to

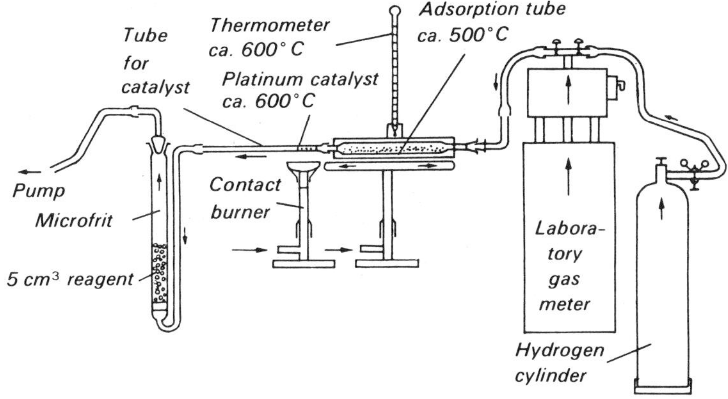

Figure 5.5.2.9
Desorption apparatus of the Stratmann instrument

500°C and a quartz contact tube (length 26 cm, inside diameter 8 mm) containing a tightly coiled 3 × 6 cm^2 platinum wire grid of 3600 mesh/cm^2, heated with a Bunsen burner to 600°C. The instrument is connected to a small fritted scrubber containing 5 ml reagent which in turn is connected to a suction pump.

The color determination is effected by means of a colorimeter with a filter of 570 mμ and a 1-cm cell.

Reagents and solutions

Purified silica gel (according to the VDI-Richtlinie a suitable preparation can be obtained from the firm Lange and Rechberg, Bochum): In order to separate traces of sulfur-containing contaminants from other commercial brands of silica gel (grain size 2–6 mm), the material is heated for several hours in concentrated HNO_3 in a water bath under a reflux condenser. It is then filtered, washed in hot water (removal of acid) and dried at 150°C. The quartz tube is filled and 100 ml of hot water are passed through to effect further cleaning. Then air is sucked through while the tube is heated by the bunsen burner until the silica gel appears to be dry. This procedure is repeated 4–7 times. A single filling serves for at least 10 analyses.

Reagent: a mixture of 2 vol. Solution A and 3 vol. Solution B.

Solution A: 0.5 g urea in 1000 ml 1 N H_2SO_4.

Solution B: 33.3 g ammonium molybdate ($[NH_4]_6Mo_7O_{24} \cdot 4H_2O$) in 1000 ml water. The solutions can be used for several weeks when stored in polyethylene flasks.

Working procedure

1. Determination of the blank. A clean adsorption tube is connected to a contact tube containing the catalyst. The latter is connected to a receiver containing 2 ml

Solution A + 3 ml Solution B. Then a hydrogen stream (4 liter/hr) is passed through, the adsorption tube is heated to $500 \pm 20°C$ and the catalyst to about 600°C (weak red heat). No appreciable cooling between the two tubes should occur. This is continued for 15 min. Then the receiver is exchanged until after another 15 min of passing through H_2 a constant extinction below 0.02 is achieved (1-cm cell, 570 mμ).

2. The sample (30 liter air) is passed through for 30 min (flow rate 1 liter/min); however, the adsorption is still complete with a 6-fold flow velocity. The charged adsorption tubes are labeled, closed and stored.

3. The adsorption tube is desorbed in the same manner as in the blank determination, i.e., hydrogen is passed through the tube for 15 min. The extinction of the washing solution is measured at 570 mμ after standing for 10–15 min (until the color maximum appears). If the solution is too dark, it can be diluted and then measured.

The calibration curve is plotted for the range from 3.12–12.8 μg SO_2 by applying 0.10–0.40 ml 0.001 *N* H_2SO_4 to the silica gel layer. The acid is reduced to H_2S as in the normal determination and measured. The blank of a fresh adsorption tube prepared in the same manner and subjected to the same procedure (reduction and measurement), but without introduction of air, must be subtracted. In this determination the Lambert–Beer law is satisfied. 0.1 ml 0.001 *N* H_2SO_4, corresponding to 3.2 μg SO_2, yielded an extinction of 0.124 ± 0.008. Thus, the content of SO_2 (μg) in the air sample used can be determined with the aid of the equation

$$\mu g\ SO_2 = (E - B) \times 29.5,$$

where E is the measured extinction of the sample and B is the extinction of the blank.

The standard deviation is $\pm$ 0.03 mg SO_2/m^3, the detection limit 0.3 μg SO_2.

Variants: Instead of determining the H_2S formed as molybdenum blue Abel and Barth [1] introduced the hydrogen sulfide into a zinc acetate solution and determined it as methylene blue by adding dimethyl-p-phenylenediamine and ferric salt (see page 145).

5.5.2.10 *Determination of* SO_2 *according to* West *and* Gaeke

On the basis of the Schiff reaction between pararosaniline, formaldehyde and sulfur dioxide, which has been used in analytical chemistry for a long time for the detection of both SO_2 and formaldehyde, several methods for the quantitative colorimetric determination of SO_2-traces were developed in air analysis. The variant of West and Gaeke [264] is the most widespread of these methods and was adopted in VDI-Richtlinie 2451. The authors employed the data of Feigl on the stability of the disulfitomercurate ion $(Hg[SO_3]_2)^{2-}$ and used a sodium tetrachloromercurate solution (from 2 NaCl + $HgCl_2$), in which SO_2 remains stable for at least 24 hr, as the absorption solution for the SO_2 in the air sample.

The structural formula of the red reaction product is given by Naumann, West, Tron and Gaeke [187] as well as Pate, Lodge and Wartburg [199].

West and Gaeke developed the following working procedure:

Reagents: 0.1 mg Na_2HgCl_4 solution: 27.2 g $HgCl_2$ (0.1 mole) and 11.7 g NaCl (0.2 mole) per 1000 ml solution.

Pararosanaline in HCl: 4 ml of a 1% aqueous solution of pararosaniline, which was allowed to stand for 24 hr, and 6 ml concentrated HCl are mixed and shaken until the brown color disappears. Then the solution is made up to 100 ml.

According to Herrmann [103], the brown color of the solution can be made as light as amber (no further decoloration is permitted) by treatment with finely ground activated carbon (no more than 5 g carbon per 1000 ml 0.04% pararosaniline hydrochloride solution).

0.2% Formaldehyde: 5 ml 40% formaldehyde are diluted to 1000 ml.

Working procedure: 10.0 ml 0.1 *M* Na_2HgCl_4 are introduced into a small fritted scrubber and a 38.2-liter air sample is passed through for 10–30 min. To the solution 1.0 ml pararosaniline hydrochloride and 1.0 ml formaldehyde are added. A blank of 10.0 ml Na_2HgCl_4 is treated in the same manner. After 20–30 min the extinction can be measured at a wavelength of 560 mμ. For the evaluation a calibration curve is plotted using sodium pyrosulfite. When 38.2 liter of air are introduced, 1 μg SO_2 corresponds to 0.01 ppm SO_2.

0.5 μg SO_2 corresponds to an extinction of 0.015. Nitrogen dioxide interferes when present in concentrations exceeding 2 ppm.

VDI-Richtlinie 2451 recommends the following variant of the West–Gaeke method. A 100-ml fritted scrubber (G2-frit, lateral access) with a ground-glass joint is used. It contains 20 ml Na_2HgCl_4. The measurement time is 30 min; during this period about 20 liter air sample are passed through. 10 ml solution are removed from the scrubber for the determination. The pararosaniline solution is prepared by dissolution of 40 mg of this reagent ($HO \cdot C(C_6H_4NH_2)_3$) in 6 ml HCl and dilution to 100 ml. The solution is stable for 10–14 days. As above, 1 ml of this solution and 1 ml formaldehyde solution are used for the color reaction. The color is measured at 540–550 mμ against a blank of the reagent solution. If the extinction exceeds 0.75, a smaller volume of the absorption solution is made up to 10 ml with Na_2HgCl_2 and again colorimetrically measured.

In accordance with this VDI-Richtlinie the calibration curve is plotted using a 0.05 *N* sodium pyrosulfite solution standardized by iodometry. For this purpose 5 ml of the above solution are diluted with fresh Na_2HgCl_4 solution to 500 ml and 5 ml of this solution are again diluted to 50 ml, yielding a 0.00005 *M* solution, each ml of which contains 1.6 μg SO_2. 2, 4, 6, 8 and 10 ml of this solution are respectively made up to 10 ml with Na_2HgCl_4. 1 ml of pararosaniline and 1 ml of formaldehyde solution are added and the extinction is measured after 30 min at 540–550 mμ. The extinctions are plotted along the ordinate axis as functions of the SO_2-concentration (in μg). When fresh pararosaniline solutions are used, a new calibration curve must be plotted.

The standard deviation is $\pm$ 0.03 mg SO_2/m^3, the detection limit 0.2 μg SO_2.

Variants

West and Ordoveza [265] recommended the addition of 0.6 g amidosulfonic acid to 1 liter Na_2HgCl_4 solution in order to eliminate the interference of high NO_2-concentrations (up to 10 ppm). Pate, Ammons, Swanson and Lodge [197] prescribed the addition of the same amount of amidosulfonic acid after introduction of the air sample but prior to the addition of pararosaniline.

Herrmann [103] proposed that the color obtained by the West–Gaeke method be compared with standard solutions at the sampling site. The standard solutions are prepared from 1 part (by wt.) of Bengalrosa (Merck) and 1.27 parts (by wt.) Supracenviolet 3 R (Bayer). All the instruments and solutions for sampling and determination can be placed into an 8-kg case.

According to Huit and Lodge [112], the presence of dimethylformamide in a 3–5 *M* solution doubles the color intensity.

Data on the selectivity of the West–Gaeke and hydrogen peroxide methods with and without membrane and glass fiber filters placed in front of the absorption solution are given by Hochheiser, Santer and Ludmann [107].

On the suitability of various commercial pararosaniline preparations see Nietruch and Prescher [189].

For a comparison of the results obtained with the Stratmann and West–Gaeke methods see Stratmann and Buck [246a].

5.5.2.11 *Determination of* SO_2 *with* Fe(III)-1,10-*phenanthroline*

The reduction of Fe(III) to Fe(II) by SO_2 is sometimes used in the determination of iron. Stephens and Lindstrom [243] employed this reaction for the determination of SO_2 in the presence of 1,10-phenanthroline which with Fe(II) yields the bright yellow color known in oxidimetry as ferroin indicator.

The absorption solution is prepared as follows: 10 ml of 0.001 *M* ferric ammonium sulfate solution are mixed with 10 ml of a 0.03 *M* 1,10-phenanthroline solution; the pH of the mixture is adjusted to 5–6 with NaOH. The mixture is diluted to 75 ml and 1.0 ml n-octyl alcohol is added as an antifoaming agent. The mixture is filled into a fritted scrubber and placed in water bath at 50°C. The air sample is passed through at a rate of 2–3 liter/min. Then the absorption solution is mixed with 2 ml 5% ammonium difluoride solution in a 100-ml volumetric flask, cooled down to room temperature and then made up to the mark. The extinction is measured at 510 mμ.

The calibration curve is plotted after injecting 5–100 μl SO_2 into a purified air stream. 27 μl SO_2 in a $\frac{1}{2}$-inch cell yielded an extinction of 0.132 ± 0.005.

Ozone and formaldehyde start to interfere when present in amounts above 60 μg; H_2S distorts the quantitative results.

When 100 liter of air are used, the detection limit is 0.05 ppm SO_2.

For automatic measurements based on this principle see Malanchuk [177a].

5.5.2.12 *Polarographic determination of* SO_2

Principle: The American Conference of Governmental Industrial Hygienists recommend the polarographic method of Kolthoff and Miller [140], especially if larger amounts of NO_2 are present. The principle of the method consists in the reduction of SO_2 to sulfoxylic acid (H_2SO_2) at the dropping mercury electrode for a half-wave potential of -0.50 V. This technique is less sensitive than the colorimetric methods but covers a larger concentration range.

Sampling: In order to determine low SO_2-concentrations at least 1 m^3 air sample is required. The gas is trapped in an impinger through which air is passed at a rate of 30 liter/min for 30 min. The impinger is filled with 100 ml (according to the original prescription), or better still, with 50 ml or even less absorption solutions to increase the sensitivity.

5–20 ml of the solution obtained are placed into the polarographic cell and pure nitrogen is bubbled through the solution and through an equal volume of acetate buffer. Both solutions are mixed under airtight conditions and nitrogen is passed for $1\frac{1}{2}$ min through the mixture. Then the nitrogen stream is interrupted and the polarographic determination is carried out between -0.35 and -1.00 V. In the same way a calibration curve is plotted by using sulfite solutions containing 3–30 μg SO_2/ml. Here the heights of the plateaus (diffusion currents) are plotted as a function of the concentration.

The lowest concentration at which SO_2 can be determined is 2 μg/ml absorption solution or 0.008 ppm in air (10 ml solution, 5 liter/min air for 30 min). The average deviation from the theoretical value is about $\pm 7\%$.

Solutions: Absorption solution 2% glycerol in 0.05 N NaOH. Acetate buffer 2.5 M CH_3COOH in 0.5 M CH_3COONa (pH = 3.85–3.95). Standard sulfite solution, 1.484 g sodium metabisulfite $Na_2S_2O_5$ or 1.625 g $NaHSO_3$ are dissolved in 1000 ml absorption solution and standardized by iodometry. 1 ml standard solution is equivalent to 1 mg SO_2.

Ciaccio and Cotsis [41] trapped the SO_2 in the air sample in 25 ml 0.3 M Na_2HgCl_4 solution. For the polarographic analysis 2.0 ml hydrazine hydrate solution are added, the mixture is agitated for 2 min, centrifuged and 10 ml of the clear solution are transferred into the polarographic cell. The cell is rinsed for 2 min with nitrogen and the pH is adjusted to 1.0 with 2.0 ml 5 N HCl. The polarographic determination is performed between 0 and -1.0 V (Ag/AgCl–KCl).

Determination of SO_2 *by emission flame photometry*

The blue coloration of flame caused by the combustion of sulfur-containing air constituents was used by Crider [49a] for the continuous automatic determination of SO_2 and other sulfur-containing pollutants (SO_3, H_2S) in the range from 0.1 to 3 ppm SO_2 or 0.17 to 5 mg H_2SO_4. For this purpose the air sample (325 ml/min) is mixed with hydrogen (640 ml/min) and then ignited upon leaving a nozzle mounted in a borosilicate glass tube (diameter 2.5 cm). The light of the flame is directed by means of a lens via a light filter of 402 mμ (corresponding to the emission band of 350–420 mμ) onto a photomultiplier cell. The induced photocurrent is compared every 2 min with the photocurrent produced by a sulfur-free air sample.

The current difference of 1.4×10^{-9} A obtained by an air mixture containing 0.37 ppm SO_3 is reproducible with a relative standard deviation of $\pm 8\%$, so that 0.1 ppm SO_2 can still be detected with some certainty. Other sulfur-containing pollutants (SO_3, H_2S) yielded in the same molar ratio to SO_2 approximately the same flame emission, while concentrations of 100 ppm CO, 1.5 ppm NO_2 and 25 ppm hydrocarbons had no influence on the measurement result.

5.5.2.13 *Automatic continuous* SO_2*-determinations*

Among the methods used for this purpose the determinations based on electrical conductivity are the most important. The relevant instruments were already described in Section 4.4.2. The instrument most frequently applied is the "Ultragas 3" of the firm Woesthoff. Its mode of operation is described in VDI-Richtlinie 2451.

SO_2*-determinations with the Ultragas* 3

Principle: The air sample is led through an acidified H_2O_2 solution and the increase in the electrical conductivity due to the formation of H_2SO_4 is measured.

The instrument is described on page 73.

The reaction solution for the determination of SO_2 contains 1 ml 0.1 *N* H_2SO_4, 0.1 ml 30% H_2O_2 and 0.2 ml Docen solution in 1 liter deionized water, whose conductivity is below 2 $\mu ohm^{-1} \cdot cm^{-1}$. The reaction solution should have an electrical conductivity of 39 $\mu ohm^{-1} \cdot cm^{-1}$ at 20°C. The gas flow rate is between 40 and 275 ml/min and the throughput rate of the reaction solution 0.4–1.5 ml/min.

To effect the determination the instrument is connected to 220 V ac mains. A constant zero point is established by means of air purified with an activated-carbon filter. The indication must be checked twice a month with a test gas whose SO_2-content corresponds to the MIK-value. Air bubbles must be carefully removed from the fluid pipes.

The most sensitive measuring range of the instrument is 0–1 ppm SO_2. The zero point varies by about $\pm 3\%$ and the sensitivity by about $\pm 5\%$ of the full-scale deflection. For a comparison with the MIK-values half-hourly average values are calculated.

Sensitivities to interfering substances

1 mg H_2SO_4	simulates	0.6 mg SO_2
1 mg H_2S	simulates	0.01 mg SO_2
1 mg NO_2	simulates	0.08 mg SO_2
1 mg HCl	simulates	0.84 mg SO_2
1 mg NH_3	simulates	−1.8 mg SO_2
1 mg Cl_2	simulates	0.8 mg SO_2

The continuous determination of the SO_2-content by means of the "Picoflux" was already described on page 74; for coulometric measurements see page 75.

Since the conductivity method is not very specific, it frequently happens that much higher SO_2-levels are found than in other, more specific procedures (see, for example, Benson, Nevill, Thompson, Terabe and Omichi [19b], as well as Stalker, Dickerson and Kramer [242b], and Stratmann and Buck [245a]).

Automatic colorimetric SO_2*-determinations with pararosaniline*

The disadvantage of the West–Gaeke method lies in the fact that when the SO_2-level of the air is unknown, we do not know whether a sufficient air volume was sampled for the subsequent colorimetric determination. Also for automatic continuous recording the method must be modified, so that the color reaction can be measured without having to deal with a long waiting time after introducing the sample (the waiting time in the conventional West–Gaeke method is 20–30 min).

Helwig and Gordon [102] prepared a suitable reagent as follows. The pH of 20 liter water is adjusted to 1.5 by adding concentrated HCl (about 80 ml). To this 40 ml pararosaniline hydrochloride solution are added. The latter is prepared by dissolving 0.4 g pararosaniline hydrochloride in 100 ml hydrochloric acid (1 vol. concentrated HCl + 9 vol. water). In continuous measurements the throughput rates of the air and solution are 250 ml/min and 3.3 ml/min, respectively. The color reaction is almost complete after 5 min.

A NO_2-concentration of 2 ppm results in an SO_2-loss of 0.1 ppm. Ozone concentrations of the same order of magnitude as the SO_2-concentration do not appreciably interfere, provided that the time of residence in the solution is short.

The Imcometer (see page 70) can also be used for automatic colorimetric SO_2-determinations.

Determination of SO_2 *by semicontinuous pH-measurements*

The "Turicum II" of Kündig and Högger [143] (page 78) was especially recommended for recording SO_2 concentrations in the atmosphere.

The air sample is first freed from ammonia and partially from HCl and SO_3 by scrubbing with 40% phosphoric acid. Then it is introduced for 10 min at a rate of 0.85 liter/min into a receiver containing 8 ml of a 1% H_2O_2 solution (pH 5.00), in which a glass electrode is immersed. At the same time, a similar air stream, previously freed from SO_2 by passing it through a tetrachloromercurate solution, is led into a receiver with a glass electrode and containing the same absorption solution. The potentials of the two glass electrodes are continuously and separately recorded by a two-channel pH recorder. After 10 min the solution is automatically renewed. When the air sample does not contain SO_2, two parallel lines are obtained; on the other hand, when it contains, for example, 0.15 ppm SO_2, the measuring electrodes indicate a periodically interrupted trend of the potential. The instrument is calibrated with

SO_2–air mixtures of suitable concentrations. The accuracy is about $\pm$ 0.03 ppm. According to the data of the manufacturer, the modified version of the instrument responds to SO_2 but not to HCl, H_2SO_4, SO_3, NO, NO_2 or NH_3.

Determination of SO_2 *via radioactivity measurements*

A very interesting method for the determination of SO_2 is described by Bersin, Bronsaides and Hommel [19a]. SO_2 reacts with sodium chlorite to form chlorine dioxide. The latter releases atoms from the quinol clathrate of ^{85}Kr, the radioactivity of which is measured.

5.5.2.14 *Bell method according to* Liesegang [163]

In the methods using cloth strips for long-term (summation) measurements Liesegang bells are the most widely employed in Germany. As a rule, the exposure time is 100 hr.

Apparatus: The Liesegang bell consists of a porcelain cylinder (height 14 cm, diameter 5 cm) with a rounded edge which can hold 50 ml absorption liquid. A cartridge of filter paper is wrapped around the cylinder and immersed into the solution. The bell is protected against rain and bird droppings by an aluminium cover. In order to investigate the influence zone and the efficiency of an immission source 8 bells are placed in 8 wind directions at distances of 500 or 1000 m and a height of 75 cm. The wind directions are continuously recorded by means of an anemometer.

The absorption solution consists of a mixture of water : glycerol : K_2CO_3 = 1 : 1 : 1 (by wt.).

Working procedure: After exposure (usually 100 hr) the paper cartridge and liquid are extracted with water. The filtrate is acidified with HCl, evaporated and sulfur dioxide is precipitated as $BaSO_4$. The result is expressed in mg S ("bell value"). Values for comparison with the conventional SO_2-levels are listed in Tables 5.5.2.14a and b.

Table 5.5.2.14a
BELL VALUES

Unpolluted country air	up to 5 mg S
Spread-out cities without industry	up to 10 mg S
Densely populated cities	up to 20 mg S
In the vicinity of some industrial plants	up to 300 mg S
Damage to vegetation not to be expected	below 30 mg S
Possible damage to vegetation	above 50 mg S
Damage to vegetation	above 100 mg S

Summation method with the lead peroxide candle

This method is widely used in Great Britain and the USA. It is described in British Standard 1747, IV, 1963 and in ASTM D 2010-62 T, 1962.

Table 5.5.2.14b
EMPIRICAL CORRELATION
BETWEEN CONCENTRATION (mg SO_2/m^3)
AND BELL VALUE (mg S)

mg SO_2/m^3	0.1	0.1–0.2	0.2–0.5	0.5–1.0
Bell value	10	10–20	20–50	50–200

Apparatus: A cylinder (height 12.8 cm, diameter 2.5 cm) made of porcelain, glass or plastic is placed in a louvered shelter for protection against rain and contamination by birds. The louvers are set at an angle of 45°, thus exposing the cylinder to the ambient air. The PbO_2 paste is applied onto a 100-cm^2 cotton gauze wrapped around the cylinder. The result is the lead peroxide candle. It is set up at the site where the pollution is to be determined.

Preparation of the paste: 8 g fine lead peroxide powder are mixed with a gum tragacanth solution to a uniform paste which is applied to cotton gauze (12.8×8 cm^2). The gum tragacanth solution is prepared as follows: 3 g tragacanth powder are suspended in 15 g ethanol. Then 150 ml hot water are gradually added while mixing until a uniform solution is formed. The cylinder with gauze is dried in SO_2-free air.

The exposure time is usually taken as 30 days. A candle can take up 15% of the SO_2-amount equivalent to 8 g PbO_2.

Processing: After exposure the gauze with the paste is allowed to stand with a solution of 5 g Na_2CO_3 in 60 ml water for 3 hours. Then it is boiled for half an hour (supplementing the water evaporated) and filtered. The filtrate is acidified with HCl (methyl orange) and the sulfate precipitated with $BaCl_2$ and weighed as $BaSO_4$. The result is expressed in mg $SO_2/cm^2 \cdot$ day.

According to Rayner [209], the SO_2 in the exposed lead peroxide can also be determined by titration with barium perchlorate using Thoron as the indicator (see page 157). The PbO_2 paste is boiled with ammonium carbonate instead of sodium carbonate. The solution (or an aliquot thereof) is adjusted to a pH of 3.2 with $HClO_4$. Thoron is added and the solution is titrated with 0.02 *M* $Ba(ClO_4)_2$ to pink. When the sulfate content is low, it is better to plot a calibration curve instead of carrying out the stoichiometric calculation.

For a comparison of the results obtained by this method with those of individual analyses (peroxide method) see Eaves and Macaulay [58a].

5.5.3 Sulfuric acid and sulfur trioxide

General

Sulfuric acid mists (H_2SO_4 or SO_3) in industrial waste gases occur when the sulfur trioxide formed from sulfur dioxide in sulfuric acid plants by the contact process is incompletely absorbed. Sulfuric acid is formed in the atmosphere by oxidation of SO_2, especially in the presence of soot, catalytically acting metal oxides from ash components and air humidity.

Such soot and smoke particles which contain free H_2SO_4 are apparently responsible for the noxious effects of flue gases in respiratory air.

The MAK-value for H_2SO_4 mists is 1 mg H_2SO_4/m^3.

Analysis

Principle: To isolate H_2SO_4 mists together with acid adsorbates on soot and smoke, the air sample is drawn through a double layer of filter paper (Mader, Hamming and Bellin [177]). The amount of acid taken up by the paper is potentiometrically titrated with 0.002 *N* NaOH.

Preparation of the filters: Sheets of Whatman No. 4 filter paper are leached with distilled water for 60 hr (the water is changed 5 times). Circular disks of 2.5 cm diameter are cut from the dried papers which must fit into a suitable suction device (see page 40). For the determination of the blank or to establish whether the filter is free of acid, two disks are macerated and stirred with a glass rod in 20 ml CO_2-free water of measured pH. The pH-value, which has to be determined several times with an accuracy of $\pm$ 0.03 units, must not differ by more than 0.1 unit from the original value.

Working procedure: Two dry clean filter disks are inserted into the suction device and the air sample is aspirated at a rate of 28 liter/min (1 cu. ft/min). For H_2SO_4-levels around the MAK-value about 50 liter are required, while for investigations in the open up to 1 m^3 air is necessary.

Before determining the amount of acid absorbed, both filter disks are macerated and crushed in 20 ml water (just as in the determination of the blank). A glass electrode is immersed into the filter pulp and 0.002 *N* of sodium hydroxide is added until the pH of the blank is attained. 1 ml 0.002 *N* of sodium hydroxide corresponds to 98 μg H_2SO_4.

In this determination all the acids adhering to the filter are determined. If only sulfuric acid (including the sulfates) has to be determined, the titration is performed with barium perchlorate using Thoron as the indicator (see page 157).

The determination of SO_3 in the presence of SO_2 in flue gases by selective absorption is described by Fielder and Morgan [67]. SO_3 is trapped by selective absorption in 4:1 isopropanol water and is then titrated with Thoron serving as the indicator.

An automatic instrument—the "Sulfatherm"—for the determination of SO_3 in the presence of SO_2 in flue gases is manufactured by the firm Siemens. SO_3 is isolated with condensing water vapor from the gas sample. In the condensate SO_2 is removed by evaporation (see Sieth [231b] and Wahnschaffe [259a]).

5.6 Nitrogen-containing air pollutants

5.6.1 Ammonia

5.6.1.1 *Presence and general description*

Ammonia occurs free or in the form of its salts as traces in air. The ammonia level in the air depends on the vicinity of natural or artificial decomposition processes, especially of organic substances, and on the time which has passed since the last precipitation (rain, snow). It is a normal end product of the decomposition (putrefaction) of nitrogen-containing organic substances. In larger concentrations it is formed by the decomposition of urea and uric acid, and thus reaches the air from sewers, lavatories, stables and manure pits (in most cases together with hydrogen sulfide).

Ammonia is widely applied in the chemical industry and thus can be encountered at various work places. In addition to the production plants of ammonia synthesis, it can be found in coking plants and processing of its side products (ammonia liquor), as well as in the production of nitric acid, fertilizers, hydrocyanic acid, urea, plastics and pharmaceutics. Ammonia is an auxiliary material for the production of soda and is used in refrigeration plants.

Physical properties

The boiling point of liquid ammonia is $-33°C$. It is strongly absorbed by water; the aqueous solution over NH_3 vapor contains 34 wt. % ammonia at 20°C. Ammonia water is marketed in the following concentrations: 25% (d = 0.91) and 35% (d = 0.88). The explosion limits for an NH_3–air mixture is 15.5–28 vol. % NH_3.

Toxicology

The odor threshold for ammonia is 20–40 mg/m^3. Levels of 100 mg/m^3 are tolerable without adverse effects for a certain time, while those of 1500–2500 mg/m^3 are highly dangerous after $\frac{1}{2}$ hr. Ammonia gas acts on the mucous membranes of the mouth and nose as well as the upper respiratory tracks as a strong irritant.

The MAK-value is at present 50 ppm = 35 mg NH_3/m^3.

5.6.1.2 *General analytical information*

In determining the ammonia level in air, the alkaline properties of this substance can be used; the acid consumption can be measured when the gas is introduced into an acid of known titer. The determination is even simpler when the air sample is bubbled through the acid until the indicator dissolved in the latter changes its color. The air volume used up to this color change constitutes a measure for the NH_3-level. This procedure is, of course, only applicable, when no other pollutants are present which can change the color of the indicator. When determining the MAK-values of NH_3

(50 ppm), such pollutants (e.g., SO_2, Cl_2, NO_2) will be rarely present in amounts interfering with the measurement. On the other hand, when low NH_3-levels (ppm-range and below) have to be determined, more specific methods will be used.

Two colorimetric methods can be applied for this purpose. Buck and Stratmann [32] used the Nessler reagent and developed an accurate procedure. A disadvantage is the interference of small amounts of H_2S, frequently occurring as substances of organic origin which accompany ammonia. Therefore, NH_3 must be separated from the alkaline solution prior to its determination. A modification of the indophenol reaction, recently reported by Leithe and Petschl [160], eliminates this difficulty. By means of the two principal photometric methods NH_3-levels in the ppb-range can be determined in 100–1000 liter air.

5.6.1.3 *Alkalimetric determination of* NH_3 *in amounts of the order of the MAK-value*

Principle: This method is used in the Osterreichische Stickstoffwerke, Linz, for the determination of the NH_3-level in air of fertilizer silos and other storage or working spaces. Diluted sulfuric acid with bromophenol indicator is prepared in an amount equivalent to the MAK-value. The air is introduced into the acid until the indicator changes its color. The volume of air introduced up to this point constitutes a measure for the NH_3-level.

Apparatus: Aspirator (see page 37) for 5 liter air, consisting of two polythene flasks with a capacity of 5 liter air each. The aspirator is tubulated or provided with a siphon, whereby volumes as small as 0.2 liter can be read off. A 150-ml fritted scrubber G I of Jena glass (free of grease) with ground-glass joint (see page 44) is employed. The supply tube for the gas must be dry down to the frit.

Reagents: 0.0005 *N* H_2SO_4 prepared by diluting 5 ml 0.1 *N* H_2SO_4 to 1 liter. One liter of the acid also contains 4 ml n-butanol and 6 ml bromophenol blue solution (0.1 g in 100 ml ethanol).

Working procedure: 20 ml of the prepared sulfuric acid solution are introduced into the scrubber. The ground-glass joint of the supply tube is inserted in such a way that the inlet opening for the air sample is first closed with a rubber stopper or by a squeezed hose end in order to prevent wetting of the supply tube by the acid within the frit and oversaturation with NH_3 prior to the aspiration of the air sample. Then the aspirator set at zero is connected, adjusted to a low throughput rate, the closure of the supply tube is opened and the aspiration effect is regulated in such a way that an approximately 6-cm high foam zone is formed corresponding to an air throughput rate of about 1 liter/min. The solution immediately begins to turn gray, the scrubber is shaken and the solution seeping through the frit is brought to the top by inclining the scrubber. The aspirator is disconnected right after the solution turns blue. The air volume in the aspirator is read off after lifting the lower flask to the level adjustment (x liter).

$$\text{Calculation: mg NH}_3/\text{m}^3 = 170\,\text{x}.$$

This procedure can be modified when other concentration ranges are to be determined. The air throughput rate can be considerably increased by means of an absorption solution without n-butanol. The alcohol is used to ensure quantitative absorption, though, according to recent investigations of Leithe and Petschl [159], this is not absolutely necessary.

5.6.1.4 *Determination of low ammonia concentrations according to* Buck *and* Stratmann

Principle: Buck and Stratmann [32] absorb NH_3 in the impinger at an air throughput rate of about 1 m^3/30 min. For NH_3-concentrations between 17 and 400 $\mu g/m^3$ the absorption is practically complete. The receiver contains in addition to 0.01 *N* H_2SO_4 also potassium permanganate in order to eliminate hydrogen sulfide and formaldehyde, which also react with the Nessler reagent. The ammonia is distilled off from alkaline solution and determined photometrically with the Nessler reagent at 450 mμ.

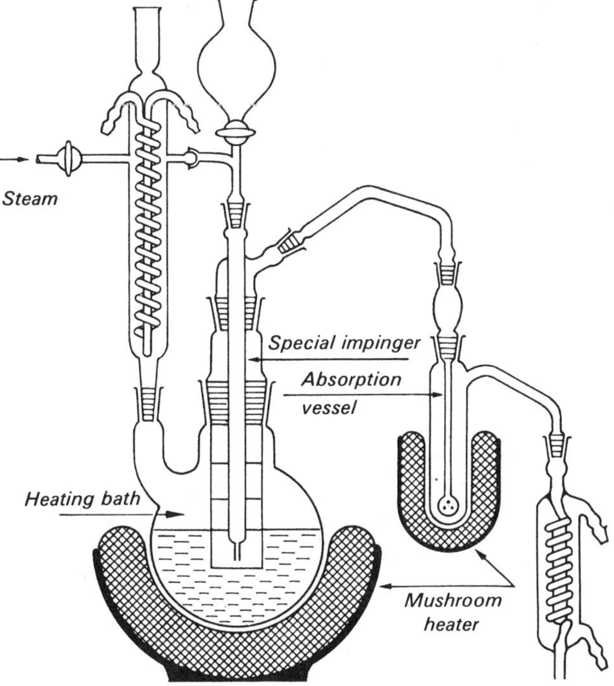

Figure 5.6.1.4
Distillation apparatus for separating ammonia from accompanying substances

Apparatus: Impinger with outer ground-glass joint from which the ammonia can be distilled (without decanting) into a suitable distillation apparatus.

Working procedure

Sampling: The 275-ml special impinger mentioned above is filled with 50 ml 0.01 *N* H_2SO_4 and 10 ml 0.4% $KMnO_4$. 0.8–1.0 m^3 air are aspirated within 30 min.

Purification by distillation: 30 ml 0.01 *N* H_2SO_4 are introduced into the receiver, 30 ml 30% NaOH are added, and liquid in the heating bath (acetic amylate, b.p. 138–140°C) is boiled. The receiver is also heated (to about 120°C), so that after NH_3 has been taken up by the receiver the excess water vapor escapes without diluting the absorption liquid, condenses outside and is removed. The process is carried out until 100 ml condensate are collected (30 min) and all the NH_3 is collected in the receiver. The content of the receiver is transferred into a 50-ml volumetric flask; 4 ml Nessler reagent (Merck, No. 9028) are added and the volume is made up to the mark. After 20 min the photometric determination is carried out at 450 mμ for a 2-cm layer thickness.

To plot the calibration curve NH_3-amounts between 10 and 100 μg are subjected to steam distillation, treated as above and measured.

5.6.1.5 *Determination of low ammonia concentrations via the indophenol reaction*

Principle: As shown by Leithe and Petschl [160], the indophenol reaction (blue coloration with sodium hypochlorite and sodium phenylate) has the following advantage over the Nessler reagent when determining low NH_3-concentrations: hydrogen sulfide does not interfere, so that distillation of the absorbed ammonia can be foregone. When carrying out the reaction and plotting the calibration curve, the prescription for the composition and the sequence of adding the reagents must be adhered to. The detection limit and reproducibility correspond to those of the Nessler reaction.

Reagents: (see also Niedermair [188]) Sodium phenylate; 62.4 g phenol are dissolved in 100 ml sodium hydroxide (270 g NaOH/liter), 8.0 ml acetone are added and the volume is made up to 1000 ml with water.

Sodium hypochlorite solution: Chlorine is introduced into 200 g 10% NaOH while cooling with ice until the reaction is neutral. Then 30 ml 3% NaOH are added. The solution is adjusted to a content of 0.7% free chlorine (the chlorine concentration is determined by iodometry), in a sample acidified with HCl, i.e., the solution is diluted 1:10. Further, its content of free alkali is determined by introducing the solution into 20 ml 3% H_2O, evaporating the excess H_2O_2 and titrating with 0.1 HCl against methyl red, and then adjusted to 1.5% free NaOH.

The solutions are prepared by using NH_3-free water (obtained with the aid of a cation exchanger and by double distillation).

$MnSO_4$ solution: 0.05 g $MnSO_4$ (anhydrous) per 100 ml solution.

Working procedure

A suitable fritted scrubber is filled with 25 ml 0.01 *N* H_2SO_4 and the air sample is introduced for 30 min at a rate of 30 liter/min for NH_3-determinations in the ppb-range. The solution is transferred into a 50-ml volumetric flask and neutralized with 2.5 ml 0.1 *N* NaOH. Then 0.2 ml $MnSO_4$, 8 ml phenylate and 2 ml hypochlorite solutions are added and the volume is made up to 50 ml. The solution is heated to 50°C, allowed to stand for 15 min (this period can be extended to 2 hours) and the photometric determination is carried out at 650 mμ in a 1- or 5-cm cell, depending on the intensity of the color, against a blank solution treated in the same manner.

The calibration curve is obtained with 0–100 μg NH_3 (as $(NH_4)_2SO_4$) in 30 ml water; addition of the reagents, treatment and measuring as for the sample. Up to about 100 μg NH_3 the curve is a straight line with a reciprocal slope of 144. Thus, for example, 144 μg NH_3 in 50 ml measuring solution (layer thickness 1 cm) yield an extinction E = 1.0; 71.5 μg NH_3/50 ml yield E = 0.500. Thus, the calculated molar extinction coefficient is 5900.

The detection limit (corresponding to the 3-fold standard deviation of the E of a blank solution) is 1 μg NH_3.

Interfering substances: Monoalkylamines, but not di- or trialkylamines, give a similar reaction with indophenol as ammonia. Formaldehyde interferes when present in an amount equal to one fifth of the ammonia content. It can only be eliminated by alkaline distillation. On the other hand, hydrogen sulfide interferes independently of the NH_3-level only when present in amounts equal to or exceeding 50 μg. Larger amounts of H_2S can be removed from the H_2SO_4-containing absorption solution by evaporation until the characteristic odor disappears, without interfering with the subsequent indophenol reaction. Nitrite and sulfite only interfere when present in amounts 100 times larger than the NH_3-content.

5.6.1.6 *Determination by direct UV spectrophotometry*

Gaseous ammonia exhibits several strong absorption bands in the range between 190 and 230 mμ. Accordingly, Gunther, Barkley, Kolbezen, Blinn and Staggs [88] carried out the determination of ammonia in air by direct UV spectrophotometry at 204.3 mμ in 10-cm quartz cells. The molar extinction coefficient at this wavelength is 2790. The limit concentration is 7 ppm.

Kolbezen, Eckert and Wilson [139] described a corresponding instrument for the continuous automatic determination of ammonia levels above 10 ppm, such as occur, for example, in storehouses for citrus fruits.

5.6.2 Determination of monoethanolamine in air

The determination of monoethanolamine ($NH_2CH_2CH_2OH$) in the presence of or together with ammonia is of importance in the air of small enclosed spaces, such as in submarines, where the carbon monoxide is continuously removed from the air by scrubbing with the above base.

According to Williams and Miller [271], the base can be determined with test tubes containing ninhydrin. These tubes contain a 2-cm long layer of 2% ninhydrin (triketohydrindene hydrate) and silica gel (Davison No. 15), an intermediate layer of 1 cm finely ground calcium carbonate and finally a 1-cm layer of finely ground boric acid, which absorbs the ethanolamine but not ammonia. The tubes are fused after filling and can be stored for several months.

First 5–10 liter air are passed through in one direction and then in the opposite direction. The color reaction takes place after the tube is heated for 3 min at 90–95°C. The length of the colored silica-gel layer corresponds to the sum of the two bases (NH_3 + ethanolamine), while that of the H_3BO_3 layer corresponds to NH_3 alone. The tubes are calibrated with test air mixtures of known composition.

Determination of primary aliphatic amines with ninhydrin

Häntzsch and Prescher [93] prefer the use of impingers for the absorption of low concentrations of primary aliphatic amines (methylamine to hexylamine) in the ppm-range by HCl diluted with water or isopropanol. When evaporating the hydrochloric acid solution with base levels in the μg-range, losses of 60% at normal pressure and 20–40% in vacuo were observed. The concentration is determined by the ninhydrin reaction as follows:

A 10-ml volumetric flask is filled with 2 ml of a freshly prepared solution containing 0.158 g ninhydrin in 100 ml isopropanol ("for chromatography"). The residue from the evaporated acid absorption solution is taken up in 3 ml isopropanol–hydrochloric acid mixture (14 ml 25% HCl per liter isopropanol) and introduced into the flask. Then 5 ml α-picoline ("for chromatography") are added, and the flask is placed in a thermostat set at 85°C for exactly 7 min. The flask is then cooled to room temperature and the extinction measured at 575 mμ against an amine-free blank solution.

5.6.3 Hydrazine

Hydrazine gained importance since its use as a rocket propellant in World War II. In the laboratory it is used because of its reducing properties as a reagent, in engineering for the removal of dissolved oxygen in the processing of high-pressure boiler supply water and recently in the plastics industry.

The toxic effect of free hydrazine is similar to that of ammonia, but enhanced by the reduction effect of this compound; a MAK-value of 1 ppm has been proposed.

Analysis

The photometric determination of small hydrazine amounts in air using the color reaction with dimethylaminobenzaldehyde has recently been described by Pilz and Stelzl [204].

A small impinger or fritted scrubber is filled with 10 ml absorption solution. The air sample (10–100 liter) is introduced at a rate of 5 liter/min. Then the solution is transferred into a 50-ml volumetric flask, 10 ml reagent are added and the volume is made up to the mark. The solution is allowed to stand 20–30 min until the color has completely developed and the photometric determination is carried out at 455 mμ against a simultaneously set up blank.

Absorption solution: 40 ml sulfur dioxide (d = 1.84) for forensic purposes are diluted to 500 ml with water.

Reagent: To 5 g dimethylaminobenzaldehyde are added 5 ml absorption solution. The mixture is then diluted with pure ethanol to 100 ml. The solution is only stable for 1 week; when it acquires a brown tint the solution can no longer be used.

The calibration curve is obtained on the basis of hydrazine sulfate solutions, corresponding to 0–25 μg $NH_2 \cdot NH_2$ in 50 ml solution according to the above procedure.

Calculation: 1 μg $NH_2 \cdot NH_2$ per ml solution = $E_{1\ cm} \times 0.510$. The molar extinction coefficient is hence 64,000. The detection limit for 100 liter air sample is about 3 ppb.

Buck and Eldridge [30a] report on the continuous coulometric determination of the asymmetric dimethylhydrazine in the ppm-range with coulorimetrically generated bromine as the reagent (sensitivity $\pm$ 0.1 ppm).

5.6.4 Nitrous oxide

5.6.4.1 *General*

Nitrous oxide (laughing gas) is of no importance as an air pollutant but is interesting because of its natural occurrence in low concentrations. In concentrations of 0.3 ppm it is a normal constituent of the unpolluted atmosphere and is dissolved in sea water in about the same amounts. Nitrous oxide is formed upon decomposition of nitrogen-containing inorganic and organic substances; it is contained in appreciable amounts in explosion clouds and the waste gases of oxidation processes with nitric acid. It is formed in small concentrations in the production of nitric acid by ammonia combustion or of fertilizers by heating ammonium nitrate. N_2O traces are also found in tobacco smoke (40 μg per g tobacco). Since there are relatively few specific methods for detecting N_2O and since this gas is readily soluble in water, it is frequently not recognized by the analytical chemist and is probably contained in some other waste gases.

Nitrous oxide is currently being applied in dental practice as well as in surgery as an anesthetic; for this purpose it is marketed in very pure form and stored in steel cylinders. Nitrous oxide is sometimes also encountered in foodstuffs, for example, whipped cream.

Toxicological effects

Because of its low reactivity at room temperature no toxic effects are known except the narcotic effect when the gas is present in high concentrations.

Boiling point $-89°C$. Solubility in water: at an N_2O-partial pressure of 760 mm Hg 1 vol. water dissolves 0.63 vol. N_2O at 20°C. The gas is colorless and practically odorless.

Remarks pertaining to the analytical methods

The analytical chemistry of nitrous oxide (which is mainly interesting because of its use in anesthesia) is characterized by a lack of specific chemical reactions at room temperature. Therefore, only physical methods (namely, mass spectrometry and IR analysis) are suitable for determining low concentrations. The IR spectrum has a strong band at 1275 cm^{-1} and several weaker ones between 2200 and 2250 cm^{-1}.

Low NO_2-concentrations in air and waste gases can be determined by gas chromatography. A gas-solid chromatographic method using silica gel or activated carbon as the fixed phase was first developed by Janak and Rusek [119]. A simpler and more sensitive method of gas-liquid chromatography developed by Leithe and Hofer [160a] will be described in the modified version used at the main laboratory of the Österreichische Stickstoffwerke.

5.6.4.2 *Gas-chromatographic determination of low* N_2O*-concentrations*

Principle: Low N_2O-concentration in the ppm-range can be concentrated over silica gel at $-70°C$ and desorbed at room temperature. Air with a N_2O-level of above 50 ppm is directly led into the separating column. For lower levels the gas should be concentrated in a silica gel column. Low CO_2-levels (up to about 400 ppm) appear as distinct peaks after the N_2O-peak; when the CO_2-concentration is higher, as, e.g., in exhaled air, it is expedient to interpose a small cartridge with soda lime.

Apparatus: Separation column (10.5 m × 6 mm). Fixed phase 19% propylene carbonate and 16.5% glutaronitrile on 0.2–0.3 mm Sterchamol. The carrier gas is helium (60 ml/min). Temperature 20°C. Sample volume 5–10 ml. Detector with thermistor probe.

Under these conditions the air peak appears after 6 min, the N_2O-peak after 13 min and the distinct CO_2-peak after 14 min.

Concentration column: U-tube (30 cm × 6 mm) filled with silica gel type 12 of the firm Davison. It is immersed during sampling (air throughput rate 500 ml/min) into a Dewar flask filled with acetone and dry ice (-70 to $-75°C$). For desorption the concentration column is first flushed with helium, immersed into water at room temperature and then connected to the separation column.

Soda lime cartridge (10 cm × 6 mm) of the firm Merck. The cartridge is inserted between the sample inlet and the beginning of the column. The detection limit for N_2O, when concentrated from a 10-liter air sample, is 0.05 ppm, without the concentration column—50 ppm.

When the silica gel is activated prior to use by heating for 5 hr at 160°C, the packing of the column can effect a complete separation of the N_2O naturally occurring in the air (in Linz, Austria, and its environment a concentration of 0.22 ± 0.025 ppm N_2O was found).

This method is also suitable for determining nitrous oxide dissolved in natural waters, whereby the gases dissolved in 5 liter water are transferred by boiling (reflux condenser) and passing air-free carbon dioxide into an azotometer containing 50% KOH.

5.6.5 Nitric oxide and nitrogen dioxide

5.6.5.1 *General remarks and occurrence*

In the earlier literature on air pollutants nitrogen monoxide (NO) and nitrogen dioxide (NO_2) were treated together under the common heading nitrous gases, whereby it was assumed that nitrogen monoxide is rapidly and almost completely oxidized into nitrogen dioxide in the presence of atmospheric oxygen or that it combines with this oxygen to N_2O_3 (the anhydride of nitrous acid).

It was shown, however, by Bodenstein and subsequently by other authors that low (ppm-range) NO-levels in the air are stable for a very long time. Accordingly, when 5 ppm NO are present in the air, it takes $1\frac{1}{2}$ hr at 20°C until 10% of the NO content is converted to NO_2. A 50% conversion under these conditions takes $10\frac{3}{4}$ hr, while in the presence of 1 ppm 10% are converted to NO_2 only after 8 hr. The reaction $NO + NO_2 \rightarrow N_2O_3$ is of no significance with regard to the concentrations occurring in the air.

Thus, NO and NO_2 must be separately evaluated according to their analytical determination as well as their toxic effects.

Nitrogen monoxide is formed in the air from nitrogen and oxygen at high temperatures. At 2000°C 1.5% NO are in equilibrium with N_2 and O_2. When the reaction product is slowly cooled, it passes through temperature ranges in which NO largely decomposes into N_2 and O_2. If, however, the cooling process is very rapid, the temperature ranges, in which the rate of decomposition is large, are traversed very rapidly and a range is reached in which the equilibrium, though very much on the side of decomposition, is so slowly established ("freezing") that NO can be considered as practically stable.

For this reason NO is a side product of all combustion and explosion processes in which an excess of oxygen is present and the combustion products are rapidly cooled.

The most important process is the formation of NO in gasoline and diesel engines; at complete combustion and top speed the exhaust contains NO in amounts up to 4000 ppm. In Los Angeles the amount of NO emitted in automobile exhaust per day reaches 600 t. The role of nitrogen oxides in smog formation was already discussed on page 17. The waste gas of gas turbines contains up to 2000 ppm, while that from coal-processing thermal power plants 200 and 1200 ppm NO, depending on the type of heating.

Hazardous concentrations may also occur in autogenic and electric welding in enclosed spaces. Also during thunderstorms NO and NO_2 are formed.

NO and NO_2 are present in dangerous concentrations (145–1000 ppm) in cigarette smoke (see Bokhoven and Niessen [23], Haagen-Smit, Brunelle and Hara [92], as well as Norman and Keith [189a]).

Furthermore, nitrogen oxides are formed upon the thermal decomposition of nitric acid or nitrates in the presence of reducing agents. They are present in the explosion clouds of nitrate- or nitrite-containing explosives, in the combustion waste gases of celluloid and in the gaseous products of superheated nitrate-containing commercial fertilizers. They are also generated in nitration processes and in the production of catalysts from nitrates, pickling of metals, copperplate engraving and in the waste gases of the lead chamber process.

Among the emissions of the chemical industry the waste gases generated in the production of nitric acid by oxidation of ammonia are the most important. Nitric oxide, formed during the reaction of NH_3 with oxygen on the platinum catalyst, reacts with air to form NO_2, which in turn is converted into nitric acid by combining with water ($3NO_2 + H_2O = 2HNO_3 + NO$), so that again NO is formed. The reaction $2NO + O_2 = 2NO_2$ proceeds more slowly as the NO-concentration decreases; since a complete separation under technically and economically feasible conditions is impossible, the waste gas of older plants in which absorption is effected at atmospheric pressure contains 0.3–0.4 vol. % NO, while the waste gas emitted by modern pressure-absorption plants contains 0.1–0.2 vol. % NO. The NO-content of automobile exhaust in Los Angeles, where 3,500,000 motor cars are registered, thus corresponds to the emission of 60 large nitric acid plants with a daily discharge of 1000 tons 100% HNO_3 each.

Some physical data

NO: boiling point −152°C; solubility in water at 20°C: 4.7 ml NO/100 ml H_2O.

NO and NO_2 are paramagnetic.

NO_2 and N_2O_4: boiling point (N_2O_4) 21°C.

5.6.5.2 *Toxicology*

Nitrogen monoxide (see also Oettel in Ullmanns Encyclopädie der technischen Chemie (Encyclopedia of Chemical Engineering) and VDI-Richtlinie 2105).

Nitrogen monoxide—in contrast to nitrogen dioxide—does not irritate the mucous membranes. Injuries are mainly due to its effect on hemoglobin, which is converted into nitrosohemoglobin or methemoglobin. Little is known about damages purely due to the nitrogen monoxide, since larger NO-concentrations are not stable in air but are converted into NO_2. At present, when evaluating the harmfulness of a mixture of nitrogen oxides of low concentration, the effect of NO is not taken into account, i.e., the NO_2-concentration is the decisive factor; this procedure appears to be correct. Frequently, the toxic effect of NO is assumed to be 20% of the toxic effect of NO_2. Here, we draw the reader's attention to the high NO-level in cigarette smoke (up to 1000 ppm), as was mentioned above.

Nitrogen dioxide

Nitrogen dioxide can be perceived by smell in concentrations as low as 0.1 ppm. However, it is possible to get accustomed to it; by slowly increasing the NO_2-concentration the odor threshold may be as high as 25 ppm. Concentrations of 20–50 ppm, in addition to their odor, irritate the eye, while 150 ppm may bring about strong local irritations, especially of the respiratory organs. The special danger of NO_2 is that after the preliminary irritation phase a temporary recovery is felt and that the lung edema, which may or may not be lethal, occurs only after 3–8 hr.

Amounts of 30 ppm NO_2 do not harm normal plants, while damages to sensitive plants is slight.

The MAK-value is 5 ppm NO_2 = 10 mg NO_2/m^3, the MIK_D-value 0.5 ppm NO_2 = 1 mg NO_2/m^3 and the MIK_K-value 1 ppm NO_2 = 2 mg/m^3. This level should not occur more than 3 times daily. The TAL gives the immission limit for "nitrous gases, calculated as NO_2" as 1 mg/m^3 (or 2 mg/m^3, once within 8 hr).

5.6.5.3 *Possibilities of decreasing the nitrogen oxide concentration in waste gases*

Discussions concerning the decreases in the emission of NO/NO_2-containing waste gases deal mostly with the gases of ammonia oxidation plants for the production of nitric acid. The nitrous gas content (expressed as NO) is restricted in Great Britain to 3.5 g NO/m^3 and in the German Federal Republic (VDI-Richtlinie 2295) to 4 g NO/m^3 for normal-pressure absorption and 3 g NO/m^3 for medium pressure absorption.

The level of nitrogen oxides can be reduced by alkaline absorption in the last absorption tower. Using soda as a scrubbing agent, a mixture of $NaNO_2$ and $NaNO_3$ is obtained from which sodium nitrite and sodium nitrate can be produced. If the absorption liquid is allowed to react with HNO_3, nitrate is obtained, whereby the nitrogen oxides released in concentrated form are led back to the absorption towers. The general application of alkaline scrubbing using soda is complicated by the fact that sodium nitrite and nitrate have limited use and are not in high demand.

Other alkalis can also be used for the absorption in the last tower. Milk of lime yields calcium nitrate after decomposition with nitric acid, which likewise is sold as a fertilizer on a limited scale only because of its hygroscopic properties.

Interesting experiments with magnesium oxide suspensions were carried out at the Osterreichische Stickstoffwerke in Linz. The magnesium nitrite solution obtained can (in contrast to the heat-resistant calcium nitrite) be converted by heating to about 140°C into magnesium oxide and magnesium nitrate. From the latter basic magnesium carbonate can be precipitated with ammonium carbonate, which together with the magnesium oxide primarily formed is led back to the absorption towers, while ammonium nitrate present in the solution is processed into lime ammonium saltpeter.

Alkaline final absorption with ammonia is currently being employed in Linz. The solution obtained in the last tower after scrubbing of the waste gases with NH_3 is not separately processed (this would involve excessively high evaporation costs because of the unavoidable dilution) but proceeds in countercurrent flow through the entire series of towers. The nitric acid, which is the final product, contains part of the ammonia (as NH_4NO_3) used in the process. This, however, does not prevent the use of this acid for the production of lime ammonium saltpeter. A part of NH_4NO_3 is lost by oxidation as nitrogen, but a considerable reduction in the emission of nitrogen oxides is attained by using final absorption with NH_3.

Frequently, methods are experimentally developed for the elimination of the nitrogen oxides in the waste gas of nitric acid plants by treatment with combustible gases (CH_4, H_2, CO) at elevated temperatures using catalysts. When the last reduction stage is NO, the effect is merely optical. When, on the other hand, the nitrogen oxides are to be reduced to elemental nitrogen, the entire oxygen content of the waste gases (about 2.5% O_2) must react, i.e., we must allow for an appreciable consumption of the abovementioned combustible constituents. This method can thus be applied when the price for such gaseous fuel is very cheap. An additional difficulty is that the gaseous fuels added, must be metered out very accurately, since otherwise the uncombusted and/or partially oxidized fractions will cause additional air pollution.

At present there is no technically and economically satisfactory solution for the elimination of nitrogen oxides from these waste gases or the destruction of NO in automobile exhaust. Thus, high smoke stacks should be sufficient to prevent the accumulation of hazardous ground-level NO_2-concentrations.

5.6.5.4 *General analytical data* (see also Leithe [154])

Nitrogen dioxide

Direct spectrophotometric determination in the absorption range of NO_2 is employed when the gas is present in higher concentrations; this procedure can also be carried out over long optical path lengths in the free atmosphere.

The gas-chromatographic determination with the electron-capture detector was already described for concentrations between 3 and 25 ppm (see page 93).

At concentrations expected to exceed 500 ppm in waste gases a stationary sample can be taken in 1–2-liter vessels; the gas is converted into nitric acid and acidimetrically determined. Dry absorption of the nitrogen oxides with simultaneous oxidation to nitrates is used for higher concentrations; the nitrate is reduced to ammonia, which is colorimetrically determined.

A stationary sample (1 to 2 liters) can also be taken for determinations in the MAK-range. The nitrogen oxides in the sample can be oxidized into nitric acid, which is then colorimetrically determined. This procedure, however, can only be applied when the oxides are cumulatively determined without differentiation. Nitrogen dioxide is most frequently determined, especially in the range below 1 ppm, with the aid of azo dyes which are colorimetrically measured. Depending on the concentration, the absorption is effected in a stationary sample (2.5 liter) or from the flowing air sample by alkali, or, as is most frequently done, directly determined as the azo dye with the aromatic amines.

Among these diazotizing agents predominantly α-naphthylamine and sulfanilic acid (Griess reagent) were mainly used in earlier determinations. This reagent is also sometimes used today. As a rule, however, the Saltzman reagent (N[1-naphthyl]-ethylenediamine and sulfanilic acid) is preferred, though here the stoichiometric conditions are still subject to discussion.

Sawicky, Miller, Stanley, Pfaff and D'Amico [228] described some interesting examples of the determination of NO_2 from the standpoint of its radical nature. It reacts with other organic reagents having a radical character. The authors observed very sensitive color reactions with a molar extinction coefficient of 620,000. The significance of these procedures in practical analysis is still an open question.

For rapid NO_2-level determinations at work places with the Dräger test tube the reaction with diphenylbenzidine (blue color) is useful. When an oxidizing zone consisting of CrO_3 is placed before the benzidine layer, the NO present in the air is determined together with NO_2.

The continuous automatic determination of NO and NO_2 has also been widely studied. Several colorimetric instruments for this purpose have already appeared on the market.

Nitrogen monoxide

The NO present in the atmosphere is converted into NO_2 by oxidizing agents and determined together with the original NO_2-content. In a second experiment the NO_2 originally present in the air is determined. The NO-content is calculated from the difference of the two results.

Nitric oxide can be oxidized to nitrogen dioxide in the liquid phase with $KMnO_4$ in H_2SO_4 or H_3PO_4, in the solid phase on surfaces impregnated with CrO_3, in the gas phase with ozone, with oxygen under pressure or by UV irradiation (possibly in the presence of olefins functioning as catalysts).

Spectrophotometric determination of NO_2 *over long distances*

As in the O_3-determination methods developed by Regener and Renzetti (see page 129) at the First Physical Institute of the University of Vienna (Moundrea [184]), light rays from a searchlight are projected onto a receiver (spectrograph with camera) at a distance of 3 km and photometrically evaluated in the absorption range of NO_2. The diurnal variation of the NO_2-concentrations (10–100 ppb) agrees with the expected NO_2-emission in vehicular traffic.

Gas-chromatographic determination of NO_2 *with the electron-capture detector*

Because of its radical character, NO_2 exhibits a high affinity for free electrons and can thus be detected by the ECD. Morrison and Corcoran [182] used a plane-parallel ECD with a tritium source of 180 mC for the indication of NO_2 in the ppm-range. The 5-m separation column contains as the fixed phase 10% SF 97 (a methylsilicone oil) applied to Fluoropak 80 (40–80 mesh = 0.18–0.35 mm). Argon serves both as the carrier gas (10 ml/min) and the scavenging gas (30 ml/min). The temperature of the detector is 200°C.

The standard deviation is ± 0.6 ppm in the range from 3–25 ppm NO_2.

5.6.5.5 *Acidimetric determination of nitrogen oxides*

In industrial practice the nitrogen oxides, for example, in the waste gases of nitric acid plants, are determined as follows: the sample is introduced into large evacuated flasks, water or dilute H_2O_2 is added and the mixture is allowed to stand for several hours. The nitric acid formed is then titrated. According to Leithe [155], the duration of the process can be reduced to a few minutes by the following modification. The gas is sampled into a 500-ml gas burette and vigorously shaken with 20 ml 1.5% neutral H_2O_2 and 2 ml of a 1% solution of neutral foaming agent (Nekal, dodecylbenzosulfate, "Pril", or similar agents) for five minutes and then titrated with 0.01 *N* NaOH against methyl red. One ml 0.01 *N* NaOH corresponds to 460 μg NO_2.

5.6.5.6 *Dry absorption of* NO *and* NO_2 *according to* Peters *and* Straschil

Peters and Straschil [202] found that sodium chlorite ($NaClO_2$) applied to alumina is a suitable absorbing agent for determining the sum NO + NO_2 in waste gases. The nitrogen compounds absorbed are reduced by Devarda alloy and can be colorimetrically determined as ammonia.

Twenty-seven gram aluminum metal are dissolved in a prescribed amount of 40% sodium hydroxide, diluted with 5 liter water and precipitated as $Al(OH)_3$ by a CO_2-stream introduced at 40°C. The precipitate is filtered and washed to weak alkaline reaction, dried at 80°C and pulverized. One part (by wt.) is mixed with a solution consisting of 0.3 parts (by wt.) of 80% $NaClO_2$ and 0.01 parts (by wt.) alizarin (dissolved in a few drops of alkali to indicate the consumption) in 2 parts (by wt.) of H_2O into a paste. The paste is dried in vacuo at 50°C and ground to a particle size of 0.5–2 mm. About 3 g are inserted into a suitable glass tube.

The air sample is drawn through at a rate of 200–500 ml/min. After sampling, the content of the tube is first slightly heated with 30 ml 30% NaOH until the individual particles decompose. Then they are dissolved in 200 ml cold water, 5 g Devarda alloy are added and a reflux condenser is attached to the vessel. When reduction is completed, the ammonia formed is distilled into a receiver, containing a slightly acid medium, and photometrically determined with the Nessler reagent or via the indophenol reaction (see page 172).

Colorimetric determination of $NO + NO_2$ *as* HNO_3

After oxidation of the nitrogen oxides into nitric acid, the acid can be colorimetrically determined with phenoldisulfonic acid. This method is very popular in the United States, but a differentiation between NO and NO_2 is not possible. This procedure is recommended by the American Conference of Governmental Industrial Hygienists for determinations at work places.

The air sample is drawn into a glass vessel of 1–2½ liter and allowed to stand with 10 ml of a solution of 1 ml concentrated sulfuric acid and 6 drops 30% H_2O_2 in 200 ml water for 12–40 hr. Then the solution is made weakly alkaline with NaOH, evaporated to dryness and the residue well triturated with 1 ml phenoldisulfonic acid. The residue is allowed to stand for 10 min and is then diluted with 10 ml H_2O. Drops of 10 *N* NaOH (400 g NaOH/liter) are introduced until a dark brown color appears and then an excess of 4 drops is added. The solution is diluted to 25 ml, filtered in a photometric cell and the determination carried out at 410 mμ against the blank of the reagents.

The calibration curve is plotted for known amounts of $NaNO_3$ (0.1–0.2 μmole) in accordance with the above procedure. The molar extinction coefficient is about 4,400 and the detection limit approximately 1 ppm in a 2.5-liter air sample.

Phenoldisulfonic acid: 25 g phenol are dissolved in 150 ml concentrated sulfuric acid. Then 75 ml of fuming sulfuric acid with 13% SO_3 are added and the reagent is heated at 100°C for 2 hr. The preparation is properly sealed and stored.

5.6.5.7 *Determination of* NO_2 *by photometry of azo dyes*

General

Principle: NO_2 is absorbed as nitrite and colorimetrically determined as azo dye. For the formation of the azo dye usually two aromatic amines are used; one forms the diazonium compound with nitric acid, while the other couples this compound to the azo dye.

Sampling: The high sensitivity of the photometric determination of azo dyes makes it possible to work with a stationary sample in a 2-liter glass vessel when the NO_2-level exceeds 1 ppm, for example, at work places with levels in the MAK-range. The vessel must be preliminarily evacuated or sufficiently rinsed with the air sample. After sampling, the reagents are introduced into the vessel and the color reactions performed.

When the NO_2-level is low (MIK-range), about 10–20 liter of air sample are drawn through small fritted scrubbers at a rate not exceeding 0.5 liter/min. The scrubbers contain either dilute alkali or the mixture of aromatic amines in acetic acid. In order to ensure complete absorption, foam formation must be enhanced; in absorption by alkali this is achieved by adding n-butanol, while in the other case the added acetic acid leads to sufficient foaming. Absorption by alkali has the disadvantage that during sampling the required air volume cannot be evaluated because no color reaction takes place. On the other hand, it has the advantage that the disproportionation proceeds stoichiometrically according to the reaction $2NO_2 + 2NaOH = NaNO_2 + NaNO_3 + H_2O$, whereby only the $NaNO_2$ formed couples to the azo dye.

Color reaction: The classical Griess reagent (a mixture of sulfanilic acid and α-naphthylamine in acetic acid) is still preferred in some laboratories. It is used in the Österreichische Stickstoffwerke and included in a prescription of the ICI Manual. However, in recent years many analytical chemists use the somewhat more sensitive and stable Saltzman reagent (sulfanilic acid and N-(1-naphthyl)-ethylenediamine). Sometimes other aromatic amines from the large number of known azo dye components are recommended for air analysis.

NO_2*-determination with the* Griess *reagent*

According to the standard procedure at the main laboratory of the Österreichische Stickstoffwerke 20 ml colorless Griess reagent are introduced into a 100 ml fritted scrubber (G III frit). Then 10–20 liter air are drawn in at a rate not exceeding 5 liter/min until a distinct pink color appears. The acetic acid solution should be covered with a 3- to 4-cm high foam layer.

After sampling, the solution is transferred into a 25-ml volumetric flask, rinsed with a little fresh Griess reagent and made up to 25 ml. After a waiting time of at least 15 min (the color solutions are adequately stable for 2 hr) the extinction is measured at 530 mμ in a 1- or 5-cm cell, depending on the intensity of the color.

The calibration curve is plotted using a solution of 0.300 g very pure $NaNO_2$ in 1 liter, diluted again 1:20. One ml corresponds to 10 μg NO_2. A series of 0–10-ml calibration solutions is diluted to 25 ml with the Griess reagent used in the determination. The molar extinction coefficient is 46,000. The calculation is based on the assumption obtained from a test series with air mixtures of known NO_2-level, namely that 2 mole NO_2 of the air sample correspond to 1 mole of the calibration series.

Griess reagent: 2 g sulfanilic acid are dissolved in 400 ml water (distilled over a little $KMnO_4$) and 100 ml glacial acetic acid. 0.5 g of α-naphthylamine is dissolved in 400 ml water and 100 ml glacial acetic acid with heating and then rapidly filtered if necessary. Both of the solutions are combined, tightly sealed and stored in a refrigerator.

According to the ICI prescription, 20 ml 0.025 *N* NaOH are introduced into a thick-walled 2.5-liter glass vessel with a hermetic stopcock. The vessel is evacuated to a residual pressure of 200 mm Hg. To prevent breakage, the vessel should be wrapped in a resistant gauze. After sampling, the vessel is allowed to stand for 15 min and occasionally shaken to facilitate complete absorption. Then the solution is transferred into a 100-ml volumetric flask, 10 ml of Griess solutions A and B are respectively added, the volume is made up to 100 ml and the extinction measured at 530 mμ after 30 min.

The result is calculated from the calibration curve plotted for dilute $NaNO_2$ solution (1 mole $NaNO_2$ of the calibration solution corresponds to 2 mole NO_2 of the air sample).

Reagents: Solution A: 8 g sulfanilic acid are dissolved in water mixed with 270 ml glacial acetic acid and made up to 1000 ml with water. Solution B: 5 g pure (recrystallized, if necessary) α-naphthylamine is dissolved in 20 ml glacial acetic acid and made up to 1000 ml with water.

Determination of NO *after oxidation to* NO_2

The oxidant is $KMnO_4$ in sulfuric acid.

The determination of nitrogen monoxide in the air is effected by first converting the oxide into NO_2 by means of the oxidant $KMnO_4$ dissolved in H_2SO_4. The gas must be rapidly passed through the oxidizing solution, since for longer residence times the oxidation proceeds up to the formation of nitric acid.

At the main laboratory of the Österreichische Stickstoffwerke the following method is used for the simultaneous determination of NO and NO_2 in the air at work places and in the free atmosphere (see also Leithe [154]).

The apparatus, which can also be set up in the open air, consists of three 100-ml fritted scrubbers, a pump and a gas meter. The first scrubber (G III frit) serves to absorb the NO_2 originally present in the air and contains (see page 184) the above-described sulfanilic acid–α-naphthylamine solution in dilute acetic acid. The second scrubber (G I or GII frit) contains 25 ml 2.5% H_2SO_4 with 2.5% $KMnO_4$. The third scrubber, which is intended to absorb the NO_2 formed from the original NO, is similar to the first scrubber and contains 20 ml Griess reagent. The air sample is introduced at a rate not exceeding 0.5 liter/min until in both scrubbers a red color appears which can be photometrically measured. The subsequent procedure is the same as that outlined on page 184.

In accordance with the experience gained at our laboratory from test series with air mixtures containing known NO-amounts, the NO_2-yield from NO is 80% in this procedure. Since about 50% of the NO_2 is converted into the azo dye (see above), 1 mole NO in the air corresponds to 0.4 mole $NaNO_2$ in the calibration solution. The error range (relative standard deviation) is $\pm$ 8–10%.

Buck and Stratmann [34] reported the complete conversion of NO into NO_2 without losses (NO-concentrations of 0.3–2.5 mg NO/m^3 in the presence of 0.04–3.5 mg NO_2/m^3) for gas throughput rates between 20 and 60 liter/hr. The compounds are absorbed in modified Muencke scrubbers using the following solutions.

1. Mixed acid: 59 ml H_3PO_4 (d = 1.71) are diluted with water to 100 ml of a 60% acid and 10 ml concentrated H_2SO_4 added.

2. Oxidation solution: 0.5 g solid $KMnO_4$ are placed in the scrubber and 20 ml mixed acid are added. The scrubber is vigorously shaken until the solution is saturated; a residue of undissolved $KMnO_4$ remains. The mixture has to be prepared

shortly before being used and must be renewed when it turns brown due to precipitation of MnO_2.

The oxidant CrO_3 *on dry surfaces*

An oxidant for the quantitative conversion of NO into NO_2 and especially suitable for continuous apparatus was described by Ripley, Clingenpeel and Hurn [214]. Glass fiber filter papers of 10 dm^2 area are impregnated with 25 ml of a solution of 2.5% $Na_2Cr_2O_7$ in 2.5% sulfuric acid, dried in vacuum at 70°C or in the drying oven at 95°C for 90 min, tightly sealed and stored. The 5-cm strips are cut and placed into a suitable U-tube. The air throughput rate is 290 ml/min. The air sample should have a relative humidity of 15 to 75%.

Other procedures for oxidizing NO *to* NO_2

In the Imcometer NO is converted into NO_2 by UV irradiation in a long quartz spiral (Fuhrmann [71]).

According to Bokhoven and Thommassen [25], this reaction in UV light (230 mμ) is catalytically accelerated in the presence of 0.05% butadiene.

The oxidation of NO to NO_2 with ozone, also described for continuous analyzers, involves the risk of excess oxidation to N_2O_5 or HNO_3.

5.6.5.8 *Absorption in alkaline solution according to* Jacobs *and* Hochheiser

Jacobs and Hochheiser [118] used an alkaline solution to absorb NO_2. The disadvantage that during sampling it cannot be established when NO_2 has accumulated to an amount sufficient for the subsequent photometric determination is compensated by the advantage that in the presence of interfering SO_2-concentrations, sulfur dioxide can be eliminated in alkaline solution by adding a drop of H_2O_2. Another advantage is the fact that NO_2 is stoichiometrically converted in accordance with the equation $2NO_2 + 2NaOH = NaNO_2 + NaNO_3 + H_2O$, so that in the subsequent diazotization only the nitrite formed (for which 2 mole NO_2 were required) participates.

For the absorption a fritted scrubber containing 30 ml absorption solution is used. To obtain complete absorption, butanol is introduced as a foaming agent into the solution. The air throughput rate is about 1 liter/min for 40 mm. The solution is transferred into a 50-ml volumetric flask and 1% H_2O_2 (decomposition of SO_2) is added. Then 10 ml diazo solution and 1 ml of 0.1% N-(1-naphthyl)-ethylenediamine dihydrochloride in water are added. The volume is made up to 50 ml, the solution allowed to stand for 30 min and the determination carried out at 550 mμ.

Absorption solution: To 1 liter of 0.1 *N* NaOH are added 2 ml n-butanol as foaming agent.

Diazo solution: 20 g sulfanilamide are dissolved in 1 liter water containing 50 g concentrated phosphoric acid.

The standard nitrite solution for plotting the calibration curve contains 10 μg NO_2 in 1 ml. It is obtained by dissolving 0.150 g $NaNO_2$ in 1 liter water and diluting the solution 1:10.

Determination of NO_2 *with the* Saltzman *reagent*

The reagent (Saltzman [217, 218]) contains instead of α-naphthylamine as the coupling component in the Griess reagent N-(1-naphthyl)-ethylenediamine. The Saltzman reagent was previously used for the determination of the sulfonamide medicament sulfanilamide and is now frequently used for the determination of nitrite, for example, in water.

In order to obtain air samples with NO_2-levels below 1 ppm, Saltzman used a small fritted scrubber containing 10 ml absorption reagent. The air sample is introduced at a rate not exceeding 0.4 liter/min; 20–30 ml fine foam should form over the solution. For NO_2-levels exceeding 1 ppm the sample should be drawn into an evacuated vessel of known volume and containing 10 ml absorption solution. The sampling must be carried out in vacuum.

The solution is allowed to stand for at least 15 min and the extinction measured with the spectrophotometer against fresh reagent at 550 mμ. The color is retained in a hermetic flask for 1 day with only 3–4% fading.

The calibration curve is plotted for dilute $NaNO_2$ solution (e.g., 10 μg $NaNO_2$/ml). Saltzman, however, did not assume that 1 mole $NaNO_2$ is equivalent to 2 mole NO_2 in accordance with the equation $2NO_2 + H_2O = HNO_2 + HNO_3$, but found in his procedure that 0.72 mole (instead of 5 mole) $NaNO_2$ gives the same color as 1 ml NO_2 in test air mixtures (see also Saltzman and Wartburg [220]). On the other hand, Gill [79] found a nitrite equivalent of 0.58 and Stratmann and Buck (see below) a nitrite equivalent of 1 [see Staub, 29 (1969) 447].

Interfering substances: A 10-fold SO_2-concentration does not appreciably interfere. When more SO_2 is expected, 1% acetone is added to the absorption solution and the determination is carried out no later than 4 or 5 hr afterward.

Ozone, interfering in concentrations above 0.2 ppm, can be removed without any NO_2 losses by passing the air through a glass wool plug impregnated with MnO_2. The plug is prepared by impregnating glass wool with an $Mn(NO_3)_2 \cdot 6H_2O$ solution and then heating in a 200°C air stream until all the nitrogen oxides have been expelled.

Absorption reagent: 5 g sulfanilic acid are dissolved in 850 ml water, 50 ml glacial acetic acid and 50 ml 0.1% N-(1-naphthyl)-ethylenediamine dihydrochloride solution are added. The volume is then made up to 1000 ml.

Stratmann and Buck [246] thoroughly studied the Saltzman procedure. Instead of the nitrite equivalent of 0.72, as found by Saltzman, they found an equivalent of 1, which, however, does not agree with the classical formulation of the formation of azo dyes. Only for NO_2-levels exceeding 500 mg NO_2/m^3 the nitrite equivalent is

0.75. We have as yet no explanation for the different nitrite equivalents of Saltzman [34], Gill [79], and Stratmann and Buck [246].

Stratmann and Buck use the absorption solution of Saltzman to which 1% acetone is added; acetone addition was also employed by Saltzman in order to eliminate the interference of SO_2. Stratmann and Buck use 25 ml of absorption solution in a fritted scrubber at an air throughput rate of 750 ml/min. A sampling time of 10–15 min is usually adequate. The extinction is measured at 550 mμ in 1- or 2-cm cells.

To plot the calibration curve NO_2-amounts of 1.5, 2.5 and 3 g are pipetted from an $NaNO_2$ solution containing 2.5 μg NO_2 per ml into a 25-ml volumetric flask. The volume is made up to the mark with the reagent solution and the determination carried out after 10 min in the 1- or 5-cm cell, depending on the color intensity.

5.6.5.9 *Continuous automatic determinations of* NO *and* NO_2

Automatic analyzers for the determination of NO_2 (or of NO after its conversion into NO_2) are mostly based on colorimetric determinations with the Saltzman reagent. The oldest arrangement is described by Thomas et al. [251]. Further, the Imcometer (see page 70), in which NO is converted into NO_2 by UV irradiation, is also suitable as is the Actalyser K 1008 of the firm Beckman.

5.7 Carbon-containing air pollutants

5.7.1 Carbon dioxide

5.7.1.1 *General information*

Boiling point −78.5°C.
Solubility in water: 100 ml water dissolve 88 ml CO_2 at 20°C and 760 mm Hg.
Dissociation constants: $K_1 = 4.3 \times 10^{-7}$; $K_2 = 5.6 \times 10^{-11}$.

Carbon dioxide is a normal air constituent which is of vital importance for the flora. Because of its ability to absorb infrared rays from sunlight it plays a significant role in the heat balance of atmospheric air.

The CO_2-level in unpolluted air is not completely uniform. It has increased appreciably owing to the consumption of fossil fuel (coal, petroleum and natural gas) since the turn of the last century from about 290 ppm to about 320 ppm on the average. It is higher on the equator (about 350 ppm) than in the arctic (260 ppm). Heating, traffic and industry can cause local CO_2-levels which exceed the average level in the open air.

Carbon dioxide of the unpolluted atmosphere can be distinguished from that originating from the combustion of fossil fuel by radiometric methods. The former is enriched with radioactive C^{14} originating from the higher air layers, while in the latter practically no radioactive isotope is present (half-life of C^{14}: 5570 years).

According to the results of Austrian scientists (Kunz, Stetter, Wagner and others [145, 244]), carbonic acid is present in the atmosphere in two states of aggregation: mostly as normal gaseous carbon dioxide, which is determined via absorption in liquids by conductometry or IR measurements, and in lower amounts (up to 30% of the total carbonic acid) as an aerosol (nucleate phase). Aerosol particles with a diameter $< 10^{-5}$ cm are present in an amount of $10^9/cm^3$ together with dust and water vapor in the air. The presence and amount of this "nucleate phase" depends on the weather conditions. It is converted at higher temperatures (50°C or higher) and by irradiation with green light (510–537 mμ) into the gaseous phase. When the weather conditions are favorable and the temperature is higher or when absorption is effected over a long layer of calcium chloride and soda lime, the sum of the gaseous and nucleate phases can be obtained in the CO_2-determination; hence, the values are appreciably higher than those obtained by conventional gas analysis.

These very interesting findings on the CO_2 nucleate phase were hitherto unfortunately neglected, despite their importance for meteorological and biological events. Their validity, however, is not subject to controversy.

In the microclimate over territories with dense vegetation and thus increased assimilational activity, the CO_2-level is subject to appreciable daily variations. It is much higher than average in the near-ground air layers above the plant cover because of decomposition of organic substances. Near mineral water sources exhalations of practically pure CO_2 may occur.

Hygienic importance of the CO_2*-level in the atmosphere*

The human organism, i.e., its respiratory functions is to a large extent independent of variations in the CO_2-level in the atmosphere. This is also expressed by the high MAK-value (0.5 vol. % = 5000 ppm). In the open atmosphere hazardous concentrations are almost never reached and consequently, such air is rarely analyzed for its CO_2-content. On the other hand, the CO_2-level of the air, especially in the microclimate, is of interest to the plant physiologist.

In enclosed spaces, however, CO_2-poisoning due to the accumulation of higher CO_2-concentrations is always possible and frequently leads to death. In the presence of pure CO_2 instant death is to be expected because of paralysis of the respiratory centers. CO_2-concentrations exceeding 6% constitute an acute danger to life. Since CO_2 is 1.5 times as heavy as air, it is accumulated on the floor of enclosed spaces, for example, fermentation cellars, storage rooms for grain, potatoes, etc. Combustion gases can contain up to 20% CO_2, depending on the O_2-content in the air. In mines CO_2-rich gas eruption may cause lethal poisoning. A CO_2-rich atmosphere is usually indicated by extinguishing of a candle when the CO_2-amounts exceed 8%. For concentrations exceeding 2% the candle flame has a reddish light. Oxygen respirators must be worn when entering spaces subject to CO_2-hazards.

The MAK-value can also be exceeded in enclosed spaces due to the presence of a large number of people. The symptoms are feebleness (sickness), to which especially odor and exhalation contribute. An adult person at rest exhales about 300 liter air/hr, the exhaled air containing 4–5% CO_2. In automobile exhaust 7–13 vol. % CO_2 are present.

The outlet of combustion gases is dangerous not because of the accumulation of CO_2 but because of the presence of carbon monoxide which is much more toxic.

In the plant kingdom CO_2-rich air (below 1%) is not harmful but leads to increased assimilation and hence to an increase in the growth rate. For this reason CO_2-rich air is sometimes introduced into the soil as a fertilizer to enhance vegetation.

Removal of carbon dioxide from industrial and other waste gases in order to reduce air pollution is of no interest. On the other hand, numerous methods for isolating CO_2 for its further utilization are described in the literature.

In living spaces of limited size, for example, in submarines or manned satellites, the removal of exhaled carbon dioxide is important. The same applies to the regeneration of respiratory air cycled through the oxygen respirators.

Determination of the CO_2*-level*

5.7.1.2 *Survey of procedures*

The approximate determination of the CO_2-concentration in the MAK-range can be readily performed with the gas-detection devices of the firm Dräger or Auer (see page 56). The Dräger tube 0.1a records 0.1–1 vol. % for 5 strokes of the pump and 0.5–5 vol. % for 1 pump stroke.

CO_2 is accurately determined by acid-base titration. The relatively high CO_2-level in atmospheric air makes it possible to use a moderate sample volume in easy-to-handle sampling vessels (e.g., 1–2 liter) and to determine the CO_2-content with titrated alkali (e.g., 0.01 *N* $Ba(OH)_2$). With all precautions taken the reproducibility is $\pm$ 1%.

When in the flow procedure larger samples are to be processed, the inertia of CO_2-absorption in aqueous hydroxides, must be taken into account. Here, absorption in fritted scrubbers is quite convenient after introducing butanol (foaming agent).

Instead of aqueous hydroxides the much more rapidly absorbing nonaqueous alkali solutions can also be used for the determination of CO_2. A disadvantage, however, is the extreme sensitivity to CO_2 in the ambient air.

Very small sample volumes, for example, for biological studies, can be statically determined by a very sensitive colorimetric procedure based on the decolorization of a phenolphthalein solution previously made alkaline. For the same purpose gas-chromatographic methods (see page 94) have been developed. For flow procedures a small absorption device for 0.5 ml absorption lye (longer residence time) is available.

The continuous automatic determination can be carried out both by relative conductometry and IR absorption.

5.7.1.3 *Determination of the CO_2-content by shaking a metered air sample*

This method, which is still being used, was developed in essence by Pettenkofer. In the modification of Treadwell [254] a dry 5-liter flask, whose volume was determined by weighing with water, is filled with the air sample by means of a bellows. One hundred ml of approximately 0.025 *N* $Ba(OH)_2$ are added, the flask is sealed and shaken for about 15 min. The turbid liquid is rapidly transferred into a dry flask. Twenty-five ml of this solution are withdrawn with a pipette and several drops of 1% phenolphthalein in ethanol are added. The solution is titrated very slowly with hydrochloric acid (see below) and continuously stirred until the solution becomes clear. The flask must be protected against respiratory air.

The titrating agent is obtained by diluting 224.6 ml 0.1 *N* HCl to 1 liter with CO_2-free water. One ml is equivalent to 0.25 ml CO_2 at NTP.

$$\text{Vol. ‰ } CO_2 = \frac{1000\,(N-n)}{V_o}$$

N = ml HCl used for titrating 25.00 ml $Ba(OH)_2$ until the titer is established

$$V_o = \frac{(V-100)\cdot(B-w)\cdot 273}{760\cdot(273+t)}$$

n = ml HCl in the titration of $Ba(OH)_2$ shaken with air

Hesse used 0.01 *N* oxalic acid instead of HCl.

Modification according to Wagner

Wagner [259] treated the air sample in a 1-liter flask, whose volume was determined by weighing with water. The air sample is shaken with 25 ml 0.01 *N* $Ba(OH)_2$ for 30 min on a vibrating shelf, 0.01 *N* oxalic acid is added directly from the burette into the flask and the solution is titrated until it becomes colorless. The flask should be made of Jena glass and must be boiled with water before use. During titration and the determination of the blank, CO_2-free air or a nitrogen stream must be passed through the flask.

Oxalic acid: 1.2605 g $(COOH)_2 \cdot 2H_2O$ are diluted in a nitrogen atmosphere with CO_2-free water to 2 liter.

$Ba(OH)_2$ solution: 1.720 g barium hydroxide are mixed with 0.4 g $BaCl_2$ and diluted to 1 liter with water. The storage flask and pipette are kept free of CO_2 with soda lime. The titer is determined with 100.0 ml $Ba(OH)_2$ against oxalic acid in a flask flushed with CO_2-free air. When no CO_2 is present, the color change produced by 1 drop (± 1 rel. %) is sharp.

$$\text{Vol. \% } CO_2 = \frac{\text{ml}\cdot 0.1113\cdot 100}{V_o}$$

CO_2-determination in a flowing air sample

Since dissolution of CO_2 in aqueous alkalis from the gas phase takes a comparatively long time, an adequate contact time and surface must be available to ensure completion of the reaction. Deckert [54] showed that impingers (see page 42) are unsuitable for this purpose. According to Leithe and Petschl [159], only 10% of the carbonic acid present is absorbed.

On the other hand, CO_2 can be completely absorbed from flowing gases in fritted scrubbers containing aqueous alkalis even if the gas is present in amounts of the order of several ppm, when a few drops of n-butanol are added to the excess alkali. The approximately 10-cm high foam layer consisting of fine bubbles and having a "lifetime" of about 5 sec ensures complete absorption when two scrubbers are connected in series.

When a twofold alkali excess is used, about 10% of the total CO_2-amount is found in the second scrubber, while with a fourfold excess the value is only about 1%; thus, in this case the second scrubber can be dispensed with.

A 100-ml fritted scrubber (G I frit) is supplied with 50 ml 0.01 *N* $Ba(OH)_2$ and air is drawn in at a flow rate of 1 liter/min, or 100 ml 0.02 *N* NaOH are introduced into a 400-ml fritted scrubber (G I frit, diameter 5 cm) and 50 liter air are drawn through at a rate of 3 liter/min.

The excess alkali is directly titrated in the fritted scrubber against phenolphthalein; in the first case 0.01 *N* oxalic acid and in the second 0.05 *N* HCl is used.

Determination according to Holm-Jensen

Holm-Jensen [110] developed a small apparatus (Figure 2.4.3f) for the absorption of CO_2 from air. 0.5 ml 0.005 *M* $Sr(OH)_2$ is injected over 7 large spiral windings (length 160 cm) as the air sample is passed through (about 15 ml/min). Because of the large surface and long residence time the absorption of CO_2 is practically complete. Thymolphthalein is added to the solution. The air sample is passed through until the solution is decolorized and the CO_2-level is determined from the volume of the air sample. The author reported a reproducibility of $\pm$ 5%.

Instead of visual indication the CO_2-content can also be determined by conductometric measurements prior to or after absorption of carbon dioxide; for this purpose a pair of electrodes is fused into the instrument.

Acidimetric determination of CO_2 *in nonaqueous solutions*

The absorption of CO_2 from a flowing air sample can also be accelerated by using alkalis in nonaqueous solutions. Blom and Edelhauser [21] used acetone or pyridine. Two spiral absorbers (see Figure 2.4.3c) with recycling are connected in series. Acetone or pyridine and thymol blue are added and each scrubber is hermetically fitted with a microburette containing 0.01 *M* sodium methylate in methanol (obtained by dissolving sodium metal in methanol). The solvent is neutralized and a small excess of alkali is added which must be renewed while the sample is passed through. Within 15 min about 2 liter of the air sample are drawn through until the indicator is just decolorized. About 3 ml 0.01 *N* titration solution are consumed. The error is about ± 10 ppm for air of the usual composition.

The absorption of CO_2 is even more effective when alkaline solvents, such as ethylenediamine or ethanolamine, are used. The absorbed carbon dioxide is then determined by titration with sodium methylate against thymol blue. Special care must be taken to prevent entry of CO_2 from the ambient air.

5.7.1.4 *Colorimetric determination of* CO_2 *according to* Spector *and* Dodge [234]

This method can be used for analyzing small air quantities but is less suitable for series analyses. It is based on the fading of the red color of a 0.0001 *N* NaOH solution containing phenolphthalein by CO_2 because of the drop in the pH. To a 0.0001 *N* NaOH solution phenolphthalein in ethanol is added in such an amount that the transmittance in a 100-mm cell of a colorimeter or spectrophotometer is 10% for a light wavelength of 515 mμ. The air sample (50–200 ml) is mixed with 100 ml of the solution and the increase in the light transmittance measured. The transmittance is governed by the Lambert–Beer law. There is a standard deviation of ± 10%.

Automatic procedures

Differential conductometry and infrared absorption are particularly suitable for the continuous automatic recording of the CO_2-determination.

The firm, Woesthoff, Bochum, has developed two types of instruments for the conductometric determination based on the decrease in the electrical conductivity of a 0.005 *N* aqueous NaOH solution caused by the absorption of CO_2 (see page 71, as well as Schmidts and Bartscher [240]). The working procedures for determining CO_2 in air are described by Malissa and Wagner [178] and the results compared with those of accurate acidimetric titration. The first type is intended to effect the continuous measurement with parallel throughputs of air and NaOH. The second type is designed for a certain air amount, for example, 1135 ml, introduced over a period of 5 min. The experimental setup is similar to that used in the determination of carbon in steel and elemental analysis. The pen is adjusted to zero when CO_2-free air is passed through the system and to a scale deflection of 100 mm for air containing 300 ppm CO_2. The sensitivity of the reading is ± 1 mm and the reproducibility ± 1%. The instruments can be calibrated from accurate titration results (e.g., according to Wagner, see page 191) or via the introduction of additional carbon dioxide. Carbon dioxide is not quantitatively absorbed using 0.005 *N* aqueous NaOH, the absorption is about 86%, but the results exhibit good reproducibility.

The use of butanol as a foaming agent to complete absorption is not advisable when subsequent conductometric measurements have to be carried out. Escape of this volatile alcohol leads to sharp rises in the electrical conductivity. On the other hand, according to the findings of the authors, a small amount of sodium caproate produces an equally effective foam.

The air sample should not contain interfering amounts of acid or alkaline pollutants which influence the conductivity of the NaOH solution.

Measurements of the IR absorption specific for CO_2, for example, with the aid of the URAS instrument (see page 66), permit continuous recording of CO_2-concentrations (most sensitive measuring range 0–50 ppm over the entire scale width).

5.7.2 Carbon monoxide

5.7.2.1 *Occurrence*

Carbon monoxide is present in fuel gases, sometimes in high concentrations (water gas 40 vol. %, coke oven gas about 5%, household gas about 5%). The compound is formed by the combustion of carbon-containing materials in the presence of air amounts which do not suffice to oxidize the CO to CO_2 and is contained in numerous flue gases from heating plants and automobile exhaust, and especially in household gas (see Table on page 11) as well as in concentrations of up to 2% in cigarette smoke.

Traces of carbon monoxide also occur in naturally polluted air, but the main source of this gas is to be sought in the household, traffic and industry.

Hygiene

Carbon monoxide in the air of production plants has attracted the attention of the industrial hygienist and the analyst because of its widespread occurrence. On the other hand, CO-pollutants in the open air have become interesting only recently, since the dense vehicular traffic has lead to a higher incidence of hazardous concentrations. When a person stays for extended periods in the open near operating gasoline engines, he is liable to contract carbon monoxide poisoning. Foresters working with engine-driven handsaws are especially exposed to this risk. Carbon monoxide and methane were the first gases to be automatically measured and monitored at work sites.

The toxic effect of carbon monoxide is associated with the ability of CO to form with hemoglobin a comparatively stable light red addition compound—carbon monoxide hemoglobin—similar to oxygen hemoglobin. The affinity of CO for hemoglobin is 200 times as strong as that of oxygen. Thus 0.1% CO in the air blocks the same amount, i.e., 50% hemoglobin as in the case of oxygen. However, the formation of carbon monoxide hemoglobin is also reversible; when sufficient air supply is ensured, it is slowly converted back to oxygen hemoglobin. A content of about 10% CO-hemoglobin in the blood does not yet endanger its respiratory function. Since carbon monoxide occurs in cigarette smoke in considerable amounts (up to 2 vol. % = 20,000 ppm), the blood of smokers contains CO-hemoglobin.

Carbon monoxide is a potent poison; it is still a matter of dispute, whether chronic diseases due to carbon monoxide exist.

The determination of the CO-hemoglobin level in blood is interesting from the medical standpoint. In most cases the compound is spectroscopically determined from the absorption band in the yellow and green parts of spectrum which is shifted more to the short-wave range than the band of oxygen hemoglobin. Sensitive instruments detect concentrations in blood as low as 10%. However, CO can also be removed from the blood by dilute sulfuric acid and then determined by one of the sensitive chemical or physical methods described below.

The MAK-value for CO is 50 ppm = 55 mg/m^3. An MIK$_D$-value of 2 and a MIK$_K$-value of 6 ppm have been proposed.

5.7.2.2 *General information for the analyst*

Carbon monoxide is the most frequently encountered toxicant in interior spaces (work sites or living quarters). Consequently, its determination in the MAK-range and below is of special importance.

Various gas-detection devices are currently available for the rapid semi-quantitative determination of dangerous CO-concentrations by means of individual analyses. The Dräger tubes, designed for specific concentration ranges (see page 56), utilize the reduction of ammonium molybdate to molybdenum blue by CO under the catalytic effect of $PdCl_2$. In the tube of the Bureau of Mines the same principle is applied. The CO-reading of a special instrument of the Drägerwerke, Lübeck, is associated with the temperature increase caused by the catalytic combustion of CO.

In factories or installations, where frequent CO-immissions are to be expected and where continuous recording of CO and warning devices are necessary, automatic instruments must be employed. The operating principles are here also based on indicating the heat of reaction in the catalytic combustion of CO. Also infrared instruments, manufactured by various firms, are employed for sensitive and largely selective determinations of the CO-level in the air. These instruments are sufficiently sensitive to determine the low CO-levels in the open air.

More accurate determinations in solutions are required for checking and calibrating automatic analyzers as well as in other special cases. These are mostly based on the selective oxidation of CO to CO_2 by I_2O_5 or Hopcalite. Spectroscopic methods based on the formation of CO-hemoglobin are very specific but rather inaccurate and therefore rarely used. Recently, some other colorimetric methods have been proposed.

Gas-collecting tubes having a volume of about 1 liter are being employed for CO analysis in the MAK-range. Because of the low solubility of CO in water, aqueous liquid seals (water or salt solutions) can be used. Sampling in valve bags made of plastic foils (content 1–10 liter) are very convenient and involve no danger.

Gas chromatography should be used for small sample volumes (see page 95). Using suitable partition columns, CO can be directly detected via a thermistor probe; in the flame ionization detector very low CO-levels can be determined after dehydrogenation to methane (Porter and Volman [204a]).

The Dräger CO-detector

The instrument manufactured by the Drägerwerke, Lübeck, is designed for the determination of intermediate CO-levels in the air. The operating principle is based on catalytic combustion and measurement of the heat liberated in this process. Hopcalite serves as the catalyst which induces combustion of CO at 100°C but not of hydrogen (provided the CO-level is below 3000 ppm). Olefins, solvent vapors, as well as chlorine- and sulfur-containing catalytic poisons are removed by preliminary absorption on activated carbon. The air sample is introduced at a rate of 2 liter/min. The temperature increase on the catalyst (a temperature increase of 60°C is caused by 1% CO) is indicated by a special mercury thermometer. The limit concentration is 10 ppm. Measuring ranges from 0–3000 ppm, 0–5000 ppm and 0–10,000 ppm (1%) can be selected.

In instruments with warning devices the temperature is measured by means of thermocouples. When the thermoelectric voltage attains a certain value, an alarm signal is triggered.

The instrument is calibrated with air mixtures of known CO_2-level which can also be obtained via the Drägerwerke.

5.7.2.3 *Determination of carbon monoxide by reduction of mercuric oxide*

The reaction of carbon monoxide with HgO into CO_2 and metallic mercury at 180°C has been frequently used in analysis.

McCullough, Crane and Beckman [174] determined the weight loss of a Pyrex absorption tube filled with 8 g red mercuric oxide after passing through the air sample. The inside diameter of the tube is 9 mm and its length 10 cm. The tube is heated in a metal box on an air bath to 200°C; it is equipped with an arrangement for removing the metallic mercury formed. In the presence of hydrogen the above procedure cannot be applied. Methane does not react to a great extent. Other higher hydrocarbons which may react with mercuric oxide can be removed by preliminary absorption in an absorption tube which has been filled with activated carbon.

The colorimetric determination of the expelled mercury vapor is much more sensitive. Beckman, McCullough and Crane [15] first free the sample from higher hydrocarbons over activated charcoal. Then the sample is passed at 175–180°C over granular mercuric oxide and then over SeS_2 paper, which is prepared by impregnating a filter paper with 0.025 *M* H_2SeO_3, 10 min exposure to H_2S and 30 min drying at 140°C. The brown colors are standardized with air samples of known CO-content (20–300 ppm).

In an even more sensitive method (Tomberg [253]) the mercury vapor is determined be measuring the atomic absorption, i.e., light absorption of the irradiated light of the Hg resonance line of 2537 Å by the free atoms of mercury vapor (see page 223). The detection limit here is 10 ppb CO in air.

5.7.2.4 *Volumetric determination of* CO *after oxidation to* CO_2

1. *With iodine pentoxide*

Principle: Iodine pentoxide at a sufficient holding time and increased temperature (110°C or higher) is a suitable reagent for the oxidation of CO to CO_2. Older methods were improved by experiences obtained from the oxygen determination according to Unterzaucher which is frequently used in elemental analysis and which is also carried out via CO and its reactions with I_2O_5.

The elemental iodine formed in accordance with the equation $I_2O_5 + 5CO = = 2I + 5CO_2$ can be directly titrated with sodium thiosulfate against starch; however, it is much more convenient to apply the following procedure because of the sixfold higher sensitivity: iodine is oxidized into iodate with bromine, potassium iodide is added and the released iodine titrated. The following equations apply here:

$$2I + 10Br + 6H_2O \rightarrow 2HIO_3 + 10HBr$$

$$2HIO_3 + 10HI \rightarrow 12I + 6H_2O.$$

Whereas in the direct titration 1 molecule CO corresponds to $\frac{2}{5}$ of an iodine atom, in the second procedure 1 molecule CO corresponds to $\frac{12}{5}$ iodine atoms. Thus, the sensitivity is 6 times as high.

When instead of the iodine the formed carbon dioxide has to be titrated, the CO_2 must be removed from the air sample prior to the reaction with I_2O_5. On the other hand, pollutants which react with iodine pentoxide but do not yield acid reaction products exhibit no interference.

Apparatus: Gas-collecting tube (volume 1–2 liter) for absorption of the air sample, leveling vessel with liquid seal (10% Na_2SO_4 or common salt solution with 22 parts (by wt.) NaCl and 78 parts water) and a drying tube filled with P_2O_5. The air sample can then be drawn from the filled plastic foil bag (see page 36) with the aid of an aspirator.

When the formed CO_2 is to be determined, the air sample has to be passed over calcium hydroxide prior to drying. If the sample contains olefins or other oxidizable substances, these are retained by a trap cooled with dry ice and acetone which lets through CO.

The I_2O_5-tube contains a 10-cm long and 1-cm thick layer of "iodine pentoxide for flue gas analysis" bounded by glass or quartz wool. The reagent has a particle size corresponding to 100–400 mesh and should be tightly packed. The heating bath ("hollow shell" according to Pregl) is filled either with boiling glacial acetic acid (b.p. 118°C) or with high-boiling oil heated to 120°C. A bead tube wetted with 10% KI and a 10-cm long absorption tube, whose surface was increased by bucklings similar to those of the Vigreux column, serve as the traps with ground-glass joint for the direct and indirect titrations, respectively. The individual parts of the apparatus correspond to those in the instrument for the determination of oxygen according to Unterzaucher. The absorption tube is moistened with dilute NaOH.

Reagents: Iodine pentoxide is completely freed from iodine and water by heating at 190°C in an air stream.

Formic acid: 98–100%.

Sodium hydroxide: 25 g NaOH dissolved in 100 ml water.

Bromine–glacial acetic acid–potassium acetate solution: 100 g potassium acetate are dissolved in 1 liter glacial acetic acid at 60°C. After cooling, 4 ml iodine-free bromine per atom are added.

Working procedure

The air sample is introduced at a flow rate of 1–2 bubbles/sec (about 50 ml/min) through the iodine pentoxide tube (heated to 120°C) and attached receiver. To ensure complete transfer into the receiver of the iodine which is produced, an additional 200 ml CO-free air or nitrogen are introduced after treatment of the air sample.

For the direct titration the iodine absorbed in the tube by the KI solution is transferred into a titration flask with a little water. Soluble starch is added and the solution is titrated with 0.01 *N* $Na_2S_2O_3$. One ml 0.01 *N* $Na_2S_2O_3$ corresponds to 0.56 ml (NTP) or 0.70 mg CO.

In the indirect procedure the content of the tube wetted with diluted NaOH is rinsed into a 200-ml flask. Ten ml bromine–glacial acetic acid are added, the solution is shaken and formic acid is introduced in drops until the brown color of the excess bromine disappears. After 3–4 min 2 ml sulfuric acid (1 vol. concentrated H_2SO_4 + + 1 vol. water) and 300 mg solid potassium iodide are added. The solution is shaken and titrated with 0.01 *N* $Na_2S_2O_3$ to light yellow. Soluble starch is then added and the titration completed.

One ml 0.01 *N* $Na_2S_2O_3$ corresponds to 0.0936 ml (NTP) or 0.117 mg CO in the indirect procedure.

Hydrogen and methane are still not appreciably oxidized at 120°C.

2. *Oxidation of carbon monoxide with Hopcalite*

(See also Lindsley and Yoe [166], as well as Salsbury, Cole and Yoe [216].)

The use of iodine pentoxide as an oxidant for CO has the disadvantage that the holding time of the air sample is long and hence the flow rate must be low. A mixed catalyst, called Hopcalite, which contains copper and manganese oxides and sometimes also silver and cobalt oxides, is much more efficient also at low temperatures. It is used in the conventional respirator for the removal of CO from the air and in automatic CO-analyzers (see page 196). In the quantitative determination the catalyst is heated to 100°C so as to ensure complete desorption of the carbon dioxide formed. When a 2 × 20 cm Hopcalite layer is used (the catalyst can be prepared from precipitated CuO and MnO_2), an air flow rate of 1 liter/min can be used. Under these conditions the carbon dioxide produced can be quantitatively transferred by foam absorption in a fritted scrubber into receivers containing 0.01 *N* $Ba(OH)_2$ or NaOH (see page 192).

The absorbed CO_2 can also be determined by potentiometric titration or conductometry.

The application of potentiometric single-point titration is very advantageous for continuous determinations. Depending on the flow rate ratios between absorbing agent and air sample, various CO-concentrations can be analyzed (Leithe and Petschl, to be published).

5.7.3 Carbon oxysulfide

Carbon oxysulfide (COS, boiling point −50°C) occurs cometimes in natural gas and sulfur waters. In industrial gases it is a component of "organic sulfur", for example, in coke-oven gas. It occurs along with carbon disulfide in the waste gases of viscose plants. COS is odorless and highly toxic. No MAK-value has been established.

Analysis

Whereas the determination of low COS-concentrations in the presence of H_2S, CO_2 and CS_2 was hitherto very complicated, a good method has now been formed on the basis of gas chromatography. For the determination of COS in natural gas Schols [240a] used a 360 cm × 5 mm column, filled with 30% N,N-di-n-butyl-acetamide, which is applied to the carrier "Kromat FB" (firm Burrell). The detector consists of a thermistor conductivity cell and helium (50 ml/min) serves as the carrier gas at 28°C. A 5-ml gas sample is used. The retention time is about 5 min. COS appears before the C_4 hydrocarbons and after CO_2 and the C_3 hydrocarbons.

The gas-chromatographic separation of COS from air CO_2, H_2S, CS_2 and SO_2 is described by Hodges and Matson [108a]. The authors applied either a 180-cm column containing 20% benzylcellosolve on firebrick, kieselguhr or chromosorb G at 30°C (carrier gas helium), or a 30-cm column filled with silica gel (Davison 08, 80–100 mesh) at 100°C (40 ml He/min). The COS-peak appears after air and CO_2 and before H_2S, CS_2 and SO_2.

A preliminary concentration of COS at −78°C over silica gel is recommended for air analysis.

5.7.4 Methane

Methane is always present in atmospheric air. It also occurs in concentrations of 1.2–1.5 ppm at places where a local pollution by natural gases, large-scale biological processes (e.g., anaerobic decomposition processes in swamps) or industrial waste gases is not expected. This permanent methane level must be taken into account, for example, in the gas-chromatographic analysis of hydrocarbons. As borne out by C^{14}-measurements, methane in the air is mostly of recent biological origin.

Methane is present in concentrations of up to 99% in natural gas as well as in damps and marsh gas. It is the anaerobic decomposition product of organic compounds. Further, methane is contained in large amounts in the waste gases of anaerobic putrefaction processes of wastes (e.g., waste water processing) and in fuel gases (e.g., in coke-oven gas up to amounts of 25–30%).

Air–methane mixtures are significant because of the possibility of explosions (explosion limits between 5 and 14% methane). Such mixtures (damps) constitute a permanent hazard in coal mines. Lower methane levels (below 1%) have no harmful effects on humans or vegetation.

Low methane levels in the air are sometimes determined for metabolism tests and for finding leakage sources. However, determinations near the explosion limits, especially in underground mines, are of special importance. For this purpose portable and rapidly recording manual devices or continuous automatic analyzers are employed which trigger alarm signals when methane is present in dangerous concentrations.

Methane can be determined in the air (in the absence of other combustible carbon compounds) by combustion with the aid of copper oxide or noble metal catalysts at red heat. The amount of carbon dioxide produced is then assessed. Any carbon monoxide which may be present can be removed by preliminary combustion at 280°C; methane is not oxidized at this temperature.

Among the earlier physical methods, portable interferometers were frequently used; the percentage of methane could be directly read off a scale. The reading, however, is not specific for methane, since the changes in the refractivity can be also due to other variations in the air composition.

Instruments with catalytic combustion on electrically heated platinum wires are easier to operate. The changes in the wire resistance caused by the temperature increase due to the combustion of methane is measured on a Wheatstone bridge. Such devices can be designed to record methane levels of 0–5%. The "Gesellschaft für Gerätebau", Dortmund, manufactured a convenient instrument of this type, whereby the power is supplied by a small storage battery. Of course, such devices do not yield specific results, since they are based on combustion processes which may occur also with other compounds.

Instruments which measure the infrared absorption are more sensitive and specific, and are especially suitable for automatic analysis. The URAS device (see page 66) has the most sensitive measuring range for methane (0–200 ppm), but can also be adjusted to higher concentrations. The UNOR I of the firm Maihak, Hamburg, is specially designed for underground tests in coal mines.

Gas-chromatographic determination of methane

This technique has proven to be very successful for the determination of methane in small air amounts. The working procedure was already outlined in Section 4.5.3. The model GC-M of the firm Beckman (double FID) is an apparatus suitable for the determination of very low methane levels in an air sample of 0.3 ml. Helium (70 ml/min) serves as the carrier gas, while the dioctyl sebacate column (2.2 m $\times$ 3 mm) is used for separation. The methane peak appears after approximately 1.6 min.

5.8 Halogens

5.8.1 Fluorine

5.8.1.1 *Occurrence*

Elemental fluorine is rarely found in industrial waste gases (fluorination processes), but fluorine compounds (below referred to as "fluorine") are often, may it be in low concentrations, constituents of industrial emissions. Fluorine occurs mostly as HF and sometimes as SiF_4; in humid air the latter is slowly converted to HF.

The most important sources of fluorine-containing industrial air pollution are the following:

1. Plants for phosphate fertilizers. The crude phosphate serving as the raw material consists mainly of fluorapatite $3Ca_3(PO_4)_2 \cdot CaF_2$ and contains up to 4% fluorine. When decomposed with sulfuric acid, part of the fluorine escapes as HF or SiF_4 (e.g., up to 20% escapes in the production of superphosphate as well as phosphoric acid). When nitric acid is used, only small amounts of fluorine are discharged into the atmosphere.

2. Plants for the production of aluminium by dissolution in fused cryolite and subsequent electrolysis. The raw material is here Al_2O_3 (purified clay), to which cryolite (Na_3AlF_6) and AlF_3 are added to reduce the melting point. At the anodes, which consist of coke and coal-tar pitch, fluorine-containing waste gases (HF and SiF_4 via reactions with the anode material) as well as CO_2 and CO are generated. The amount of fluorine produced is about 6–8 kg per ton aluminium.

3. Disturbing amounts of HF are sometimes discharged into the atmosphere by ironworks, when CaF_2 (fluorite) is added to the scrap in order to produce low-boiling slags (this occurs, for example, in the Siemens-Martin furnace). Sometimes the iron ores themselves also contain appreciable amounts of fluorides (e.g., in Utah, USA).

4. Small amounts of calcium fluoride are frequently found in the raw materials used in the production of bricks and ceramics, or as additive in enamel production. Fluorine is released as HF at high burning temperatures.

5. Dead, slate-containing gangues of coal frequently contain fluorine. Consequently, the fluorine level in the atmosphere is appreciably increased during the winter months.

Hygiene of fluorine and damage to vegetation

We distinguish between two types of toxic effects of fluorine-containing waste gases: *acute* and *chronic*; the former occur in industrial accidents. Fluorine etches the skin and the wounds heal very slowly.

Chronic injuries can be caused by gaseous pollutants (HF or SiF_4) or by fluorine-containing dusts. The symptoms are weakening of the bone substance (osteomalacia) and teeth (spotty dental enamel and embrittlement) which is caused by disturbances in the calcium metabolism.

On the other hand, small amounts of fluorides (up to 1 mg per day) are administered to children and juveniles (e.g., in the USA and occasionally in Germany) or are added to the drinking water in order to increase resistance against tooth decay. This fluorine amount, which is certainly not toxic, would be present in the daily respiratory air volume of about 10 m^3 and correspond to an amount of 0.1 ppm F^-.

The MAK-value for gaseous fluorine compounds (calculated as HF) is 2 ppm and for dust-like fluorine compounds 2.5 mg/m^3. The limits for sparingly soluble fluorine compounds, for example, CaF_2 or apatite, have not yet been established. Here, the solubility of the compounds in gastric juice has to be allowed for.

A MIK-value, which would take into account damage to vegetation has not yet been established.

Fluorine compounds in the air may lead to appreciable damage to vegetation, even when present in very low amounts. In gladiolus, concentrations as low as 2 $\mu g\,F^-/m^3$ may lead to leaf damage. Also, certain stone fruits and conifers are very sensitive in this respect, while numerous other agricultural plants, for example, wheat, are comparatively resistant to fluorine. A synergetic effect is assumed when F^- and SO_2 are simultaneously present.

The capacity of plant organs to accumulate fluorine is remarkably high. Whereas plants grown in clean air contain only about 5–20 ppm F^- in the dry substance, the fluorine level after long-term exposure to fluorine-containing waste gases was found to increase to levels of 100–2500 ppm. Such green fodder enriched with fluorine may lead to serious chronic poisoning (growth retardation, diseases of the skeleton) in pasture animals which are not very sensitive to direct inhalation of fluorine immissions. The animal fodder should not contain more than 1.8 mg F^- per day and kg body weight.

Removal of fluorine compounds from industrial waste gases

Separation of fluorine compounds from waste gases is not only important for air pollution control but it is also interesting from the economic standpoint, since the isolated substances can be widely used.

The waste gases of the superphosphate plants, especially those immediately formed upon mixing of the crude phosphate with H_2SO_4, are removed by suction and scrubbed in special installations (spraying towers, Ströder scrubbers; see also page 16). The prescribed reduction of the F^- level to less than 50 mg HF/m^3 (in the regulation of the British Alkali Inspectorate the upper limit is 115 mg/m^3) can be readily achieved. On the other hand, the removal of the malodorous organic pollutants formed in low concentrations from the crude phosphate, which, though toxicologically harmless, constitute an annoyance in neighboring residential areas, is somewhat more complicated. The same applies to the large amounts of highly diluted air escaping from the cellars which are used for maturing and storage of the superphosphate. The alkaline scrubbing liquids can be processed into fluosilicates or AlF_3 and other fluorides.

The situation is similar for the anode waste gases in the production of aluminum. Here, as well, the scrubbing of the concentrated furnace gases is less difficult than the removal of the highly diluted gases from the storage rooms. The gases are scrubbed with soda solution, whereby NaF is formed, or with sodium aluminate solution for conversion to cryolite.

The removal of fluorine from the hot waste gases of iron works can be effected by dry procedures (treatment with lime dust).

Determination of fluorine

5.8.1.2 *Survey*

Though the toxic effects of fluorine on plants have been considered down to the ppb-range ($\mu gF^-/m^3$ air), the MAK-value is as high as 2 ppm. For this reason several methods are required for this wide sensitivity range and air amounts of up to 4 m^3 are sometimes used.

The earlier procedures are based on the effect of the fluorine ion on color lakes of zirconium or thorium with alizarin. The metals are removed from the color lake to form the more stable metal fluoride. The lake is decolorized and the weaker color can be observed on a filter paper or measured with the aid of a spectrophotometer. Larger fluorine amounts can be titrated; alizarin indicator is added to the fluoride solution and the latter is then titrated with thorium or zirconium salt solution until the yellow color of alizarin changes into the red color of the lake.

A more recent method of Belcher, Leonard and West is based on the measurable blue color which develops when the fluorine ion reacts with the red lake consisting of alizarin complexone (3-aminomethylalizarin-N,N-diacetic acid) and ceric or lanthanum salts.

For the separation of the fluorine-containing toxicants from the air sample VDI-Richtlinie 2452 prescribes an impinger (see page 42) with a throughput of 30 liter/min or the procedure of Buck and Stratmann with a sorption tube (60 liter/min). However, Leithe and Petschl found that in the analysis of the (dust-free) air at the production sites of superphosphate and aluminum only one third of the fluorine amount absorbed in the fritted scrubber in the presence of n-butanol (foaming agent) and at an air flow rate of 1.5–2 liter/min is trapped in the impinger. In the fritted scrubber not only HF but also other fluorine compounds are absorbed because of the long residence time in the foam. In the impinger, however (where the residence time is short), HF is preferentially trapped.

From the alkaline absorption liquids fluorine is, as a rule, distilled by steam after separation of the interfering substances and acidification with sulfuric or perchloric acid. However, there also exist separation procedures using ion exchangers.

Usually, the gaseous component of fluorine in the air is determined, but it may happen that analysis of the dust component is also necessary. This especially applies to enclosed spaces in plants for phosphate fertilizers. The dust is collected in accordance with the procedure described on page 101. The filter is first impregnated with milk of lime, ashed in a nickel crucible in a muffle furnace at 600°C, fused a few minutes with 2 g NaOH and dissolved in water. If necessary, a few drops of H_2O_2 are added to destroy sulfites and the solution is transferred into the distillation flask.

A Dräger tube can be used for an orientative determination in the MAK-range.

A long-term determination of the ground-level F concentration can be effected by a simple method using cloth strips. This is especially useful for evaluating damages to agricultural plants when the ground-level concentration is only local and its small extent renders high costs for the analysis unfeasible. A similar aim can be achieved by determining the fluorine content in plant materials from the areas in question.

Continuous recording of the fluorine level in the atmosphere is possible by means of the "Mini-Adak" and the Imcometer.

5.8.1.3 *Method of* Belcher, Leonard *and* West [16] (*VDI-Richtlinie* 2452)

Sampling

A 300-ml impinger (sampling time 1 hr), 70 ml 0.1 *N* NaOH and 5 ml 3% H_2O (to oxidize sulfite, sulfide and nitrite), is connected to a suction pump and a gas meter via an empty wash bottle and a pressure gage. The suction rate is 1.7 ± 0.1 m^3/hr. The underpressure of the gage should be 80 ± 10 mm Hg.

Distillation for the removal of interfering substances

The absorption liquid is transferred into the flask of the steam-distillation apparatus (Model, Hohenheim*) shown in Figure 5.8.1.3. Instead of the asbestos plate used in the model of the "Landesanstalt", a 140°C oil bath is employed, but overheating of the solution in the flask must be avoided. To remove the excess H_2O_2, 2 g ferrous sulfate dissolved in 10 ml H_2O and a few scraps of glass frits are added, followed by the introduction of 120 ml 72% H_2SO_4 via the dropping funnel. The content of the flask is heated to 130°C and 250 ml water are distilled off. When the fluorine level

* A suitable distillation flask is also manufactured by the firm E. Bühler, Tübingen.

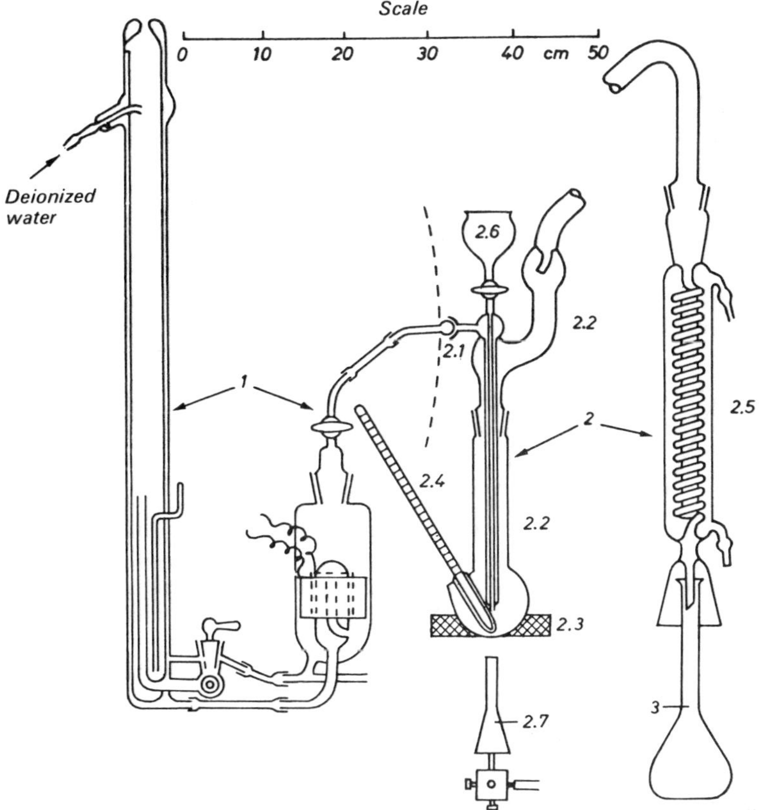

Figure 5.8.1.3
Steam distillation, model "Hohenheim"

1	"Buchi" steam generator	2.5	Spiral cooler for distillate
2	Distillation apparatus "Hohenheim"	2.6	Dropping funnel for H_2SO_4
2.1	Supply tube for steam	2.7	Bunsen burner
2.2	Distillation flask	3	Volumetric flask for distillate (100 and 250 ml, resp.)
2.3	Asbestos plate		
2.4	Thermometer		

is below 5 μgF^-, 100 ml of distillate is sufficient. The steam generator contains distilled water and a NaOH pellet.

Determination

From the 250- and 100-ml distillates in the volumetric flask 50 ml are respectively withdrawn and transferred into a 100-ml volumetric flask. Then 2 ml buffer solution,

10 ml alizarin complexone solution, 10 ml lanthanic nitrate solution and 25 ml acetone are added. The volume is made up to 100 ml with water and the solution is allowed to stand for $1\frac{1}{2}$ hours in the dark. The spectrophotometric determination is carried out at 610–620 mμ in a cell with a layer thickness corresponding to the color intensity. The reference solution (zero value) is a red blank solution prepared from the reagents.

When the extinction of the aliquot (50 from 250 ml) is so low that it lies within the range of the reading error, the residue of the distillate (200 ml) is made just alkaline with NaOH. It is then evaporated to about 50 ml on a platinum or quartz dish, neutralized with 0.1 *N* $HClO_4$ and the measurement repeated. The detection limit is in this case about 0.6 μg F^-/m^3.

Evaluation

The measured extinctions are evaluated on the basis of a calibration curve (range 1–30 μg F^-) which is plotted for a calibration solution containing 1 μg F^-/ml. Further, a blank distillation was to be carried out with subsequent spectrophotometric determination and the blank is subtracted from the determination result. When the calibration curve is a straight line and passes through the origin, its slope can be immediately used for the calculation of the result in accordance with the following formula:

$$\mu g\ F^-/m^3 = \frac{E - E_o}{Kv} \times \frac{a}{x},$$

where E is the extinction of the distillate relative to the blank, E_o is the extinction of the blank distillate without sample, K is the slope of the calibration curve, v is the column of aspirated air sample (m^3), a is the total volume of the distillate (ml), and x is the volume of the distillate aliquot (ml).

Preparation of reagents

Fluorine-free sulfuric acid, 72% (d = 1.635): 1 liter of 24 *N* (72%) H_2SO_4 is heated with some glass frits to 120°C in a 2-liter round-bottom flask with a thermometer and condenser. Then steam is introduced from a steam generator which contains water and a pellet of NaOH. About 1 liter liquid is distilled over at 150°C. Then the distillation flask is more strongly heated and the distillation interrupted when the thermometer in sulfuric acid indicates 171°C.

Alizarin complexone solution, 0.0005 *M*: 193 mg 3-aminomethyl-alizarin-N, N-diacetic acid (alizarin complexone of Merck) are suspended in a little volume of water and dissolved with the addition of the smallest possible amount of NaOH. The pH is adjusted to 5.0 (color change from violet to red) with dilute HCl and the solution made up to 1 liter.

Lanthanic nitrate solution, 0.0005 *M*: 216.5 mg lanthanic nitrate hexahydrate per liter.

Buffer solution: 105 g sodium acetate trihydrate and 100 ml acetic acid are diluted with water to 1 liter (pH = 4.3).

NaF calibration solution: 442.1 mg NaF are dissolved in 1 liter water. 10 ml of the solution are made up to 2 liter with water. 1 ml = $\mu g\ F^-$.

Twice distilled water has to be used in all cases. The solutions (except for sulfuric acid and acetone) should be stored in plastic containers. Quartz containers are recommended for sulfuric acid.

The response sensitivity (50 ml from 250 ml distillate) at a layer thickness of 4 cm is 3 $\mu g\ F/m^3$, the standard deviation in the 10 $\mu g\ F^-/m^3$ range is $s = \pm 0.5\ \mu g\ F/m^3$.

Variant used in the Österreichische Stickstoffwerke for fluorine amounts exceeding 0.01 mg/m^3

The foam absorption procedure is applied to effect complete trapping of the fluorine. Further, an aliquot of the distillate is orientatively titrated with thorium nitrate in order to select the most favorable amount ratio for the colorimetric determination.

The absorption is carried out in two series connected 250-ml fritted scrubbers (G 1 frit). Each scrubber contains 100 ml of approximately 0.1 *N* NaOH and 3 drops of n-butanol as a foaming agent. The air throughput rate is 1.5–2 liter/min. For air analysis a sample volume of 100 liter, for air in rooms in the MAK-range 25–50 liter and for waste gases a correspondingly smaller sample is taken.

The alkaline content of the scrubbers is first evaporated to about 20 ml and transferred into the steam-distillation apparatus. A spatula of silver sulfate and about 0.2 g finely powdered quartz sand are added. Then 20 ml 60% $HClO_4$ are introduced via a dropping funnel and 200–250 ml are distilled off. The distillate in the receiver is maintained alkaline by adding 0.1 *N* NaOH (phenolphthalein). The weakly alkaline distillate is evaporated on a platinum dish to about 80 ml, transferred into a 100-ml volumetric flask and made up to the mark.

For the determination of the approximate F^--level (in the spectrometric determination no more than 25 $\mu g\ F^-$ must be present) 10 ml of the distillate are preliminarily titrated in the following manner (see also page 209); 10 ml of the distillate are adjusted in a colorless wide-mouth bottle to pH 7 with 0.1 *N* HCl against phenolphthalein. Then 0.4 ml indicator solution are added and the solution is again adjusted with 0.01 *N* HCl until it just turns yellow. The solution is titrated in accordance with the procedure outlined on page 210.

0.1 ml of 0.01 *N* $Th(NO_3)_2$ corresponds to about 10 $\mu g\ F^-$.

Using the result of the preliminary titration, the distillate is pipetted into a 100-ml volumetric flask in such an amount that the highest value (about 25 $\mu g\ F^-$) of the calibration curve is not exceeded. Then 0.05 *N* $HClO_4$ is added until the phenolphthalein color disappears; the reagents are added (as outlined on page 206) and the above described procedure carried out.

The extinction is measured against the reagent mixture without the sample. The final result is obtained by subtracting the blank of the chemicals for the absorption, distillation and color measurements.

If the perchloric acid used contains interfering amounts of fluorine, the latter can be eliminated by blowing off with steam at 155°C.

Determination in the presence of large amounts of NO_2

When waste gas samples (e.g., from plants in which crude phosphates are decomposed by HNO_3) contain large amounts of NO_2 which interfere in the distillate, this substance can be most readily eliminated by treatment with amidosulfonic acid. The alkaline absorption solution is first neutralized by sulfuric acid against phenolphthalein and 1–2 g solid amidosulfonic acid are added with continuous mixing until the nitrite reaction with the Griess reagent becomes very weak. Then dilute NaOH is added, the solution evaporated and the analysis continued as above.

5.8.1.4 *Variant for small* F^-*-levels according to* Buck *and* Stratmann

For the determination of small F^--concentrations (1–2 $\mu g/m^3$) with a relative standard deviation of ± 5% Buck and Stratmann [33] developed the following procedure:

1. Fluorine is absorbed in sorption tubes (silver tubes, wall thickness 1 mm, inside diameter 12 mm, length 100 mm, containing silver beads of 3 mm diameter, wetted with 20% Na_2CO_3 and dried at 200°C) at a flow rate of 4 m^3/hr and a sampling time of 10–15 min. F^- is desorbed with steam generated from 30 ml water. For the distillation 50 ml 24 *N* H_2SO_4 (free of fluorine) are used.

2. To prevent the accumulation of excessive amounts in the steam distillation, the distillate is trapped in a quartz receiver, which contains 35 ml water + 1 ml 0.1 *N* NaOH (the receiver is heated to 145°C on a heating bath). The distillate is transferred into a 100-ml volumetric flask, 0.05 *N* HCl is added up to a very weak acid reaction (phenolphthalein), the reagents are introduced, the solution is made up to the mark and the extinction measured, as described on page 206.

3. The calibration curve (5–25 μg F^-) is plotted after distillation and addition of all the reagents. Further, an averaged total blank is determined and taken into account.

This procedure was tested by the authors on prepared mixtures of air and HF. It is, however, rather doubtful whether other fluorine-containing toxicants can be completely absorbed (Leithe and Petschl [159]).

5.8.1.5 *Titration of fluorine with thorium nitrate and alizarin sulfonic acid*

This procedure is somewhat older, but is especially suitable for determining higher F^--levels. The principle of this reaction is as follows: thorium does not form a red color lake with alizarin sulfonic acid in the presence of F^-, since a more stable colorless thorium–fluorine complex is formed. When titrating a fluorine-containing solution with a diluted thorium salt solution in the presence of alizarin sulfonic acid as indicator, first the colorless thorium fluoride is formed. The initial drops of excess thorium form the pink color lake. Care must be taken to adjust the pH to a certain optimum value, so that the color change in the titration with 0.01 *N* $Th(NO_3)_4$ can be distinctly seen. The change (red to yellow) can also be observed in back titration. Interfering components are removed by the procedure described above, i.e., by means of steam distillation according to Willard and Winter [270], but here perchloric acid is preferably used instead of H_2SO_4. On the basis of the experiences of Gericke and Kurmies [77] the following procedure is applied in the laboratory of the Österreichische Stickstoffwerke.

A small 2 × 8 cm colorless wide-mouth bottle with a flat bottom serves as the titration vessel. In order to distinctly observe the end point of the titration first a reference solution is prepared. To 10 ml water 1 drop of phenolphthalein solution and 1 drop of 0.1 *N* NaOH are added. The distinctly red solution is accurately decolorized by carefully adding 0.1 *N* NaOH. Then 0.4 ml alizarin indicator solution is added, after which the color of the solution changes from orange to red. 0.1 *N* HCl is again carefully added until the solution just turns yellow. Finally, 0.5 ml hydrochloric acid–thorium nitrate solution is added (this serves as the standard solution) and a small amount (0.015–0.02 ml) of the 0.01 *N* thorium nitrate solution is introduced; the color changes from yellow to light pink. This solution serves as the standard for the color at the end point.

Calibration curve

A calibration curve is necessary, since the consumption of the titration solutions does not quite stoichiometrically correspond to the fluorine amount present. An NaF stock solution is prepared by dissolving 0.221 g NaF in 1 liter of water. 100 ml of this solution are again diluted to 1 liter after adding a few NaOH pellets (1 ml = 10 μg F^-). The volumes corresponding to 10, 20, 30, 40 and 50 μg F^- are respectively diluted to 10 ml and titrated with 0.01 *N* thorium nitrate solution, as in the determination of the color standard. The calibration curve is plotted from the consumption of the titration solutions and the corresponding F^--concentrations. Since for F^--amounts exceeding 50 μg in 10 ml the end point color cannot be distinctly seen, it is recommended to adjust the sample to amounts not exceeding 50 μg/10 ml; for this purpose an aliquot of the distillate can be preliminarily titrated. The value obtained in the blank titration must be subtracted from the result to give the actual F^- concentration.

Working procedure

The sample is withdrawn and distilled as described on page 208. The distillate is evaporated to 15–20 ml, introduced into a 25-ml flask and the volume made up to the mark. 10 ml are pipetted and the determination is carried out in the same way as for plotting the calibration curve.

Reagents: Alizarin indicator solution: 0.125 g sodium alizarin sulfonate are dissolved in 1 liter water.

Hydrochloric acid–thorium nitrate stock solution: 20 ml of 0.01 *N* thorium nitrate and 75 ml *N* HCl are diluted with water to 1 liter. 0.01 *N* thorium nitrate solution: 1.3804 g $Th(NO_3)_4 \cdot 4H_2O$ are dissolved and diluted to 1 liter with water. The content of water of crystallization can be determined by igniting the nitrate to ThO_2.

Determination with the gas-detection device

Rapid semiquantitative determinations of F^- in the MAK-range can be effected by means of the Dräger gas-detection device (see page 56) with the test tube 0.5 a. When 10 pump strokes are applied, fluorine amounts between 1 and 15 ppm can be measured. The violet color of the zirconium alizarin lake becomes yellow with increasing F^--amounts in the individual zones. When a pump is used whereby a 10 liter air sample can be aspirated, the measuring range can be reduced to 0.1–1.5 ppm.

Cloth-strip methods for long-term determinations

We describe here a test series carried out at the Ranshofen aluminum plant as an example of a long-term cloth-strip method for characterizing the quantitative distribution of F^--immissions originating from a single emission source.

Cotton strips (area 3 dm^2) wrapped around a wooden cylinder and protected against rain by metal screens are set up at 32 points at a distance of 2 km from the emission source. The strips are impregnated with baryta water, which contains glycerol to retain the moisture.

After an exposure of 1 month the strips are ashed and the fluorine is determined by the conventional method. The fluorine levels are between 0.1 and 5 mg F^-. The evaluation also allows for the wind directions recorded at a central point and serves as a basis for compensation of damages claimed by neighboring residents.

Fluorine analysis in vegetable matter

Fluorine is accumulated in plant matter (see page 202) and can be determined, for example, in the needles of conifers. This also permits a cumulative evaluation of fluorine immissions.

The air-dry material is first finely disintegrated in a mixer and homogenized; 2–4 g of the powder are moistened with 20 ml milk of lime (10 g fluorine-free CaO/liter) in a platinum dish, evaporated to dryness on a water bath and demoistened in a drying oven at 150°C. Then the sample is placed in a cold muffle furnace, slowly heated to 550°C and weakly calcined for another 30 min. The ash is disintegrated and ignited for 3–4 hr until its color is light gray. It is then transferred into a distillation flask, rinsed with a small volume of dilute perchloric acid and distilled as described above. In analogous fashion a blank of the reagents is determined, using the same amount of milk of lime and the same evaporation, calcination and distillation conditions.

Automatic determination of the atmospheric fluorine level

An automatic recorder for very small fluorine concentrations (1–35 $\mu g/m^3$) in the atmosphere is described by Adams and Koppe [1a] under the name Mini-Adak. The operating principle is the spectrophotometric recording of the decolorizing reaction of the color lake of zirconium Eriochrome Cyanine by fluorine, which was first described by Megregian. In order to reduce the consumption of reagents in periods of low fluorine ground-level concentrations, the reaction product of the dye solution and air sample is drained after the spectrophotometric determination through a flow cell into the absorption part of the apparatus until a certain decolorization is reached; only then is the absorption liquid renewed.

The Imcometer (see page 70) can also be used for recording low fluorine concentrations (Fuhrmann [71]).

5.8.2 Chlorine

5.8.2.1 *General* (see also VDI-Richtlinie 2106)

Elemental chlorine is one of the most important chemicals used in industry. The annual production of chlorine reached 600,000 tons in the German Federal Republic in 1959. It is mainly used for the production of chlorinated organic products, including plastics (especially polyvinyl chloride), insecticides (DDT, Lindane, Aldrin, etc.) and the herbicides of the 2,4-dichlorophenoxyacetic acid. It is also used as a bleaching agent in the cellulose and paper industry, as well as in laundry plants, and as a disinfectant for drinking and bathing water.

Because of its varied application it can occur at numerous work sites, where its irritating effect (even in low concentrations) is a cause for concern. It rarely occurs as a toxicant in the free atmosphere, unless it escapes as a result of industrial accidents.

Toxicology

The odor threshold is about 0.05–0.1 ppm. Irritating symptoms occur below 1 ppm and are dangerous at concentrations exceeding 3 ppm.

The threshold value for damages to vegetation is about 0.1–1 ppm.

MAK-value 0.5 ppm = 2 mg/m^3

MIK_D-value 0.1 ppm = 0.3 mg/m^3

MIK_K-value 0.5 ppm = 1.5 mg/m^3 (once within 8 hr)

The immission threshold specified by the "Technische Anleitung" is 0.3 mg/m^3 and is permitted to increase to 0.6 mg/m^3 once in 8 hr.

5.8.2.2 *General information for the analyst*

Analyses for the determination of elemental chlorine are performed at the work sites in plants. Because of the strong irritating effect of chlorine, low concentrations between 0.1 and 1 ppm are of special interest.

The colorimetric methods used for this range of chlorine levels are based on common oxidation reactions which are not only specific for chlorine but also for other oxidants, such as NO_2 and ozone. Since, however, these studies are performed in plants where we have certain information on the toxicants which may occur, this should not be too great a disadvantage.

The method most frequently applied uses o-tolidine (a reagent which is widely employed in water analysis). Amounts as low as 0.2 μg chlorine can be detected with this reagent. The compound 3,3′-dimethylnaphthidine of similar structure is even more sensitive but less specific.

Determination with o-tolidine

For the determination of elemental chlorine in the air the colorimetric method with o-tolidine is used. It is simple and sensitive, but has the disadvantage that it is not specific when other strong oxidants, such as bromine, nitrogen dioxide and ozone, are also present. The method can be used only when there is no risk of confusing chlorine with these interfering substances. The air sample is directly introduced into the acidified reagent. The resulting yellow color is either evaluated by comparing it with standard solutions made of $K_2Cr_2O_7$ or the $K_2Cr_2O_7$–$CuSO_4$ mixtures used in water analysis, or by photometry at 440 mμ.

The air sample is introduced into a small fritted scrubber, containing 10 ml tolidine reagent, until a distinct yellow color appears which can be photometrically evaluated.

Tolidine reagent: 1 g of o-tolidine is triturated with 10 ml HCl (d = 1.18), rinsed with 95 ml hydrochloric acid into a 1-liter volumetric flask and made up to the mark with water.

The calibration curve for the photometric determination is obtained from a calibration series of 0–8 μg Cl_2. The element is present in the form of 10 ml diluted chlorine water, obtained by dilution of an iodometrically analyzed stock solution. The calibration factor ("reciprocal slope" see page 54), obtained on the basis of data of Freier [70] is 16 for 10 ml solution and a layer thickness of 1 cm. Thus, for E = 0.1, 1.6 μg chlorine are contained in 10 ml solution.

For a visual comparison 4 standard solutions are prescribed in the ICI Manual. These are obtained by respectively diluting the $K_2Cr_2O_7$ solutions (1 g/liter) shown in Table 5.8.2 with water to 10 ml.

Table 5.8.2

Standard No.	ml standard $K_2Cr_2O_7$ solution per 10 ml	Corresponding amount of Cl in μg
1	5	47
2	3	28
3	2	18
4	1	9

Determination of chlorine in the MAK-range with dimethylnaphthidine

The British Factory Inspectorate prescribes in Booklet No. 10 of its collection the much more sensitive 3,3′-dimethylnaphthidine, which yields a light violet color with chlorine. For the determination in the MAK-range a 360-ml air sample is adequate. The reaction, however, is just as unspecific for chlorine as that with tolidine.

A 360-ml air sample is aspirated by a pump (120 ml per stroke) within 2 min through a small wash bottle with a narrow inlet tube or fritted scrubber. The scrubber contains 4 ml of reagent solution. After sampling, the developed color is immediately compared with the standard colorations (see below).

Reagent solution: 0.01 g fine 3,3′-dimethylnaphthidine powder (Merck) is dissolved in 5 ml glacial acetic acid and immediately diluted to 200 ml with water. The solution is supersaturated, but is stable for several weeks when stored in the dark.

Reference solution: 4 mg safranin are dissolved in 100 ml water and 4 mg gentian violet are dissolved in another 100 ml water. (The dyes can be obtained in "microscopy" quality at Merck and at the firm Gurr Ltd., London.) 7 ml of the safranin solution and 2.0 ml of the gentian violet solution are mixed and diluted to 100 ml with water. This solution corresponds to the color produced by 2 ppm Cl_2 (in 360 ml air) in the determination with 3,3′-dimethylnaphthidine. The 1:1 or 1:3 diluted solution corresponds to 1 and 0.5 ppm Cl_2, respectively.

5.8.3 Hydrogen chloride

General information

Hydrogen chloride is used mainly in the form of hydrochloric acid as etching, pickling and cleaning agent for metallic and ceramic surfaces, and occurs sometimes at the work sites of such plants. In the chemical industry hydrogen chloride is a frequently occurring secondary or waste product of the chlorination of organic compounds carried out on a large scale for the production of plastics and pesticides. It is, however, rarely emitted by the plants, since it can be easily removed from the waste gases by scrubbing and utilized as hydrochloric acid or converted by electrolysis into chlorine. On the other hand, hydrogen chloride is being formed in ever increasing amounts upon incineration of refuse consisting of chlorine-containing plastic wastes (especially PVC) and must be taken into account in this case.

Gaseous hydrogen chloride is a strong acid and thus irritates the mucous membranes; however, it has no special toxic effects (0.1 *N* hydrochloric acid is present in the gastric juice). Concentrations of up to 50 ppm in the air can be tolerated for a short time without injury to health.

The MAK-value is 5 ppm, the MIK_D-value 0.5 ppm = 0.7 mg/m^3 and the MIK_K-value 1 ppm = 1.4 mg/m^3 (this is permitted to occur once in 2 hours).

Analysis

Hydrogen chloride can be completely and rapidly absorbed from the air by 0.01 *N* NaOH in any scrubber.

The determination of the chlorides in the washing liquid can be effected on the basis of the various classical procedures, depending on the amount to be expected. The most widespread method is precipitation with $AgNO_3$. The end point of the titration can be determined with thiocyanate (Volhard), fluorescence indicators (Fajans), potentiometrically or nephelometrically. The procedure of Mohr with potassium chromate was modified by Leithe [153] (oxidimetric determination of the excess chromate) to give a detection limit of 1 μg Cl^-.

Iwasaki, Utsumi, Hajino and Ozawa [116] developed a convenient and sensitive colorimetric procedure with mercuric thiocyanate. It was modified in the laboratory of the Österreichische Stickstoffwerke for air analysis and is carried out as follows.

The air sample is passed through 30 ml 0.01 *N* NaOH in the wash bottle (fritted scrubber, Drechsel bottle, impinger) at a rate of 30 liter/min. The absorption liquid is transferred into a 50-ml volumetric flask, 3 drops of 2 *N* HNO_3, 4 ml of solution of 1 g mercuric thiocyanate (Merck) in 100 ml methanol and 8 g ferric ammonium alum in 100 ml 6 *N* HNO_3 are added. The volume is made up to the mark and the extinction is measured at 460 mμ against the reagent blank in a 1- or 5-cm cell, depending on the color intensity.

The calibration curve is plotted for NaCl solutions containing 10–20 μg Cl^-/ml in the range between 0 and 200 μg Cl^- in 50 ml reaction solution. 10 μg/Cl^- can be detected with certainty. 100 μg Cl^-/50 ml yield an extinction of 0.540 in a 5-cm cell; thus, the "reciprocal slope" is 192.

The determination is disturbed by the presence of other halides, cyanogen and sulfide.

5.8.4 Determination of bromine vapors with fluorescein paper

The formation of the pink eosin (tetrabromofluorescein) from fluorescein is used for a simple determination of bromine vapor in the air (see ICI Manual). The air sample is drawn through a filter paper impregnated with fluorescein until the developed color corresponds to a standard filter paper impregnated with eosin and fluorescein. The volume of the air sample serves as a measure for the bromine level.

Sampling: The impregnated filter paper is inserted into an aspirating device, whereby a circular filter area of 1 cm diameter is exposed for the passage of the air stream and observation. The air sample is aspirated at a rate of 800 ml/min, for example, by means of a bellows pump of 120 ml stroke volume (see page 40) and a suction velocity of 10 sec per stroke. Air is aspirated until a pink color appears which corresponds to the color of one of the two test papers.

Preparation of the impregnated filter paper.

Strips of Whatman filter paper No. 541 (width 5 cm) are immersed for 1 minute into a fluorescein solution consisting of 0.1 g fluorescein sodium in a mixture of 90 ml H_2O and 10 ml glycerol, and then left to dry in clean air. The paper is cut into 2.5 × 5-cm strips which are stable for 1 month when stored in closed containers.

Preparation of the standard test papers: Paper A: strips of Whatman filter paper No. 541 are immersed for 1 min into a solution of 0.05 g eosin (tetrabromofluorescein sodium) in 200 ml water and subsequently dried in clean air.

Paper B: Filter paper strips of the same sort are dipped in the same manner into a mixture of 0.05 g fluorescein sodium in 50 ml water and 80 ml of the above eosin solution and then dried.

The bromine level of the air sample is found according to Table 5.8.4 from the sample volume required to produce a color identical with that of the standard test papers.

Table 5.8.4

AIR VOLUME IN ml REQUIRED TO ACHIEVE A COLOR IDENTICAL

with paper A	with paper B	ppm Br_2
120		10
240		5
480		2.5
960		1
	120	2
	600	0.4
	1200	0.2
	2400	0.1

5.8.5 Phosgene

Filter paper method for the determination of phosgene (*British Ministry of Labor*)

The method is described in Booklet No. 8 of the British Ministry of Labor. For concentrations between 0.25 and 0.5 ppm, which correspond to the old MAK-value of 1 ppm, 120 ml air sample are aspirated, while the sample volume for concentrations between 0.025 and 0.5 ppm, corresponding to the range around the new MAK-value of 0.1 ppm, is 1200 ml. The brick red colors which appear in the presence of phosgene are compared with standards.

Preparation of the impregnated filter papers

Into a benzene solution containing 2% 4,(4′-nitrobenzyl) pyridine and 4% N-benzylaniline 2.5 × 6.5 cm strips of Whatman filter paper No. 1 are immersed, the excess is allowed to drip off and the strips are then dried in air. If they are stored in tightly closed containers, they remain stable and it is possible to use them for several months.

Working procedure

A strip is placed into a filter unit of at least 1 cm inside diameter and 120 ml of air are drawn at a rate of 120–180 ml/min. For the evaluation a color scale prepared from own test mixture or one obtainable via the H.M. Stationery Office is used. The firm Tintometer Ltd., Salisbury, Wilts., produces standards on glass for the Tintometer comparator.

The brick red color is largely specific for phosgene. In the presence of chlorine the element can be eliminated by inserting a dry filter paper impregnated with sodium iodide and sodium thiosulfate at the air inlet. Acetyl chloride weakens the phosgene reaction. A 100-fold amount of benzoyl chloride produces an orange spot which disappears after 3 min, while the phosgene spot remains stable. Benzyl chloride, trichloroethylene and chloroform in concentrations of up to 300 ppm do not interfere with the reaction.

For the automatic determination of phosgene by colorimetry in the range between 0.02 and 10 ppm see Linch, Lord, Kubitz and De Brunner [165b].

5.9 Determination of arsenic with silver diethyldithiocarbaminate

In contrast to the Gutzeit test (appearance of brown color due to AsH_3 on paper impregnated with $HgBr_2$) the determination of AsH_3 with silver diethyldithiocarbaminate (AgDEDTC) developed by Vasak and Sedivec [255] makes a normal colorimetric assessment in aqueous solution possible. The procedure is described in the information bulletin of Merck, Darmstadt, where the necessary reagents are made available. See also the Manual of Analytical Methods (ACGIH) and Booklet No. 9 of the British Factory Inspectorate.

Principle: Arsenic is expelled as AsH_3 with zinc from acid solution. The reaction product is introduced into AgDEDTC solutions in pyridine. The red colored complex formed is spectrophotometrically determined.

Working procedure

Sampling: If arsenic is present as dust, a suitable air volume is drawn through a paper or membrane filter (see page 200). To ensure simultaneous trapping of gaseous and dust-like arsenic compounds the air sample can be aspirated at a rate of up to 30 liter/min through an impinger containing 0.1 *N* NaOH.

Distillation: The neutralized scrubbing liquid of the dust sample suspended in 40 ml water is allowed to stand in a 100-ml wide-necked flask with 10 ml $SnCl_2$–hydrochloric acid, 5 ml KI solution and 1 ml $CuSO_4$ solution for 15 min. The flask is closed with a stopper, through which an U-shaped gas-discharge tube with a spherical extension is mounted. A cotton plug impregnated with $Pb(COO)_2$ is placed in the extension and dried in vacuo. The other end of the tube is drawn into a capillary of 1 mm internal diameter and immersed to the bottom of a small reaction tube, which serves as the trap. The trap contains the amount of AgDEDTC in pyridine required to fill the 1-cm cell.

After a standing time of 15 min, 8 g pure arsenic-free zinc are added. The mixture is allowed to stand for 60 min. During this period the gases are distilled into the trap. The red color is photometrically determined at 538 mμ.

Reagents: AgDEDTC solution. 1.0 g of this reagent (Merck) is dissolved in pure pyridine, so that the final volume is 200 ml. When stored in the dark, the solution is stable for several weeks.

Potassium iodide solution: 15 g KI are dissolved in 100 ml water.

Cupric sulfate solution: 2 g $CuSO_4 \cdot H_2O$ are dissolved in 100 ml water.

$SnCl_2$–hydrochloric acid: 0.33 g $SnCl_2$ are dissolved in 100 ml HCl d = 1.16 (32%).

The (linear) calibration curves are plotted for amounts of 1–15 μg As (As_2O_3 is dissolved in small amounts of NaOH), which are expelled by hydrogen in the manner described above, trapped in the same volume of AgDEDTC solution and measured. The molar extinction coefficient is about 13,000.

The detection limit is 0.2 μg As, the relative standard deviation $\pm$ 5%.

The corresponding antimony compound has an absorption maximum at 510 mμ and can be determined in the same way.

Hydrogen sulfide and phosphine which interfere are eliminated by cotton impregnated with lead acetate.

5.10 Determination of the lead content in air

5.10.1 General

Lead occurs in polluted air not only in dust but also in the form of gaseous compounds. This element is widely applied (e.g., because of its resistance against sulfuric acid) as a structural material in the chemical industry, in the manufacture of storage batteries and for coatings (water pipes and cable sheathing). It is used both in pure metallic form and in alloys (antimonial lead and metals for soldering, anti-friction and printing). Lead compounds are also used as pigments in coating materials (white lead, rust protection with minium and lead cyanamide) and in the glass, ceramic and enamel industries. Although lead (lead arsenate) is no longer important as a pesticide, it is nevertheless frequently found in stabilizers for plastics.

The frequent occurrence of lead in the dust at work sites and its slow-acting toxic effects, which lead to chronic diseases, justify the interest of the industrial hygienist in the analytical chemistry of the metal. The MAK-value for lead in dust is currently set at 2 mg/m^3.

The pollution of the outside air by tetraethyllead, which is used as an additive in motor gasoline to improve its antiknocking property, is particularly important. The MAK-value of tetraethyllead (expressed as Pb) is 0.75 mg/m^3. Thus, the lead content of the air in streets with heavy traffic is subject to constant surveillance.

Survey of the analytical methods

To effect the separation of lead occurring in smoke or dust, the air is drawn through a suitable paper filter. The metal is then released with dilute HNO_3 (sometimes H_2O_2 must be added) and either determined colorimetrically, as the red dithizonate, or polarographically.

Lead-containing vapors (e.g., tetraethyllead compounds) are either absorbed in iodine solution or ClI and determined as above. The direct determination of lead in the air sample by atomic absorption is particularly convenient.

Isolation of lead from dust

The lead-containing dust is collected by drawing the air sample through a circular paper filter (e.g., Whatman No. 42) inserted into a suction device (see page 40). Depending on the sensitivity required, 15–200 liter air can be drawn at a rate of 3–25 liter/min.

The dust-laden filter is extracted by treatment for 20 min with 100 ml 10% lead-free nitric acid on a water bath. The filtrate and washings are evaporated to dryness on the water bath.

5.10.2 Polarographic determination in accordance with the ICI Manual
(See also Wilson and Hutchinson [275])

After evaporating the HNO_3 solution, the residue is again dissolved in 5 ml 1 *N* NaOH. Then 0.2 ml of 1% gelatin solution is added, the volume made up to 10 ml and an aliquot of 5 ml is transferred into a polarographic cell. A weak hydrogen stream is passed through for 3–4 min in order to remove the oxygen present. The polarographic determination is carried out between -0.5 and -1.0 V (referred to

SCE). The half-wave potential is -0.75 V. The graphically determined plateau height is converted with the aid of a calibration curve to the lead content of the sample. The curve is plotted for 2–20 ml of standard solution containing 0.01 mg Pb per ml in the form of lead nitrate. The standard solution must be treated in the same way as the sample. Further, the blank of a filter paper without lead-containing dust must be determined.

This method permits determining 0.01–1 mg Pb/m^3 in 200 liter air.

5.10.3 Determination with dithizone as a field method (ICI Manual)

(See also Dixon and Metson [57])

The filter dusted with 15 liter of air (1 cm diameter of the dusted surface) is vigorously shaken and disintegrated into a slurry with 2.5 ml dilute nitric acid (5 ml HNO_3 d = 1.42 and 0.2 ml 30% H_2O_2 to 100 ml) in a glass-stoppered tube (length 12.5 cm, inside diameter 2 cm). 5 ml of the NH_3–sulfite–cyanide–citrate solution are added, the mixture is again shaken, 5 ml dithizone solution are added, the mixture is shaken for 1 min and the red dithizonate solution allowed to settle.

The color is measured by means of a spectrophotometer at 510 mμ or by comparing the color with cobalt sulfate solutions prepared in glass tubes of the same type from a stock solution (7.2 g $CoSO_4 \cdot 7H_2O$ to 100 ml aqueous solution). For the preparation of the $CoSO_4$ reference solutions, corresponding to certain lead concentrations, the color tint corresponding to the blank is assessed as follows: an unused filter paper of the size and quality used in sampling is treated with the same reagents, extracted and processed as in the sample determination. The color intensity of the obtained tetra layer is now adjusted with 9 ml water, while adding the required volume (*a* ml) of the $CoSO_4$ stock solution. This volume, which corresponds to the blank, is now introduced into each of the 5 glass tubes. Then 0.7, 1.35, 2.75, 5.75 and 7.15 ml of $CoSO_4$ stock solution are respectively added and the volume is made up to 10 ml. These tubes respectively correspond to 0.1, 0.2, 0.4, 0.8 and 1.0 mg Pb/m^3 for a sample volume of 15 liter.

Preparation of the solutions

Dithizone solution: 40 mg dithizone are purified, if necessary, by reprecipitation (dissolution in 0.2 *N* NH_3 water and precipitation with 5 *N* HCl) and dissolved in 1 liter carbon tetrasulfide. The solution is shaken with aqueous 0.1 *M* SO_2 and the aqueous layer discarded. The carbon tetrasulfide solution is stored underneath a thin layer of aqueous SO_2 in a dark flask.

KCN solution: 10 g potassium cyanide per 100 ml aqueous solution. To remove lead traces, the solution is shaken with 10 ml of a solution of 40 mg dithizone in 1 liter $CHCl_3$ in a separating funnel until the extracts are colorless. The aqueous layer is filtered.

Sodium metabisulfite solution: 10 g $Na_2S_2O_5$ are dissolved in 60 ml water, adjusted to a pH of 9.0 with NH_3 water and the lead is removed by shaking with the dithizone chloroform solution, just as in preparation of the KCN solution.

Ammonium citrate solution: 25 g ammonium citrate are dissolved in 60 ml water, adjusted to a weakly alkaline reaction (litmus) with NH_3 water and the lead is removed just as from the KCN solution.

NH_3–sulfite–cyanide–citrate solution: To 325 ml NH_3–water ($d_{20} = 0.880$) 30 ml of the KCN solution, 60 ml of the sodium bisulfite solution and 20 ml of the ammonium citrate solution are added. The mixture is diluted with water to 1 liter. The solution remains stable for several months.

Working procedure of Lahmann *and* Möller

Lahmann and Möller [147b] collected the lead-containing dust by drawing the dust-laden air through a glass fiber filter No. 9 of the firm Schleicher and Schüll (filter area 18 × 23 cm), using for this purpose the High Volume Sampler of Staplex, New York, with a throughput rate of 80 m^3/hr. The lead is brought into the solution by boiling with HNO_3. The dithizone–chloroform solution is added and the mixture is agitated in the presence of hydroxylamine hydrochloride, cyanide, ammonia and tartrate. The color of the lead dithizonate in chloroform is measured after processing and purification at 510 mμ. The detection limit is 1.5 g Pb/m^3 air for a sampling time of 1 hour and the reproducibility, expressed in terms of the standard deviation, is $s = \pm 0.1$ μg Pb/m^3.

Colorimetric determination of tetraethyllead vapors in air

A procedure employing 1 *N* iodine solution is described in the collection of methods of the IUPAC. A Drechsel wash bottle, whose upper part has a doughnut-shaped thickening or is provided with a glass nipple to take up about 1.5 g glass wool, is used as the absorption vessel. The iodine vapors are removed by passing the air after leaving the wash bottle through a wide absorption tube with activated carbon.

Preparation of the 1 *N* iodine solution: 250 g KI are dissolved in 750 ml distilled water. Ammonia is added to alkaline reaction (thymol blue) and the solution is shaken with portions of dithizone–chloroform solution (4 mg in 100 ml) until the extract is no longer colored (see page 220). Then dilute HNO_3 is added to the solution until the indicator turns green. 125 g pure iodine are dissolved in the solution and diluted to 1 liter with Pb-free water.

The Drechsel wash bottle is supplied with 10 ml of the 1 *N* iodine solution and a receiver containing 20 g activated carbon is connected in order to remove the iodine vapors. At least 100 liter air are passed through at a rate not exceeding 25 liter/min.

Then the absorption liquid is treated with a solution in which the lead traces (see page 220) have been removed (10 g KCN, 100 g $Na_2S_2O_5$ and 20 g ammonium citrate are dissolved in 550 ml water, lead is removed and the solution is diluted with 1950 ml NH_3–water, d = 0.900). The solution is decolorized and agitated in the separating funnel with 10 ml of the dithizone–chloroform solution (4 mg/100 ml). The color of the chloroform layer is either measured by means of a spectrophotometer at 510 mμ and evaluated via a calibration curve plotted for ethyl nitrate and corresponding to 10–50 μg tetraethyllead, or compared with the $CoSO_4$ standard solutions as outlined on page 219.

Moss and Browett [183] effected the absorption of the tetraethyllead vapors with the aid of iodine monochloride solution instead of the iodine solution and the dithizonates obtained are colorimetrically determined.

5.10.4 Determination of lead-containing vapors by absorption flame photometry (atomic absorption)

Principle (see also Leithe [156])

Whereas in normal emission flame photometry the atoms of the elements are thermally excited in the gas flame, so that they emit light at certain wavelengths (spectral lines), the intensity of which is utilized for quantitative analysis, in absorption flame photometry the atoms of these elements are in the ground state. They absorb light at certain wavelengths (resonance lines) which is emitted by special lamps (hollow-cathode lamps). The initial intensity of the light is thus attenuated in proportion to the concentration of the elements in question.

A wide range of atoms can be determined by absorption flame photometry (atomic absorption). This method has become increasingly popular and is especially suitable for trace analysis, since it is rapid, specific and sensitive. Hollow-cathode lamps for more than 40 elements and corresponding apparatus manufactured by various firms are available on the market.

As far as air analysis is concerned, a corresponding method for mercury is described on page 222. Recently, Thilliez [249] developed a method for the determination of vapors of lead compounds. The air sample is first blown into a flame where free lead atoms are released. The atoms also remain stable for a short time outside the flame. They can be determined after emission of the Pb-resonance line (283.3 mμ) from a Pb-hollow-cathode lamp by measuring the intensity drop of the light which has passed the zone occupied by the lead atoms. The measurement is carried out with the aid of a spectrophotometer.

Apparatus: Thilliez used a special burner in order to extend the optical path through the zone occupied with lead, thus considerably increasing the sensitivity. After the flame leaves the platinum burner head, the hot vapors are led into a T-shaped tube of transparent silica. In the upper, 1-m long horizontal part of this tube the vapors are irradiated by the hollow-cathode lamp. The light is then spectrally resolved and the intensity at the given wavelength measured by the spectrophotometer.

Hydrogen or NH_3-synthesis gas ($3H_2 + N_2$) are used. The air sample is introduced at a rate of 3–14 liter/hr directly into the burner. For the connections no rubber or PVC hoses should be used.

Lead concentrations between 1 and 500 μg Pb/m^3 can be determined in this way. To determine higher lead concentrations, the pathlength of the T-tube has to be extended or conventional burners have to be used.

The apparatus is calibrated by metering out small amounts of tetraethyllead, as outlined in Chapter 3.1. The author reported a reproducibility of $\pm$ 5%. The agreement with results obtained by chemical methods is good.

Taking into account the significance of lead determinations in the free atmosphere, we hope that suitable auxiliary instruments based on this principle will soon appear on the market.

The apparatus of Thilliez was first used and calibrated to determine the level of alkyllead vapors in the air. However, it is in principle possible to collect first lead from Pb-containing dusts on filter paper in the event that they cannot be directly introduced into the flame. The paper could be extracted with HNO_3 and the extract immediately injected into a suitable flame.

5.11 Determination of mercury

The frequent application of mercury in physical instruments and sometimes in bath solutions leads to occasional pollution of the air at work sites. Furthermore, numerous mercury preparations are used as medicaments as well as pesticides, so that mercury vapors may be formed during preparation and application.

The MAK-value is 100 μg/m^3. Individual determinations are usually performed by colorimetry (as dithizonate or with selenium sulfide). Direct determination of mercury vapors by atomic absorption with the aid of special instruments is convenient and sensitive.

5.11.1 Determination of mercury vapor by absorption

A very sensitive method for determining gaseous elemental mercury in the air is based on the principle that the light emitted by a mercury vapor lamp is absorbed by free Hg atoms outside the lamp. This phenomenon (below referred to as atomic absorption) is frequently used in absorption flame photometry for the determination of a wide variety of metals.

The firm Beckman manufactured an instrument designed to determine mercury amounts between 5 and 3000 μg/m^3; it has two measuring ranges (5–100 μg/m^3

and 30–3000 $\mu g/m^3$). The light of a mercury vapor lamp (filtered out resonance line 237 mμ) is incident on two photocells. The photocurrents induced in Hg-free air are balanced. If Hg-containing air is introduced into one (extended) of the optical paths, the relevant photocell is irradiated at a lesser intensity and will yield a weaker current. The difference between the two signals is amplified, indicated on a measuring device or continuously recorded.

The determination is very specific; the instrument, however, is sensitive to certain substances at the above wavelength. The following cross sensitivities apply (0.1 scale width = 0.1 Hg/m^3 is indicated by the concentrations of interfering substances listed in the table below):

Aromatics	500–1000 ppm
SO_2	800 ppm
Ozone	20 ppm
NO_2	2000 ppm

An arrangement, which is much more sensitive (detection limit 0.3 nanogram) and selective, is described by Ling [167a].

5.11.2 Colorimetric determination with dithizone

Determination of mercury as vapor and in compounds in accordance with the ICI Manual

Principle: The air sample is drawn through a $KMnO_4$ solution in H_2SO_4. In this process Hg vapor, inorganic Hg compounds and most of the readily decomposable organomercury compounds are absorbed and decomposed. In the presence of methylmercury compounds, which are less readily decomposable, the permanganate solution is heated for 30 min on a boiling water bath. Mercury is determined by titration with dithizone solution to a certain color tint, which is adjusted by parallel titration by adding the required volume of a Hg calibration solution.

The sampling is carried out by introducing 30 ml of the absorption solution into an impinger; 40 liter air sample are drawn at a rate not exceeding 2 liter/min.

Determination: The $KMnO_4$ solution is decolorized by adding 5 ml hydroxylamine hydrochloride and 1 ml of EDTA solution is introduced. Then 10 ml acetate buffer are added and the solution is agitated with a little chloroform until a small excess of $CHCl_3$ sinks to the bottom. Small portions of dithizone solution are then added from a 5-ml burette until a mixed color (orange-yellow Hg-dithizonate and green dithizone excess) appears. A parallel titration with the Hg calibration solution is carried out until the same color is reached and the Hg-concentration in the solution is determined from the Hg-content and the amount of calibration solution used.

Reagents: Absorption solution: To 160 ml $KMnO_4$ solution (4 g in 100 ml solution) a mixture of 20 ml water and 20 ml concentrated H_2SO_4 is added.

Hydroxylamine hydrochloride solution: 20 g of the reagent in 100 ml solution.

EDTA solution: 2 g of the disodium salt of ethylenediaminetetraacetic acid in 100 ml solution.

Acetate buffer: 100 g sodium acetate trihydrate in 100 ml water.

Dithizone stock solution: 20 mg dithizone in 100 ml carbon tetrasulfide (stored in the refrigerator).

Hg calibration solution (2 μg Hg/ml): 0.027 g $HgCl_2$ are dissolved in 1 liter hydrochloric acid (50 ml concentrated HCl per liter). Then 10 ml are diluted with the same hydrochloric acid to 100 ml.

Method of the American Conference of Governmental Industrial Hygienists

For sampling an impinger is recommended which contains an aqueous solution of 0.25% iodine and 3% KI.

The absorption solution is prepared by adding 5 ml of an ammonium citrate solution, 1 ml hydroxylamine hydrochloride solution (or sodium bisulfite solution) and 2 drops of phenol red (indicator). The pH is adjusted to 8.5 (maximum color intensity of the indicator) with ammonia water. Then the solution is extracted with a solution containing 20 mg dithizone in $CHCl_3$ in portions of 5 ml until no color change is observed in the last portion. The collected chloroform extracts are shaken with a mixture of 50 ml 0.1 *N* HCl, the aqueous phase is twice washed with 5 ml $CHCl_3$ and discarded. The whole chloroform phase is shaken with 50 ml 0.1 *N* HCl and 10 ml 40% KBr, and mercury goes over into the aqueous layer. After washing the aqueous layer with a small amount of chloroform, the chloroform layer is discarded. The aqueous layer of an aliquot thereof is adjusted with 10 ml buffer to pH 6. The aqueous phase is shaken with a chloroform solution containing 10 mg dithizone/liter and the latter is filtered through a cotton plug inserted into the stem of the separating funnel into a colorimetric cell. The photometric determination is carried out at 485 mμ.

The calibration curve is plotted for $HgCl_2$ solutions corresponding to amounts of 0–15 μg Hg. In the sample or an aliquot thereof at least 3 μg Hg should be present. The influence of other heavy metals is eliminated by pH adjustment (see above) or by addition of KBr.

Reagents: Buffer solution: 300 g anhydrous Na_2HPO_4 and 75 g K_2CO_3 in 2 liter solution.

Chloroform, which has been purified by shaking with hydroxylamine hydrochloride solution.

Ammonium citrate solution: 40 g citric acid monohydrate are mixed with 20 ml water and ammonia water is slowly added until the solution exhibits an alkaline reaction to phenol red, and is then made up to 100 ml. Heavy metals are removed by shaking with dithizone–chloroform solution.

Hydroxylamine hydrochloride, 20%.

Colorimetric determination with selenium sulfide

(Booklet 13 of the British Work Inspectorate); see also Sergeant, Dixon, and Lidzey [229] as well as page 196.

Principle: Mercury vapor is absorbed at a rate of 50 liter/20 min by iodized activated carbon. It is then desorbed in a hot medium in a tube while adding iron as a reducing agent and sodium oxalate as a carbon monoxide generator, and passed over a filter paper impregnated with selenium sulfide (prepared from selenious acid and thioacetamide). The resulting brown color is compared with a calibration scale obtained in the same way.

6

Analysis of Organic Air Pollutants

6.1 Hydrocyanic acid

6.1.1 General

Hydrocyanic acid is very often applied in organic technical synthesis and as a pesticide (against insects and rats in closed premises). Cyanide is used in large amounts in leaching of gold and electroplating.

Anhydrous hydrocyanic acid boils at 26°C.

Toxicology

The main danger of cyanide and the soluble simple cyanides is the sudden resorption of concentrated doses (a drop of liquid hydrocyanic acid leads to death in a few seconds), but the danger is less acute when small concentrations are inhaled over longer periods. Nevertheless, 100–200 mg HCN/m^3 inhaled in $\frac{1}{2}$ to 1 hr are lethal. As soon as the typical almond smell is perceived, immediate precautions must be taken.

The MAK-value is 10 ppm = 11 mg/m^3.

The first-aid antidote in poisonings by HCN is inhalation of amyl nitrite from a 0.3 ml ampoule, which is broken in a handkerchief and held against the mouth and nose.

6.1.2 Analytical methods

In addition to Dräger test tubes for hydrocyanic acid the toxicant can also be determined in the air by the Prussian blue reaction on a reagent-impregnated paper prepared by the chemist. At present, small HCN-concentrations in air are determined via cyanogen bromide with pyridine–benzidine or benzidine–barbituric acid reagents. The red color is photometrically determined.

Determination of cyanogen bromide formed from hydrocyanic acid with the pyridine–benzidine reagent

Principle: HCN is absorbed in the receiver and converted to cyanogen bromide. After removal of excess bromine by As_2O_3, pyridine and benzidine are added. The red color formed by the decomposition of pyridine with benzidine is measured by the spectrophotometer.

Working procedure: 25 ml 0.1 *N* NaOH are placed into a fritted scrubber (the alkali must contain 3 drops of n-butanol as a foaming agent) and the air sample is introduced at a rate of 1–3 liter/min. To 10 ml of the absorption solution are added 1 ml acetate buffer and 0.5 ml saturated bromine water, the mixture is agitated and allowed to stand for 2 min. The excess bromine is removed by adding 0.5 ml arsenious oxide solution. Then 6.0 ml pyridine–benzidine reagent and 4.0 ml acetone are added. The mixture is allowed to stand for 30 min and the extinction measured at 530 mμ against the reagent blank in a colorimetric cell of suitable layer thickness.

The calibration curve is plotted for HCN-concentrations of 1–6 μg (obtained from KCN) in 10 ml solution. The result is referred to 25 ml absorption solution. 5 μg HCN in 10 ml absorption solution yield an extinction of 0.74 in 1 cm layer thickness after the addition of the reagents. The calibration curve is linear.

Reagents: Acetate buffer (pH = 5): 82 g sodium acetate are dissolved in 60 ml H_2O, 59 ml glacial acetic acid are added and the volume is made up to 200 ml.

Arsenious oxide solution: 8 g As_2O_3 are dissolved in a solution containing 4.4 g Na_2CO_3 in 48 ml water under heating, filtered, if necessary, and the volume is made up to 242 ml.

Pyridine–benzidine reagent: 1.0 g benzidine hydrochloride is heated in 15 ml 0.2 *N* HCl and mixed with 15 ml pyridine (for analysis). The reagent must be freshly prepared.

Calibration solution (1 ml = 1 μg HCN): A stock solution containing 650 mg KCN in 250 ml solution is diluted to a concentration of 50 mg HCN/liter (titration with 0.1 *N* $AgNO_3$). 10 ml of this solution are diluted to 500 ml.

According to Grigorescu and Toba [82], 5 ml pyridine–barbituric acid reagent (without acetone) can be added instead of the pyridine–benzidine reagent. The reagent is prepared by mixing 55 ml freshly distilled pyridine with 20 ml hydrochloric acid (1 vol. concentrated HCl + 2 vol. water) under cooling. Then 30 ml 0.8% barbituric acid are added. The color is measured at a wavelength of 570 mμ.

Procedure in the MAK-range with iron hydroxide paper strips

(British Factory Inspectorate, Booklet No. 5)

Apparatus: The impregnated reagent paper strips (7.5 × 10 cm) are placed into a suction device in such a way that a circular area of 10 mm diameter is exposed. 360 ml air sample (3 pump strokes) are aspirated within 6 minutes.

Preparation of the reagent paper: Strips of filter paper (Whatman No. 50, 7.5 × 10 cm) are dipped into a solution containing 10 g $FeSO_4 \cdot 7H_2O$ in 100 ml for 5 min and dried over a radiator. A 2-cm strip is removed from the bottom. The rest (7.5 × 8 cm) is cut into 3.5 × 2.5 cm rectangular pieces, each of which is immersed for 15 min into a solution containing 20 g NaOH in 100 ml. The strips are predried with filter paper, rapidly transferred into a vacuum desiccator and completely dried. The color is gray-green to pale brown. The pieces are separately stored in evacuated and fused test tubes.

Evaluation: After aspirating 360 ml air sample, the pieces are immediately placed into a white dish with H_2SO_4 (30 ml concentrated acid in 100 ml solution). While wet, the Prussian blue color is compared with a color scale given in the British prescription or prepared by the chemist. In 360 ml sample 2.5 ppm HCN give a very pale blue-gray color and 10 ppm a bright sky-blue. The colored papers can be stored for long periods when thoroughly rinsed and dried.

6.2 Determination of carbon disulfide at the work site

(Booklet No. 6 of the British Factory Inspectorate)

With a small suction pump of known stroke volume (e.g., the pump described on page 40) or the hand-operated bellows pump of the Drägerwerke (see page 56, stroke volume 100 ml) 10–2000 ml air (1–20 strokes) at a rate of at least 600 ml/min are introduced into 14 ml absorption solution in a suitable wash bottle (impinger or fritted scrubber) until a distinct yellow color appears.

In the presence of hydrogen sulfide the air sample is passed through a filter unit provided with a lead acetate paper (see page 40) before being introduced into the wash bottle.

After a standing time of 15 min, the yellow color is compared with 4 standard solutions. The air volume used (number of pump strokes) and the color classification yielded the CS_2-concentrations listed in the table below.

ppm CS_2 PER STANDARD COLOR

Number of strokes	ml air		I		II		III		IV	
1	126	100	166	210	330	420	660	840	1320	1680
2	252	200	83	105	165	210	330	420	660	840
3	378	300	55	70	110	140	220	280	440	560
5	630	500	33	42	65	84	130	168	270	336
20	2520	2000	8	10.5	16	21	33	42	110	84

Preparation of the solutions

Diethylamine solution: 2 ml diethylamine are dissolved in 100 ml benzene.

Copper acetate solution: 0.1 g copper acetate is dissolved in a small amount of ethanol under heating and made up with the same solvent to 100 ml.

Absorption solution: To 10 ml absolute ethanol in the gas wash bottle are added immediately before use 2 ml diethylamine solution and 2 ml copper acetate solution.

Calibration solution: 1.0 ml CS_2 is diluted with absolute ethanol to 100 ml. 1.0 ml of this solution is diluted with absolute ethanol to 50 ml (0.02 vol. % solution).

Standard colors: 0.25 ml (I), 0.5 ml (II), 1.0 ml (III) and 2.0 ml (IV) are diluted with absolute ethanol to 10 ml and each standard is mixed with 2 ml diethylamine solution and 2 ml copper sulfate solution. After a standing time of 15 min, the solutions can be compared with the sample solutions.

6.3 Aldehydes

6.3.1 General

The aldehyde group frequently occurs in the air as a pollutant. In the free atmosphere aldehydes are present as partial oxidation products of combustion and explosion processes, and, consequently, they are normal constituents of automobile exhaust.

At various work sites we must also expect the occurrence of aldehydes in the air. These compounds are frequently used as intermediate products in the production of plastics, auxiliary substances for the textile industry and other organic synthetic products. Formaldehyde is used as a disinfectant and preserving agent, a means of combating pests and, on a large scale, for the synthesis of plastics (urea or melamine resins, phenol-formaldehyde resin, Delrin, etc.) as well as for their curing.

Toxicology

Formaldehyde strongly irritates the mucous membranes; it is perceived in concentrations of 0.2 ppm and above. Its effect is annoying rather than toxic. Acrolein formed by overheating of fats has an even stronger irritating effect. With increasing chain length the irritating effect decreases but the toxicity increases.

The MIK_K-values are not permitted to be attained more than once in 4 hr.

Remarks concerning the analysis

Since the reactivity of aldehydes is high and the number of known specific reactions is large, there exists a wide variety of methods for the determination of low aldehyde concentrations. There are also methods which are specific for individual important and dangerous members of this series (formaldehyde, acrolein); other methods enable the determination of the whole group (sometimes including the ketones).

MAK- AND MIK-VALUES OF THE ALDEHYDES
(VDI-Richtlinie 2306)

	MAK-value		MIK_D-value		MIK_K-value	
	ppm	mg/m^3	ppm	mg/m^3	ppm	mg/m^3
Formaldehyde	5	6	0.02	0.03	0.06	0.07
Acetaldehyde	200	360	2	4	6	12
Acrolein	0.1	0.25	0.005	0.01	0.01	0.25
Furfural	5	20	0.02	0.08	0.06	0.25

At present, colorimetric, but also titrimetric procedures are most frequently used. A procedure based on a specific catalytic effect is very sensitive. Of course, gas chromatographic or colorimetric methods can also be employed.

6.3.2 Determination of formaldehyde according to Schryver

Principle: The phenylhydrazone of formaldehyde yields a red color after adding potassium ferricyanide. This reaction is characteristic for formaldehyde and is not disturbed by other homologous aldehydes. The method is described in the ICI Manual and by Jacobs [116a].

Working procedure in accordance with the ICI Manual

Sampling: The sample is drawn at a rate not exceeding 1.5 liter/min through an impinger or fritted scrubber. The absorption vessel contains 10 ml phenylhydrazine hydrochloride solution and is provided with 2 marks indicating contents of 15 and 25 ml, respectively.

Determination: The inlet tube is rinsed with a little water. Then the vessel is filled with water to the first mark (15 ml) and 1 ml of the $K_3Fe(CN)_6$ solution is added. After 4 min 4.0 ml concentrated HCl (d = 1.18) is added and the mixture is allowed to stand for another 5 min. The volume is made up to 25 ml and the extinction measured in a 5-cm cell at 515 mμ against a reagent blank solution prepared in the same manner.

The calibration curve is plotted in the range from 0–12 μg for a solution containing 4 μg formaldehyde per ml. The calibration solution was obtained after appropriate dilution of a 40% formaldehyde solution.

Solutions: Phenylhydrazine solution: 1.0 g phenylhydrazine hydrochloride (for analysis, the color must not be darker than light pink) is dissolved in 80 ml H_2O and 2 ml concentrated HCl (d = 1.18) are added. The reagent solution is filtered and made up to 100 ml.

Potassium ferricyanide solution: 2 g $K_3Fe(CN)_6$ per 100 ml; the solution must be freshly prepared every day.

Jacobs [116a] passed the air sample through 1.25% aqueous KOH. Before the analysis the solution is neutralized with HCl. To a 15 ml aliquot are successively added 1 ml 5% $K_3Fe(CN)_6$, 4 ml concentrated HCl and 2 ml 1.4% phenylhydrazine hydrochloride. The determination is carried out after 15 min.

6.3.3 Iodometric determination of aldehydes

The American Conference of Governmental Industrial Hygienists recommends the iodometric bisulfite procedure of Goldman and Yagoda [80] for the joint determination of the aldehydes. However, the titration of the aldehydes with 0.01 *N* iodine solution is not very sensitive.

Principle: Formaldehyde is trapped in an aqueous $NaHSO_3$ solution. It is retained as a nonvolatile addition compound which is stable against iodine. The excess free bisulfite is completely removed with iodine solution, the sulfite is freed by adding alkali and titrated with very dilute iodine solution.

Working procedure

10 ml 1% $NaHSO_3$ are introduced into two respective impingers or small fritted scrubbers and a minimum of 25 liter air sample is passed through at a rate of 1–3 liter/min. At least 95% of the aldehydes are trapped.

Analysis: The solutions are rinsed into a titration flask. The titration is performed after adding 1 ml soluble starch (to dark blue) and 0.01 *N* iodine is introduced until the indicator turns pale blue. Then 25 ml soda–acetate buffer are added and the solution is titrated with 0.01 *N* iodine to the same pale blue color.

One ml 0.01 *N* iodine corresponds to 5 ppm formaldehyde in a 25-liter sample. An error in the titration of $\pm$ 0.1 ml thus yields an error of $\pm$ 0.5 ppm in the determination.

SO_2, alcohols and free halogens do not interfere; the lower aldehydes are trapped and determined. To effect complete trapping and determination of the aldehydes up to C_4 the absorption solution is cooled in ice (Wilson [274]).

6.3.4 Determination of formaldehyde with fuchsin

Principle: The known indication of aldehyde with the Schiff reagent (fuchsin-sulfurous acid) permits a very accurate determination of low formaldehyde concentrations (lower limit 5 ppb). Higher aldehydes only slightly interfere. The working procedure outlined below was reported by Rayner and Jephcott [210]. The long time required for the color to develop is a disadvantage.

Working procedure

The air sample is drawn at a rate of 28 liter (1 cu ft)/min through an impinger containing 60–75 ml 0.005 *N* HCl. For formaldehyde concentrations below 50 ppb 1.7 m^3 air is required with a sampling time of 1 hr. Under these conditions the authors found a formaldehyde yield of 72%.

Analysis: An aliquot (up to 19 ml) is mixed in a 25-ml volumetric flask with 1 ml acetone and 5 ml Schiff reagent. A test series using a standard solution containing 5 μg HCHO/ml is carried out in the same manner. The color is allowed to develop for 3 hr, or, when the formaldehyde level is low, overnight. Then the extinction is measured at 569 mμ. The authors used a 1-inch cell (2.5 cm). 2 μg formaldehyde yield an extinction of 0.05. This corresponds to a formaldehyde content of 5 ppm in 1.7 m^3 air.

Calculation: The result is to be multiplied by 1.39 because of the incomplete absorption during sampling.

Schiff reagent: 0.5 g basic fuchsin (British Drug House) are triturated with a little water, dissolved in 500 ml water and filtered into a 500-ml reagent bottle. 4.69 g sodium bisulfite are added, the solution is allowed to stand for 15 min and then 17 ml 6 *N* HCl are added. Before use the solution is allowed to stand overnight.

Interfering substances: Acrolein gives the same reaction as formaldehyde. The reactions of the higher aldehydes become much less intense during the standing time of 3 hr, so that they appreciably interfere only when present in concentrations considerably higher than the formaldehyde level. SO_2 and NO_2 interfere only when present in a 100-fold excess.

Variants: Lyles, Dowling and Blanchard [172] used as reagents a mixture of dichlorosulfitomercurate and bleached pararosaniline in HCl, in analogy to the determination of SO_2 by West and Gaeke. The color is also measured at 560 mμ.

Determination of aldehydes with 3-methyl-2-benzothiazolone-hydrazone (MBH)

Principle: Sawicky, Hauser, Stanley and Elbert [225] used MBH as a very sensitive reagent for the aldehyde group. The reagent yields a bright blue color with ferric chloride and formaldehyde which is suitable for the photometric determination of aldehyde. This reaction can be performed on a filter paper wetted with the reagents, on silica gel in a test tube and by passing the sample through an aqueous reagent solution. A very sensitive procedure was described by Hauser and Cummins [101].

A small gas wash bottle is provided with 35 ml of the 0.05% reagent solution. The air is passed through at a rate of 0.47 liter/min for 24 hr. To an aliquot of 10 ml are added 2 ml oxidant solution, the mixture is allowed to stand for 12 min and the extinction is measured at 628 mμ against the reagent blank.

The calibration curve is plotted for 0–1.7 μg HCHO. The molar extinction coefficient is 65,000.

Solutions: Absorption reagent: a 0.05% aqueous MBH solution (obtainable at the British Drug House and Aldrich Chemical Co., Milwaukee, USA).

Oxidation solution: 1.6 amidosulfonic acid and 1.0 g $FeCl_3$ in 100 ml.

6.3.5 Catalytic procedure of Hughes and Lias

Principle: A very sensitive test tube procedure for the semiquantitative determination of aldehyde traces (0.1 ppb and above) was developed by Hughes and Lias [111]. It is based on the spot test of Feigl; the oxidation of p-phenylenediamine with hydrogen peroxide to black oxidation products (Bandrowski's base) is strongly accelerated by the catalytic effect of the aldehydes.

Working procedure

A vertical glass tube (70 × 5 mm) is half filled with purified silica gel and the air sample drawn in at a rate of 700 ml/min for 10–15 min. Then the remainder of the tube is filled with fresh silica gel (precautions must be taken to avoid mixing of the layers) and the tube is closed with a cotton wad. The tube with the inlet end down is directly immersed into the freshly prepared developing solution which rises through capillary action. The solution first passes through the pure silica gel, where it is freed from traces of dark contaminants. In the presence of aldehyde a brown-red band appears within 30 sec at the layer interface; the concentration of aldehyde can be evaluated from the color intensity.

For the calibrations 5 standard tubes with air mixtures containing the following aldehyde concentrations are prepared:

Standard tube	I	II	III	IV	V
Moles of aldehyde	10^{-7}	10^{-8}	10^{-9}	10^{-10}	10^{-11}

Since the color is stable for a short time only, the tints are copied with color pencils for evaluation or determination of the concentration. It is possible to interpolate values between two orders of magnitude.

Purification of silica gel according to Shepherd [230]

Silica gel of the firm Davison, Baltimore, is treated with HNO_3, d = 1.42, on a boiling water bath for 1 week. Then the acid is removed by suction and the silica gel is leached at 100°C with water which is frequently renewed. Leaching is terminated when the pH of the washings is less than 5 after a standing time of 48 hr. The gel is then first dehydrated at 100°C and then for 3 days at 320°C. The material is ground to a particle size of 0.3–0.5 mm (30–50 mesh).

Developing solution: A saturated aqueous p-phenylenediamine solution is mixed with 3% H_2O_2 in a ratio of 1:2. Three ml of the mixture are prepared immediately before it is drawn through the test tube.

Sensitivity: In 9 liter air the detection limit is 0.1 ppb formaldehyde corresponding to the color of tube V. The sensitivity for aliphatic aldehydes is about the same, while the reaction for aromatic aldehydes is less sensitive by one order of magnitude. Nitriles react in the same manner but the reaction is less intense.

6.3.6 Determination of acrolein with 4-hexylresorcinol

Cohen and Altshuller [47] used the reaction developed by Rosenthaler for the determination of acrolein in foodstuffs for the analysis of air and automobile exhaust. The reaction is sensitive (molar extinction coefficient 10,000) and specific.

Working procedure

The air is sampled at a flow rate of up to 2 liter/min in two small fritted scrubbers connected in series. Each scrubber contains 10 ml ethanol and is cooled with ice.

After sampling the solutions are combined and to every 2.5 ml of the alcoholic solution are added 2 drops of the 4-hexylresorcinol solution, 0.1 ml $HgCl_2$ solution and 2.5 ml trichloroacetic acid. The mixture is heated for 15 min at 60°C, cooled for 15 min and the extinction measured at 605 mμ. A blank is treated in the same manner. The extinction coefficient (1 μg/ml · cm) is 0.35 at 605 mμ. The calibration curves are plotted for pure acrolein in alcoholic solution after dilution.

Reagents: 4-hexylresorcinol solution: 5 g of the reagent are dissolved in 5 ml 96% ethanol and made up to 10 ml.

Mercuric chloride: 3 g HCl_2 in 100 ml 96% alcohol.

Trichloroacetic acid: 100 g of the purest product are dissolved in 10 ml water on a water bath.

Air samples with a very low acrolein content are directly introduced into two fritted scrubbers, each containing 10 ml of the reaction mixture (0.1 ml 4-hexyl-resorcinol solution + 0.2 ml 3% $HgCl_2$ + 5 ml trichloracetic acid + 5 ml ethanol).

Interfering substances: The reaction is very specific for acrolein. Low saturated aliphatic aldehydes and ketones, mono- and diolefins (butadiene), benzene, SO_2 and NO_2 do not interfere.

6.4 Determination of acetone in air

Principle: In accordance with Buchwald's procedure, acetone is removed from the air sample by silica gel, desorbed by NaOH and titrimetrically determined via the iodoform reaction.

Working procedure: The air sample is drawn at a rate of 1–5 liter/min through an absorption tube containing 10 g silica gel. The absorbing agent was activated by preliminary heating for 4 hours at 150°C. The silica gel is then treated with 50 ml 1 *N* NaOH. Twenty ml of this solution are allowed to stand with 2–20 ml 0.1 *N* iodine solution (depending on the acetone concentration to be expected) for 15 min with intermittent agitation. The yellow iodoform is precipitated. Then 5 ml 6 *N* H_2SO_4 are added and the released iodine is titrated with 0.01 *N* sodium thiosulfate. A blank is determined in the same manner.

One ml 0.1 *N* I corresponds to 0.97 mg acetone.

The reaction is not specific for acetone, but takes place with all compounds containing the group $CH_3 \cdot CO$ (e.g., ethanol).

6.5 Aliphatic chlorinated hydrocarbons

6.5.1 General information (see Table 6.5)

The aliphatic chlorinated hydrocarbons form a systematic closed group of compounds which exhibit similar physiological effects. Some of their uses and especially the methods of determination of these compounds are closely related.

The members of the groups which mainly occur as air pollutants are those having a small number of carbon atoms (C_1 C_3) with 1 4 chlorine atoms. These are very

Table 6.5

PROPERTIES AND APPLICATION OF SOME CHLORINATED HYDROCARBONS

	MAK		MIK_D		MIK_K		Boiling	Application
	ppm	mg/m^3	ppm	mg/m^3	ppm	mg/m^3	point	
Methyl chloride	50	105	5	20	—	—	−24°	In refrigerators and for methylation
Methylene chloride	500	1750	5	20	15	55	40°	Fat solvent
Chloroform	50	240	2	10	6	30	61°	Fat solvent, earlier used for anesthesia
Carbon tetrachloride	10	65	0.5	3	1.5	10	77°	Solvent, fire-extinguishing liquid, starting material for freons
Ethyl chloride	1000	2600					12.5°	Coolant, medicine, production of tetraethyllead
Vinyl chloride	500	1300					−14°	Production of PVC
1,2-Dichloro-ethane	100	400	2	8	6	25	84°	Solvent, pesticide, intermediate product
1,1,1-Trichloro-ethane	200	1080	5	30	15	90	74°	Solvent
Trichloro-ethylene	100	520	5	30	15	90	87°	Fat solvent and cleaner for clothes
1,1,2,2-tetra-chloroethane	1	7					146°	Intermediate product
Tetrachloro-ethylene C_2Cl_4	100	670	5	35	15	110	121°	Solvent

rarely encountered in appreciable amounts in the open atmosphere but occur frequently at work sites because of their widespread application in engineering, medicine and sometimes in the household.

Certain members of the group are used as coolants, fire-extinguishing agents, anesthetics and alkylation agents (see Table 6.5). However, most of these compounds are employed as solvents for fats, oils, etc., in chemical cleaning and in the household as a cleaning agent and spot remover. They are less inflammable than the benzine hydrocarbons but more poisonous.

The toxicological effect is anesthetic, but the substances with low MAK-values cause primarily serious injuries to the liver.

The MAK-values, which were established some time ago, exhibit large differences. The MIK-values have recently also been established; for example, for methylene chloride this value is 1/100 of the MAK-value. The MIK-value is permitted to occur once in 4 hours.

6.5.2 General data for the analyst

Since chlorinated hydrocarbons are always used as individual substances and not in mixtures, procedures which are specific for only a single substance of this group are of little interest. If in special cases mixtures occur, it is possible to isolate and determine the individual substances; gas chromatography can, of course, also be used for the determination of pure substances. Insofar as no further differentiation is required, the organically bound chlorine of all the pollutants can be converted into the chloride ion and determined by the usual methods. The Fujiwara reaction (red color with NaOH in the presence of pyridine) common to all aliphatic chlorinated hydrocarbons can be applied in some variants to determine the substances in this group and even a certain specificity can be achieved over a given range.

Since the MAK-values of the alkyl and alkylene chlorides vary over a range from 1 to 1000 ppm, the sample size must be correspondingly selected.

6.5.3 Determination of the organic chlorine as the chloride ion

Principle: A general procedure for determining halogen-containing organic pollutants by conversion into the halogen ions can be also applied to alkyl and alkylene chlorides. The organic chlorine is split off by sodium metal. The formed Cl^- ion can be determined by nephelometry, titration or colorimetry. The following procedure is listed in the Manual of ICI-Practice and can be applied for chlorine amounts between 50 and 350 μg.

Working procedure

The air sample is drawn at a rate of 175 ml/min through a fritted scrubber or impinger made of heat-resistant glass. The absorption vessel contains 2-aminoethanol.

After sampling the supply tube is removed from the wash bottle, the gas outlet is closed and a reflux condenser is connected to the ground-glass joint. Three ml of dioxane are added and 0.3 g sodium metal is thrown in small pieces through the condenser and the solution is heated to moderate boiling. After dissolution of the metal the process is repeated with another 0.3 g Na. When the process is repeated for a third time, the solution is boiled a little more strongly for about 30 min. The solution is cooled, a few ml are introduced in drops and 1:1 HNO_3 is added. The solution is then cooled, filtered (if necessary), transferred into a 100-ml Nessler cylinder and 1 ml 0.1 *N* $AgNO_3$ added. The volume is made up to 100 ml and the solution is allowed to stand in the dark for 5 min.

The reference series is obtained with the same reagents, using 0.1–1.0 ml 0.001 *N* NaCl, which corresponds to 35–350 μg Cl^-. Instead of a visual comparison, a nephelometer or turbidimeter can be used.

The following variant is recommended by the American Conference of Governmental Hygienists.

The chlorinated hydrocarbon is extracted from the air sample by drawing the latter at a rate of 2 liter/min through 15–20 g dried silica gel in a glass tube. The loaded silica gel is allowed to stand with 50 ml sec. butanol or isopropanol for 1 hr. An aliquot is decanted into a flask with a reflux condenser. After adding 1.7–2.0 g sodium metal (cut into 5 or 6 pieces), the solution is moderately boiled for 1 hr under the condenser. Now 15–20 ml 95% ethanol are added through the condenser to the hot solution and boiling is continued until the sodium is completely dissolved. Then 50–60 ml water are added. The solution is subsequently acidified with 6 *N* HNO_3 and, if two layers form, 5–10 ml ethanol are added to homogenize the solution. The titration is best carried out potentiometrically with 0.05 *N* $AgNO_3$, using a silver–silver chloride electrode.

The detection limit is 1–5 ppm in 60 liter air, depending on the sensitivity of the titration.

6.5.4 Colorimetric determination of chlorinated hydrocarbons

Principle: Chlorinated hydrocarbons are most frequently determined colorimetrically via the Fujiwara reaction, i.e., a red color is produced when the hot solution is treated with NaOH in the presence of pyridine. The reaction can be used for quantitative determinations when all experimental conditions are satisfied. The individual substances of this group exhibit some deviation in their behavior and the color reaction. This is explained by the amounts of alkali selected and quantities of water added, and the conditions. The following working procedure was taken from the ICI Manual. For experiences gained with this method see Hunold and Schühlein [114], as well as Lugg [171].

Chloroform, trichloroethylene and tetrachloroethane require the mildest conditions of alkalinity and heat supply. These individual compounds can be distinguished by the absorption maxima which appear at different wavelengths. Carbon tetrachloride is determined in the presence of somewhat less water with longer heating; dichloroethane requires more water and alkali, while tetrachloroethylene must be strongly heated.

Working procedure for chloroform, trichloroethylene and tetrachloroethane

A small fritted scrubber is provided with 10 ml pyridine and the air sample is led through at a rate not exceeding 1 liter/min. In order to trap the concentrations corresponding to the MAK-value, 500 ml air are introduced for chloroform and trichloroethylene, while for tetrachloroethane, which has the very low MAK-value of 1 ppm, 20 liter sample must be passed through.

Then exactly 2.0 ml aqueous 0.02 *N* NaOH are added. The vessel is immersed into a boiling water bath for 5 min, 5 ml water are added and the solution is quickly cooled to room temperature. At the same time a blank determination is carried out. The blue-reddish color produced by the chloroform is determined at 525 mμ and the yellow-red color associated with trichloroethylene and tetrachloroethane at 440 mμ in 1-cm cells.

When plotting the calibration curves, the concentrations and volumes of the solution listed below are to be selected and treated in the same manner as the sample in order to determine the toxicant concentrations in the MAK-range.

	MAK-value mg/m^3	Volume of air sample	Stock soln. μg/ml	Volume of stock soln. in 10 ml pyridine measuring soln.
Chloroform	240	500 ml	75	0.25–3.0 ml
Trichloroethylene	520	500 ml	58	1.0 –8.0 ml
Tetrachloroethane	7	20 liter	64	1.0 –8.0 ml

Specificity: Carbon tetrachloride yields only a very weak color reaction under the above conditions; HCl and Cl_2 interfere only when present in concentrations exceeding 200 ppm. At 525 mμ 100 μg $CHCl_3$ yield the same extinction as 500 μg trichloroethylene, while at 440 mμ 100 μg trichloroethylene or tetrachloroethane yield the same extinction as 300 μg chloroform.

Working procedure for carbon tetrachloride (*MAK-value* 65 mg/m^3)

A 2-liter air sample is led in 2 min through a wash bottle containing 10 ml pyridine. Then exactly 0.40 ml aqueous 0.1 *N* NaOH is added. The mixture is immersed for 15 min into a boiling water bath, 5 ml are added and the mixture is rapidly cooled to room temperature. The photometric determination is carried out at 525 mμ in a 1-cm cell.

Calibration curve: 1–6 ml stock solution (80 μg CCl_4 per ml pyridine solution) are filled to 10 ml with pyridine and treated in the same way as the sample solution.

Specificity: The extinction produced by 100 μg CCl_4 corresponds to that respectively produced by 80 μg chloroform, 600 μg trichloroethylene and 700 μg tetrachloroethane; HCl and Cl_2 interfere when present in amounts exceeding 50 ppm.

Working procedure for 1,2-*dichloroethane* (*MAK-value* 400 μg/m^3)

An air sample of 1 liter is led in 1 min through a wash bottle containing 20 ml *aqueous* pyridine (120 ml water made up to 1000 ml with pyridine). Then the solution is immersed for 15 min into a boiling water bath, cooled immediately with cold water, 0.8 ml aqueous 1 *N* NaOH is added and the solution again cooled for 5 min in cold water. The solution is then diluted with ethanol to 30 ml. A blank determination is carried out in the same way. The solution is measured in a 2-cm cell at 415 mμ.

Calibration curve: 0.2–2.0 ml stock solution (500 μg $C_2H_4Cl_2$ per ml aqueous pyridine solution) is made up to 10 ml with aqueous pyridine.

Calculation: Here, we must allow for the fact that only 80% of dichloroethane are absorbed by the aqueous pyridine solution.

Specificity: CCl_4, $CHCl_3$ and trichloroethylene do not produce any interference in the above procedure. Cl_2 interferes when present in concentrations exceeding 3 ppm.

Working procedure for tetrachloroethylene C_2Cl_4 (*MAK-value* 670 mg/m^3)

A 15-liter air sample is drawn in 15 min through a wash bottle containing a mixture consisting of 15 ml pyridine and 5 ml aniline (both substances must be colorless). The mixture is rinsed into a 100-ml flask and heated for 10 min to moderate boiling. Then 0.3 ml of KOH in CH_3OH (5 g per 100 ml) are added and heating is continued for 45 min. The mixture is cooled to room temperature and made up to 50 ml with methanol in a volumetric flask. The blank is determined in the same way. The photometric determination is carried out at 400 mμ in a 1-cm cell.

Calibration solution: 1–5 ml stock solution (2 mg tetrachloroethylene per ml pyridine solution) are introduced into a 100-ml flask and 15 ml pyridine and 5 ml aniline are added. The mixtures are treated as above and measured.

Specificity: The extinction produced by 10 mg tetrachloroethylene corresponds to that produced by 2 mg trichloroethylene, 11 mg chloroform or 12 mg CCl_4.

6.5.5 Gas chromatographic determination of chlorinated hydrocarbons

The gas-chromatographic determination of chlorinated hydrocarbons is not difficult because of the differences in their boiling points. We do not recommend the use of a flame ionization detector, since it is not sufficiently sensitive to chlorinated hydrocarbons. On the other hand, the electron-capture detector is highly sensitive to this group of compounds, but quantitative analysis is complicated by the fact that the effect varies strongly according to the individual substances.

Let us now describe a working procedure developed and tested in the Österreichische Stickstoffwerke in which a mixture of chloroform, carbon, tetrachloride and trichloroethylene in the MAK-range is analyzed.

An air sample (10–20 ml) is introduced directly into the separating column without preliminary enrichment. The separating column consists of a glass U-tube (total length 1600 mm, inside diameter 6 mm, fixed phase 28% paraffin oil on Sterchamol 0.3 mm, carrier gas H_2 90 ml/min, temperature of the column 77°C, paper feed rate 8 mm/min, thermal conductivity cell).

The peaks, which are distinct and can be readily measured, appear in the following sequence: chloroform, carbon tetrasulfide and trichloroethylene after 6, 10 and 13 min, respectively. The limit concentration is about 20% of the MAK-value.

6.6 Ethylene oxide

(MAK-value 50 ppm–90 mg/m^3)

Oxidation with periodic acid and determination as formaldehyde

This procedure was first described by Gage [72] and is reported in the ICI Manual in a simplified form. Whereas the ICI Manual prescribes the simple absorption of ethylene oxide in water, Gage avoids the losses occurring in this absorption by employing silica gel. In both prescriptions the absorbed ethylene oxide is oxidized with periodic acid to formaldehyde under heating. The excess of periodic acid is removed with the aid of arsenite and the yellow color produced by the reaction of formaldehyde with acetylacetone at pH 6 is photometrically determined. One mole ethylene oxide yields 1 mole formaldehyde.

Sampling: An air volume, whose ethylene oxide level should not exceed a value of about 40 μg (500 ml corresponds to the MAK-value), is drawn at a rate not exceeding 500 ml/min through a small fritted scrubber or impinger containing 10 ml water. When a larger volume is not required, the air sample can be introduced into a gas-collecting tube (this may or may not be evacuated) and agitated with 10 ml water which is pushed in.

Gage led the air sample through a tube with 2 g silica gel (particle size 0.25–0.35 mm).

Oxidation and color development: After rinsing the supply tube, the contents of the sampling vessel are mixed with 5 ml water and 1 ml HIO_4, heated on a boiling water bath and loosely closed for 40 minutes. Two ml sodium arsenite solution and 2 ml acetylacetone solution are added and the mixture is heated for 3 min on a boiling water bath. After cooling, the solution is diluted to a total volume of 20 ml and the photometric determination carried out at 420 mμ against a blank with reagents treated in the same manner.

The calibration curve is obtained in the same way from a stock solution containing 14.1 μg ethylene glycol (corresponding to 10 μg ethylene oxide) per ml. A calibration series from 0–5 ml stock solution (each of the standards is diluted to 15 ml with water) is treated and measured in the same way as in the sample determination.

The color can be visually compared with a potassium chromate solution containing 1 mg K_2CrO_4/ml. The color of this solution corresponds to that obtained by 10 μg ethylene chloride in the above method.

Substances, which after being oxidized with periodic acid yield formaldehyde (e.g., methanol), and formaldehyde itself interfere with the determination.

When the sample is absorbed according to the procedure of Gage by silica gel, the content of the tube is poured into 10 ml water, treated as described and filtered prior to the photometric determination. The calibration curve is plotted with the addition of 2 g silica gel, because of the amount of colored substance adhering to the absorption agent.

Solutions: Periodic acid, 2 g/100 ml; sodium arsenite, 6.5 g/100 ml.

Acetylacetone solution. Twenty-five gram ammonium acetate, 3 ml glacial acetic acid and 0.2 ml freshly distilled acetylacetone are dissolved in water and diluted to 100 ml.

6.7 Epichlorohydrin

The compound epichlorohydrin $ClCH_2—CH—CH_2$ (with O bridging CH and CH_2), is widely applied in the organic synthesis and plastics industry. Owing to its strong irritant effect it has a low MAK-value (5 ppm) and can be determined as in the case of ethylene oxide after oxidation with periodic acid as formaldehyde. (Daniel and Gage [52]; see also the ICI Manual.)

The air volume (1–2 liter) is aspirated by means of a suitable suction device of known stroke volume at a rate of about 0.5 liter/min through two series-connected fritted scrubbers or impingers, each containing 8 ml water. The two scrubbers are separately processed. In the first wash bottle about 80% of the epichlorohydrin present is absorbed.

Oxidation and color development

The volume of the solution is made up to 10 ml in each bottle by rinsing the supply tubes. The loosely closed vessels are transferred onto a water bath and heated for 1 hr. Then 1 ml periodic acid is added to each of the bottles, which are loosely closed and heated for another 20 min. Then 2 ml sodium arsenite and 2 ml acetylacetone solutions are introduced into each of the two bottles, which are heated for another 3 min on a boiling water bath. After cooling, the solutions are diluted to 15 ml and the extinction measured at 412 mμ against a reagent blank.

The calibration curve is obtained from a calibration series of 0–5.0 ml of an epichlorohydrin stock solution (20 μg epichlorohydrin/ml), which is treated and measured in the same way as the sample.

A potassium chromate solution (1.00 mg K_2CrO_4/ml) can be used for a visual comparison. The color produced by 1 mg K_2CrO_4 corresponds to that produced by 20 μg epichlorohydrin, processed in accordance with the above method, when both reagents are dissolved in the same volume (15 ml).

The concentration of the reagents is the same as for the determination of ethylene oxide (see page 240).

6.8 Acrylonitrile

(MAK-value 20 ppm–45 mg/m^3)

Determination with lauryl mercaptan

Acrylonitrile attaches 1 mole lauryl mercaptan to its double bond. Whereas lauryl mercaptan is converted by iodine solution into the disulfide, the addition product with acrylonitrile is stable against iodine. When a known amount of lauryl mercaptan is used, the excess can be quantitatively determined with iodine. The use of this reaction for the analysis of air is described by Haslam and Newlands [98] (see also the ICI Manual).

Two fritted scrubbers or impingers are connected in series, cooled in ice, 10.0 ml lauryl mercaptan are introduced and an air sample of at least 15 liter (for determining 5-ppm levels) are aspirated at a rate of 1 liter/min. Then to each bottle is added 1 ml KOH in ethanol (5 g/100 ml). The mixtures are allowed to stand for 4 min and then 2 ml glacial acetic acid are added to each. The solutions are separately transferred with a little amount of isopropanol into two 50-ml volumetric flasks. Then a certain amount of 0.03 *M* iodine solution equivalent to the mercaptan solution is added and the volume made up to the mark with isopropanol. The color of the solution is photometrically determined at 425 mμ against the solution used in the determination of the equivalent volume.

To determine the volume of 0.03 *N* iodine solution equivalent to 10.0 ml lauryl mercaptan, 10 ml of the latter are mixed in a 50-ml volumetric flask with 1 ml KOH in ethanol. After the solution is allowed to stand for 4 min, 2 ml glacial acetic acid are added. The mixture is titrated with 0.03 *N* iodine to a pale yellow end point. The volume of the consumed iodine solution is recorded and then the flask is filled with isopropanol to the mark. This solution serves as a reference in the photometric determination of the sample solution.

The calibration curve is determined by mixing 10.0 ml lauryl mercaptan solution with a series of 0–5 ml acrylonitrile calibration solutions (0.25 mg/ml). These are treated in the same manner as in the analysis of the sample and are measured against the comparison solution.

Lauryl mercaptan solution: 1 g of the reagent/100 ml isopropanol. 0.03 *N* iodine solution: 75 ml aqueous 0.1 *N* iodine solution are diluted to 250 ml with isopropanol. In air mixtures of known composition 90% of the acrylonitrile was detected.

Determination via hydrocyanic acid

The British Factory Inspectorate describes a procedure (Booklet No. 16) whereby acrylonitrile is first converted to cyanide by alkaline permanganate solution. The latter is converted via cyanogen bromide with pyridine–aniline into a reddish-yellow compound, whose concentration is determined by comparison with standard colors.

Working procedure

An air sample of 120 ml (1 stroke of the pump) is drawn for 3 min through a small wash bottle containing 2 ml alkaline permanganate solution. The $KMnO_4$ excess is eliminated by adding 1 ml sodium arsenite solution; 0.5 ml HCl are added and the mixture is allowed to stand for 30 sec until it becomes colorless. Then 0.5 ml saturated bromine water is added, the solution is mixed and excess bromine is removed by 1 ml sodium arsenite. Finally, 1 ml pyridine–aniline reagent is added and the color is allowed to develop for 4 min. The color is then compared with standard colors or a colorimetric determination is performed.

Preparation of the solutions

Alkaline $KMnO_4$ solution: 3.5 ml of a 0.1 *N* $KMnO_4$ solution are added to 100 ml 0.1 *N* NaOH.

Sodium arsenite solution: 0.75 g sodium arsenite per 100 ml.

Pyridine–aniline reagent: 40 ml pyridine "for analysis" are diluted with 60 ml water and 2 ml freshly distilled aniline are added.

Standard colors: $CoSO_4$ solution. 10 g $CoSO_4 \cdot 7H_2O$ are dissolved in 85 ml water. $K_2Cr_2O_7$ solution: 0.1 g $K_2Cr_2O_7$ in 100 ml.

Standard 10 ppm: 4.0 ml $CoSO_4$ soln. + 12.0 ml $K_2Cr_2O_7$ soln. + 50 ml H_2O.
Standard 20 ppm: 4.5 ml $CoSO_4$ soln. + 18.0 ml $K_2Cr_2O_7$ soln. + 40 ml H_2O.
Standard 40 ppm: 5.0 ml $CoSO_4$ soln. + 35.0 ml $K_2Cr_2O_7$ soln. + 20 ml H_2O.

Six ml of the standard solutions are compared with the sample solutions. The above ppm data apply to an air sample of 120 ml.

Acetonitrile up to 80 ppm does not interfere.

6.9 Colorimetric determination of esters of lower fatty acids

Esters of lower fatty acids, especially acetates, are being applied as readily volatile solvents, for example, for nitrocellulose lacquers. In the air they can be determined by the bright red color of their reaction products with hydroxylamine, the hydroxamic acid esters, and ferric chloride. Below we shall describe a working procedure listed in the ICI Manual.

A small fritted scrubber or impinger is provided with 5 ml of purified ethanol and 600–1200 ml air sample are passed through at a rate of 500 ml/min. Then the solution is rinsed with 2 ml ethanol in a 25-ml volumetric flask and 2 ml hydroxylamine hydrochloride solution and 2 ml 5 *N* NaOH are added. The solution is allowed to stand for 1 min and 2 ml 5 *N* HCl and 5 ml $FeCl_3$ solution are added; the mixture is then diluted to 25 ml with water. The solution is mixed and the photometric determination is carried out after 30 min against a blank reagent from 7 ml ethanol, treated in the same way.

The calibration curve is plotted for a measurement series of 0–5 ml of alcoholic solutions of pure esters (0.16 mg ester/ml), treated and measured as the sample solution.

Reagents

Ethanol: 100 ml absolute ethanol are boiled for 2 hr with 10 wt. % sodium hydroxide under a reflux condenser. The alcohol is distilled off and 10% is discarded as first and 10% as last runnings.

Hydroxylamine hydrochloride solution: 20 g hydroxylamine hydrochloride are made up to 100 ml with water. The solution remains stable in a refrigerator for about 14 days.

Ferric chloride solution: 12 g $FeCl_3 \cdot 6H_2O$ are dissolved in 0.1 HCl to 100 ml.

6.10 Aromatic hydrocarbons

The aromatics benzene, toluene and the xylenes are frequently encountered as air pollutants. They occur in motor fuels, and thus enter the open atmosphere via automobile exhaust; these substances are also found at work sites, since they are

Table 6.10
MAK- AND MIK-VALUES
(mg/m^3)

	Benzene	Toluene	Xylenes	Styrene
MAK-value	80	750	870	420
MIK_D-value	3	20	20	20
MIK_K-value (once within 4 hr)	10	60	60	65

often used as solvents for fats, varnishes, floor cleaners, pesticides, etc. As individual substances they are mainly used as raw materials for chemical processing.

The toxicological effect of the individual substances of this series varies; benzene is roughly ten times more toxic than the xylenes. In addition to the MAK-values, MIK-values have also been established for benzene, toluene and the xylenes, as well as for styrene and naphthalene. The MIK_D-values are 1/20–1/40 and the MIK_K-values (not permitted to occur more than once in 4 hr) are 1/8–1/15 of the MAK-values.

In order to assess the air pollution caused by aromatics from the hygienic standpoint, a determination of the individual homologs is necessary. For this purpose gas chromatography is the most suitable method, although UV and IR spectrometry are also useful.

When it suffices to determine the sum of benzene and homologs, the simple formolite reaction should be used.

6.10.1 Determination of benzene, toluene and the xylenes by gas chromatography

These compounds can be determined by the general methods outlined on page 81. The preliminary column is filled with silica gel of the firm Davison (0.075–0.15 mm) and the separating column with Silicone Elastomer E 301 in a weight ratio of 7:3. The temperature of the column is 65°C.

The ICI Manual recommends 10-fold amounts of the quantities (see Table 4.5.2b) stated by Cropper and Kaminsky (50-ml gas sample volume and 10-fold concentration of the calibration solution).

The peaks appear in the following sequence: benzene–toluene–m + p-xylene–o-xylene. If further separation of m-xylene from p-xylene is necessary, special separation phases must be applied.

6.10.2 Determination of aromatics by the formolite reaction

The so-called formolite reaction, i.e., the formation of a yellow-brown reaction product with formaldehyde in concentrated sulfuric acid, indicates the presence of benzene, toluene and xylenes. In accordance with Booklet No. 4 of the British Factory Inspectorate the determination is carried out as follows.

The air sample is drawn at a rate not exceeding 60 ml/min through a receiver containing 10 ml of a solution prepared by mixing 0.5 ml 40% formaldehyde with concentrated sulfuric acid to 10 ml and cooling to room temperature. The air sample is introduced (e.g., by means of a hand bellows pump with known stroke volume) until a color identical to the standard color described below has developed.

Standard color: 1 ml 0.1 *N* $K_2Cr_2O_7$ solution and 24 ml 0.1 *N* $CoSO_4$ solution (14.05 g $CoSO_4 \cdot 7H_2O$ to 1 liter solution) are made up with water to 200 ml in a volumetric flask.

The content of aromatics, expressed as benzene in the air sample, is obtained in accordance with the table below from the volume used until the color corresponds to the standard color.

ml Air sample	ppm	mg/m^3
100	144	480
200	72	240
300	48	160
400	38	125
600	24	80
800	18	60

6.10.3 Styrene

Styrene vapors occur at work sites where styrene and polystyrene are produced, but can also be found in the form of depolymerization products in the thermal processing of polystyrene.

Styrene as an unsaturated benzene derivative has a marked absorption capacity in the UV range. The absorption maximum in alcoholic solution (see e.g., Ley and Dirking [161]) occurs at 245 mμ (log $\varepsilon = 4.1$). However, it is better to measure at 290 mμ (log $\varepsilon = 2.8$), since in this range many other UV-active vapors, for example, benzene and acetone, do not interfere.

According to another method, described in the ICI Manual, styrene is removed from the air sample by introducing the latter into carbon tetrachloride and agitating the resulting solution with mixed acid (1 ml concentrated H_2SO_4 + 1 ml HNO_3, d = 1.42) for 10 minutes. This nitration mixture is extracted with water, the aqueous solution made up to 50 ml and the extinction measured at 400 mμ. The calibration curve is obtained in the same manner by nitration of 0–5 mg styrene in carbon tetrachloride solution in accordance with the above prescription, extraction with 50 ml water and subsequent photometric determination.

6.11 Phenol

Phenol is an air pollutant which occurs at many production plants. It is found as a secondary product in coking plants and in low-temperature carbonization plants for lignite, and thus in plants processing tars. Phenol is further found in the production and processing of phenol-based plastic molding materials and in large amounts in organochemical plants as a synthetic or an intermediate product. Phenol traces are contained in automobile exhaust and in tobacco smoke, and occasionally in the air of hospitals.

The MAK-value is 5 ppm (19 mg/m^3), the MIK_D-value 0.05 ppm (0.2 mg/m^3) and the MIK_K-value 0.15 ppm (0.6 mg/m^3, once in 4 hours).

Since MIK-values are low, large air volumes are required for the determination of phenol in the free atmosphere. Lahmann [147] reported a rather sensitive procedure for this range.

Determination as an azo dye according to Lahmann

Principle: Phenol is absorbed in dilute NaOH, coupled with diazotized p-nitroaniline solution into an azo dye, whose extinction is measured at 530 mμ.

For sampling in the MIK-range 300-ml impingers are used. These contain 75 ml 0.1 *N* NaOH; the air throughput rate is 30 liter/min. In the 30-min measurement period 800 liter air are aspirated. With this arrangement (1 impinger) 92% of the phenol could be collected. For tests in the MAK-range 3–5 liter of air are sufficient; a fritted scrubber is the most appropriate means of trapping the phenol.

In the subsequent analysis the contents of the wash bottle are transferred into a 100-ml volumetric flask, neutralized with 1 *N* HCl and the volume is made up to 100 ml. In another 100-ml volumetric flask 20 ml p-nitroaniline solution are decolorized with 1–2 drops of saturated sodium nitrite solution, and 30 ml *N* Na_2CO_3 and 50 ml absorption solution from the first volumetric flask are added. After 20 min the yellow-red color which appears is photometrically determined against a reagent blank solution at 530 mμ in a 5-cm cell.

p-Nitroaniline solution: 1.38 g p-nitroaniline are dissolved in 310 ml *N* HCl (on an agitator) and made up to 2000 ml.

The calibration curve is plotted for a series containing 10–150 μg phenol in 50 ml aqueous solutions, which are processed and measured as above.

The "calibration factor" (reciprocal slope, see page 54)

$$\frac{\mu\text{g phenol/50 ml}}{E_{5\ \text{cm}}} \text{ is } 142.9.$$

From the scattering range of the blank a detection limit of 3.5 μg/50 ml analysis solution, i.e., 0.01 mg phenol/m^3 for a sample volume of 800 liter, was found. This corresponds to 5% of the MIK_D-value.

This procedure, however, is not specific for phenol. Its homologs and all the aromatic compounds which are coupled with diazonium solutions to the dye interfere.

Determination of phenol with aminoantipyrine

Smith, McEwen and Barrow [233] (see also Jacobs) modified the determination of phenol by using aminoantipyrine for atmospheres heavily polluted with flue gases and automobile exhaust. The authors treated only the steam-volatile fractions of these pollutants. The determination is rendered more sensitive by extracting the colored compound with chloroform.

The air sample is drawn through impingers or fritted scrubbers containing 0.1 *N* NaOH. The scrubbing fluid is diluted to 100 ml. One ml 10% $CuSO_4$ is added, the solution is acidified with phosphoric acid against methyl orange and distilled. First 90 ml and after diluting the residue 10 ml of the distillate are trapped.

Fifty ml of the distillate are mixed with 2 ml 5% NH_4Cl and with ammonia to a pH of 10.0 ± 0.2. Then 1.0 ml aminoantipyrine (2 g per 100 ml solution) and 1 ml potassium ferricyanide solution (8 g per 100 ml solution) are added. The mixture is allowed to stand for 3 min and is then extracted with 3 × 5 ml chloroform in a separating funnel. The extracts are adjusted to a volume which is appropriate for the colorimetric cell and the color is then photometrically determined at 460 mμ.

The calibration curve is plotted for a series of 0–20 μg phenol in 100 ml water. The standard solutions are treated and measured in the same way as for the sample solution.

6.12 Benzidine

Benzidine is an important product of the dye industry and is also used as a reagent. From the toxicological point of view the carcinogenic effect plays an important role. Since even low amounts are hazardous, no MAK-value has been established.

The determination of benzidine is based on the bright yellow color produced when an oxidant is added (e.g., chlorine). The ICI Manual prescribes Chloramine T sodium p-toluenesulfonchloramide. The substance which produces the yellow color is extracted with chloroform and photometrically determined. The lower detection limit is 0.5 μg benzidine. Since the colored compound is sensitive to light, the work should be carried out in the dark.

Since very small traces have to be determined, a large air sample (about 300 liter) is drawn through at a rate of 5 liter/min through a fritted scrubber or an efficient impinger containing 25 ml 0.5 *N* HCl.

The contents of the wash bottle are transferred into a small separating funnel and rinsed with 15 ml 0.5 *N* HCl and 10 ml water. The funnel is cooled with ice water, and 1.0 ml Chloramine T solution (10 g/100 ml) are added. The solution is allowed to stand for 5 min, after which the colored compound is first extracted with 5 ml and then with 2 × 2 ml chloroform. The chloroform layer is made up to 10 ml and then dehydrated with a few grains of sodium sulfate. Then the color is photometrically determined at 445 mμ against a reagent blank solution in chloroform.

The calibration curve is plotted for a stock solution of benzidine in 0.001 *N* HCl, 1 ml of which corresponds to 10 μg benzidine. A calibration series of 0.2–3 ml stock solution dissolved in about 40 ml 0.2 *N* HCl is processed as above and photometrically measured.

Interfering substances: The method is a standard for p-p′-diaminodiphenyl derivatives. Diphenylamine, aniline and naphthylamine do not interfere. The interference of other amines and phenols present must be checked in each particular case.

6.13 Determination of aniline in the air as an azo dye

There are numerous methods for determining small amounts of aniline and other aromatic amines by diazotization with nitrous acid and coupling with a suitable phenol. The azo dye formed is then photometrically determined. The following working procedure for determining aniline traces in air essentially corresponds to that developed by Clipson and Thomas [45]. The phenolic component is the disodium salt of 2-naphthol-3,6 -disulfonic acid (R-salt).

1. *Working procedure with comparison solutions*

Sampling: 6 liter of the air sample are drawn at a rate of 1.5 liter/min through a fritted scrubber by means of an aspirator or a small hand pump of 120 or 100 ml capacity. The fritted scrubber contains 5 ml diluted hydrochloric acid.

The solution is rinsed with 2 × 2 ml dilute HCl in a 10-ml volumetric flask and the volume is made up to the mark with dilute HCl. Five ml are withdrawn, 0.5 ml $NaNO_2$ solution, 2 ml Na_2CO_3, 0.1 ml R-salt solution and finally 4 ml ammonia solution are added. The color is compared with that of the four mixtures prepared from the standard solutions A and B. For the determination a comparator is used.

The corresponding aniline concentrations can be taken from the table which appears on page 250.

Solutions

Dilute hydrochloric acid: 5 ml concentrated HCl, d = 1.18, are diluted with water to 100 ml.

$NaNO_2$ solution: 3.5 g $NaNO_2$ in 100 ml water (stable for 1 month).

Na_2CO_3 solution: 10 g anhydrous soda in 100 ml water.

R-salt solution: 0.8 g R-salt (disodium salt of 2-naphthol-3,6-disulfonic acid) are dissolved in 100 ml boiling water. The pH is adjusted by adding 1 *M* soda solution to a pH of 7.5–8.5. The reagent solution is cooled and filtered.

Ammonia solution: 20 ml of ammonia water (d = 0.88) are diluted with water to 100 ml.

Stock solutions for the color standard

Solution A: 1.00 g $K_2Cr_2O_7$ in 1 liter water.

Solution B: 70.26 g $CoSO_4 \cdot 7H_2O$ in 1 liter water.

Comparison solutions

Corresponding aniline concentration, ppm	Solution A, ml	Solution B, ml	Water, ml
1	2	1.8	23
2.5	4.0	6.0	19
5	5.2	8.0	16
10	10	10	4

2. *Photometric determination*

Ten ml dilute HCl are introduced into the fritted scrubber and 5–10 liter of air are aspirated as described in 1. The contents of the scrubber are rinsed with a little water into a 25-ml volumetric flask, the double amount of the reagent reported in 1 is added and the volume made up to the mark with 25 ml. The photometric determination is carried out in a 1- or 5-cm cell at 490 mμ.

The calibration curve is plotted for 0–200 μg from a stock solution containing 20 μg aniline/ml (2 g freshly distilled aniline are dissolved in dilute HCl and made up to 1 liter with water; 10 ml of this solution are made up to 1 liter). Each of the solutions of the calibration series is made up to 10 ml with dilute HCl. The solutions are diazotized, coupled, made up to 25 ml and the color is photometrically determined at 490 mμ.

Determination of aniline in the MAK-range

The British Factory Inspectorate prescribes the following procedure in Booklet No. 11 of its collection of methods and procedures.

An air sample of 1200 ml (10 × 120 ml) is introduced for 2 min into a small gas wash bottle with a constricted inlet tube or glass frit. The wash bottle contains 10 ml 1% HCl. The color reaction is produced by transferring the absorption solution into a test tube and adding 2 drops of hypochlorite solution. The mixture is allowed to stand for 5 min, heated to boiling in order to remove the excess chlorine, 5 ml phenol reagent are added and the solution is allowed to stand for another 15 min. The mixture is compared in a Nessler tube with the color standards described below.

Reagents

Hypochlorite solution: 5 g calcium chloride (at least 25% active chlorine) are stirred with 100 ml water at 50–60°C and filtered in the hot. Also a sodium hypochlorite solution can be diluted to 0.3 *N* (titration with $Na_2S_2O_3$).

Phenol reagent: 5 g phenol are dissolved in 100 ml dilute ammonia (50 ml ammonia water, d = 0.880, per 1 liter).

Color standard: 30 g copper sulfate ($CuSO_4 \cdot 5H_2O$) are dissolved in water to 100 ml and 1.2 ml of a solution of 0.3 g $KMnO_4$ in 100 ml water are added.

Then 28.8 mg toluidine blue (Merck, quality for microscopy) are dissolved in 100 ml water to a 0.028% stock solution. Ten ml of this solution are diluted to 100 ml and the color intensity is adjusted to that of the $CuSO_4$–$KMnO_4$ solution. This is Standard III. Standard II corresponds to a dilution of 7 ml stock solution to 100 ml and Standard I to a dilution of 3.5 ml stock solution to 100 ml. The aniline concentration is found from the number of pump strokes (1 pump stroke = 120 ml) and by matching the color with Standards I to III in accordance with the following table.

Number of pump strokes	Standard I (26 μg aniline)	Standard II (53 μg aniline)	Standard III (75 μg aniline)
3		33 ppm	50 ppm
5	10 ppm	20	33
10	5	10	20

Apart from the $CuSO_4$–$KMnO_4$ color standard, toluidine blue Standard III can also be matched with the reaction product of a solution containing 75 μg aniline + + hypochlorite + phenol reagent.

In accordance with the ICI Manual, the colorimetric determination can be effected by diluting the absorption solution with 5 ml water and reacting the excess acid with 1 *N* NaOH to acid reaction against methyl orange. Then 0.1 ml of 0.3 *N* NaOCl solution are added; the mixture is allowed to stand for 5 min and the chlorine excess is eliminated by adding 0.5 ml 0.1 *N* $Na_2S_2O_3$ solution. Then 2 ml phenol reagent are added, the mixture is diluted to 25 ml and allowed to stand for 20 min. Finally, the extinction is measured in a 4-cm cell at 620 mμ.

The calibration curve is plotted from a series of solutions containing 0–30 μg aniline dissolved to 10 ml with dilute HCl (25 ml of the acid of density 1.18 are diluted to 1 liter) by the procedure used in the determination of the sample.

Gas-chromatographic determination of aniline

The determination is carried out in accordance with the data of Table 4.5.2b, page 86.

6.14 Determination of nitrobenzene according to the ICI Manual

Principle: This substance is removed from the air sample by introducing 2-ethoxyethanol (Cellosolve). Nitrobenzene is reduced by zinc amalgam and hydrochloric acid to aniline, which is determined after diazotization and coupling with R-salt to the azo dye in accordance with the prescription for the determination of aniline.

Sampling: 5 ml 2-ethoxyethanol (Cellosolve) are introduced into a small fritted scrubber or impinger and 1.5 liter air are drawn through with the aid of a suitable device (aspirator or metal handpump of 100–120 ml capacity) in 10 min.

Reduction: The contents of the wash bottle are transferred into a 25-ml volumetric flask, rinsed with a little water and the volume is made up to 25 ml with water. An aliquot (10.0 ml) is moderately boiled with 3 ml 3 *N* HCl and 0.2 g moist zinc amalgam for 5 min in a round-bottomed flask provided with a ground-glass joint and a reflux condenser. Then the condenser is rinsed with 1 ml ethoxyethanol, boiling is continued for another 10 min and the vessel washed with 2 × 5 ml water. The liquid is transferred for the visual color comparison into a 50-ml Nessler cylinder and for the photometric determination into a 50-ml volumetric flask. Two ml 1 *N* HCl and 1 ml 0.5 *N* $NaNO_2$ solution are added. After a waiting time of 3 min the solution is poured into a mixture of 10 ml 1 *M* Na_2CO_3 solution and 0.5 ml R-salt solution, mixed and poured back into the Nessler tube or into the volumetric flask. After 10 min, 5–10 ml 3 *N* NH_3 solution, which dissolves the precipitated zinc hydroxide, are added and the volume is made up to 50 ml with 3 *N* NH_3 solution.

The standards are visually compared with a series of standard colors obtained from the aniline stock solution (0–10 ml diluted to 14 ml with water, mixed with 2 ml 1 *N* HCl, 5 ml $ZnCl_2$ solution and 5 ml ethoxyethanol, diazotized and coupled in accordance with the above prescription, and diluted to 50 ml).

The photometric determination is effected by measuring the contents of the volumetric flask against a reagent blank solution at 490 mμ. The blank solution is obtained by mixing 5 ml ethoxyethanol, 6 ml water, 2 ml 1 *N* HCl, 5 ml zinc chloride solution, and adding nitrite, soda, R-salt solution and ammonia as described above.

Solutions

R-salt solution (about 0.05 *M*). About 18 g R-salt (disodium salt of 2-naphthol-3,6-disulfonic acid) are dissolved in 500 ml hot water. The pH is adjusted to 7.5–8.5 with 1 *M* Na_2CO_3. The solution is filtered into a 1000-ml volumetric flask and the volume made up to the mark.

Zinc amalgam: Pure zinc shavings are covered with saturated sublimate solution, thoroughly agitated and allowed to stand for 10–15 min. The solution is filtered, the precipitate thoroughly washed with water and stored, covered with water in a glass-stoppered flask.

Zinc chloride solution: 4 g zinc chloride in 100 ml water, which was acidified with a few drops of concentrated HCl.

2-Ethoxyethanol: Commercial "Cellosolve" is distilled and the fraction 133–136°C (760 mm) used.

Aniline stock solution: (10 g aniline/ml) 1.00 g freshly distilled aniline is dissolved in 1 liter water; 10 ml of the solution are diluted with water to 1 liter.

The calibration curve is plotted for standard series of 0–14 ml aniline stock solution (10 μg/ml), which is processed into 50 ml color solution in a volumetric flask, as described in the visual comparison. The photometric determination is conducted in a cell against the reagent blank according to the above procedure.

6.15 Benzopyrene and other polycyclic-aromatic hydrocarbons

6.15.1 General

It has been known for some time that frequent handling of bituminous tar, soot and similar materials lead to a higher incidence of carcinogenic effects on the skin. Accordingly, Barry, Cook, Haslewood, Hewett, Hieger and Kennaway [12] found about 30 years ago a number of polycyclic aromatics, among which benzo(a)pyrene (3,4-benzopyrene), dibenzanthracene (1,2,5,6) and some dibenzopyrenes lead to skin cancer when applied to mice with a brush. Further, malignant tumors on various organs of experimental animals were also found after application of these aromatics (see also Oettel [194] and Butenandt [36]).

Statistical considerations showed that skin cancer occurs more often in residents of urban areas and in smokers than in rural residents and in nonsmokers. Analytical chemists have found small amounts of the so-called "carcinogenic" hydrocarbons (for example, 20–200 μg benzopyrene/1000 m^3 city air) in the air of towns with heavy vehicular traffic as well as in tobacco smoke. On the other hand, there are exceptions to this correlation; the incidence of lung cancer has shown an increasing trend in areas such as Iceland and the city of Venice, where little or no vehicular traffic exists.

Consequently, it was rather the attempt to find a solution to the cancer problem than to establish the causal relationship between cancer incidence and concentration of the above hydrocarbons which has stimulated chemists in their search for polycyclic aromatics in all materials with which people come frequently into contact, such as foodstuffs (see, for example, Grimmer [83]) and commodities. In the framework of this book we are particularly interested in the dust of the industrial and city atmospheres. It is current practice to look for the polycyclic aromatics of the air in dust, though we know that these substances have an appreciable vapor pressure and must therefore be also present in the gas phase (see Rondia [214a]).

The analytical task is rendered more complicated by the fact that these hydrocarbons, which are present in low concentrations, occur together with large numbers and amounts of organic substances of other types. In recent years all the known separation methods, especially the chromatographic methods (first column and paper chromatography, later thin-layer and gas chromatography), have been employed. However, so far no definite, commonly recognized analytical procedure has been established. Very often it is not adequate to merely isolate and determine 3,4-benzopyrene as the representative of this group, i.e., the principal task is to separate the mixture of polycyclic aromatics into about 10–15 individual hydrocarbons (see Table 6.15). Polycyclic aromatics are characterized by their fluorescence, which is often very intense when these compounds are irradiated with UV-light.

Table 6.15
POLYCYCLIC AROMATICS

	Biolog. activity (Fischer [68])	μg in 1000 m^3 air (Bonn)	Al_2O_3-eluate ml	Fract. (see page 260)
Anthracene 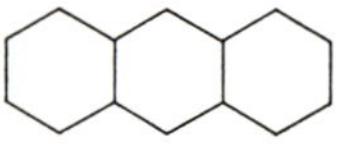	—	23	20–35	I
Phenanthrene 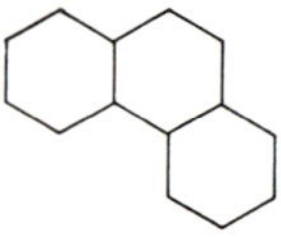	—	21	19–34	I
Pyrene 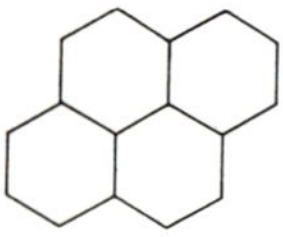	—	153	34–55	II
Fluoranthene 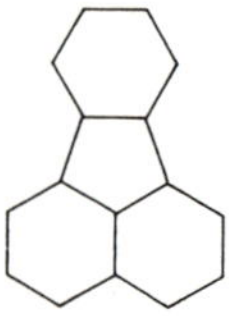	—	197	40–70	II
1,2-Benzoanthracene 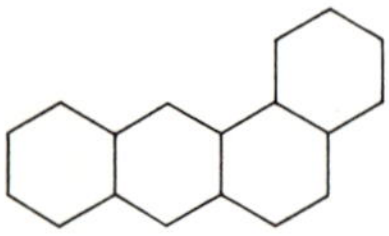	+	187	108–150	III
Chrysene 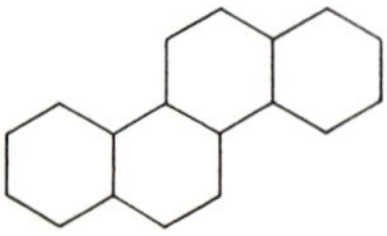	+	173	106–154	III

Table 6.15 (continued)

	Biolog. activity (Fischer [68])	μg in 1000 m^3 air (Bonn)	Al_2O_3-eluate ml	Fract. (see page 260)
1,2-Benzopyrene Benzo(a)pyrene 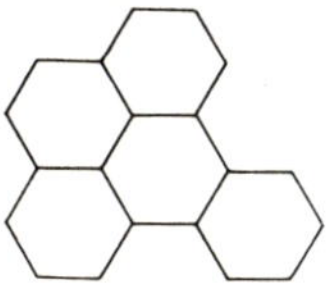	+	107	235–350	IV
 3,4-Benzopyrene Benzo(a)pyrene	+ + +	133	205–320	IV
Perylene 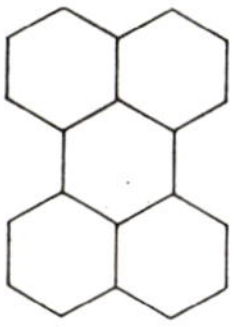	—	23	240–360	IV
1,12-Benzoperylene 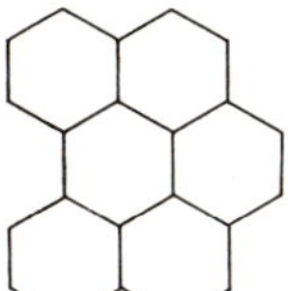	—	53	490–680	V
Anthranthrene 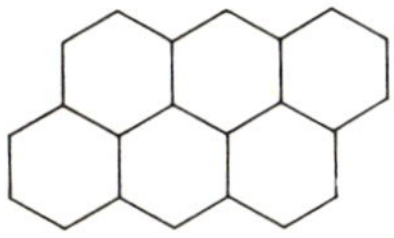	—	11	430–600	V

Table 6.15 (continued)

	Biolog. activity (Fischer [68])	μg in 1000 m^3 air (Bonn)	Al_2O_3-eluate ml	Fract. (see page 260)
Coronene	—	19	1150–1620	VII
1,2,5,6-Dibenzanthracene	+ + +	19	680–890	VI
Dibenzopyrene				
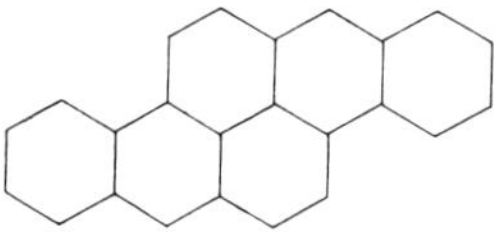	+ + +			
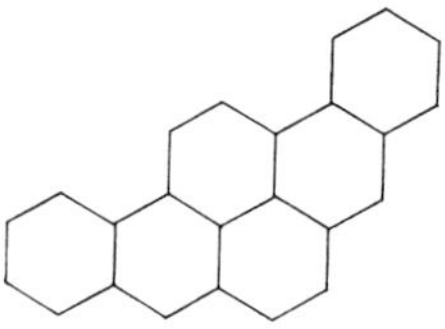	+ + + +			
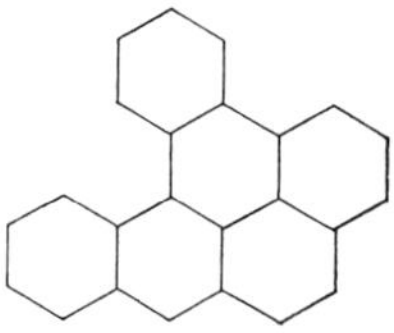	+ + +			
3-Methylcholanthrene				

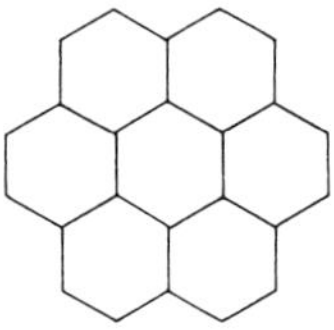

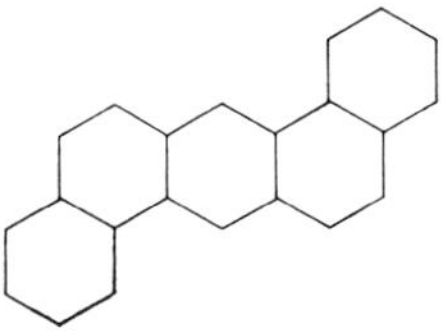

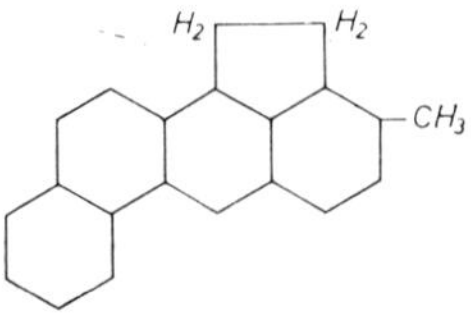

H_2 H_2 CH_3

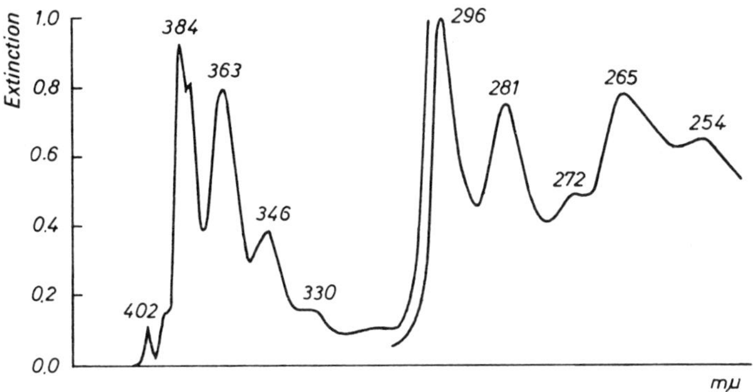

Figure 6.15.1a

Absorption curve of benzo(a)pyrene, 6.52 μg and 3.26 μg per ml cyclohexane, respectively; layer: 1 cm (Fischer [68])

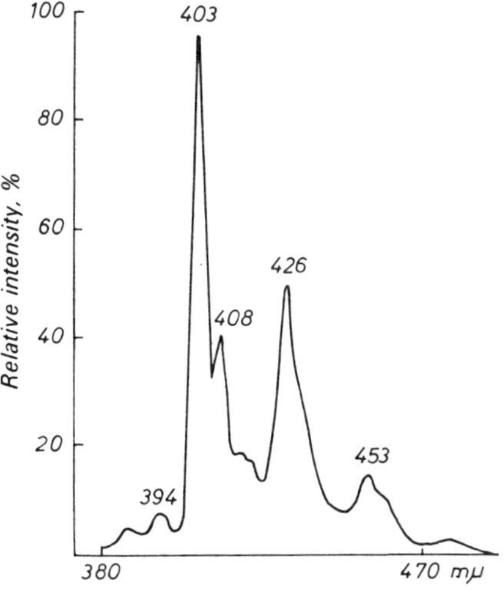

Figure 6.15.1b

Fluorescence curve of benzo(a)pyrene in cyclohexane (365 mμ) (Fischer [68])

6.15.2 Isolation and determination of benzopyrene contained in dust

In order to collect the dust (usually 1–2 g dust is required) we need an efficient device by means of which several 100 m^3 air can be drawn through a filter; glass fiber filters, for example, No. 8 Schleicher–Schüll, are frequently used for this purpose.

The next step is the extraction of the organic substances from the dust. Here, a Soxhlet cartridge, which is previously extracted with acetone until it exhibits no fluorescence, is used. Acetone or benzene and the subsequent application of methanol or cyclohexanone are used as extraction agents. Dusts with a high soot content and which firmly retain these hydrocarbons must be extracted for several hours.

Completion of the extraction process can be established by checking the fluorescence induced by irradiation with UV-light (quartz lamp); this irradiation, however, should last only for a short time so as to prevent losses by decomposition of the polycyclic aromatics. Further, precautions must be taken during evaporation; the process should not be carried out to dryness, since the hydrocarbons sought are quite volatile at higher temperatures.

After extraction the polycyclic hydrocarbons must be concentrated. Frequently, liquid–liquid extraction used. The extract is distributed, for example, between 50 ml cyclohexane, 50 ml methanol and 5 ml water; the methanol–water phase is washed several times with cyclohexane and then discarded. The polycyclic aromatics can be selectively extracted by shaking the concentrated cyclohexane phase 5 times with 5 ml nitromethane. The extractant (b.p. 101°C) is removed by careful evaporation in vacuo.

Another effective concentration process is described on page 259; Greiner and Hildebrandt, who developed this procedure, used a short silica gel column with cyclohexane.

Other separation methods based on thin-layer chromatography are described by Sawicky, Stanley, Elbert and Pfaff [227]. The authors used cellulose with 1:1 dimethylformamide–water as the mobile phase, acetylcellulose (Schleicher and Schüll, acetyl level 21%) with 17:4:4 ethanol–toluene–water as well as aluminium oxide with 19:1 pentane–ether.

The above procedure was further simplified by Stanley, Morgan and Meeker [242c]. The authors used aluminium oxide, activated at 100°C for 30 min and stored at 45% relative humidity, as the partition layer. Pentane serves as the mobile phase. For the localization and quantitative assessment of the benzo(a)pyrene spots a test solution containing 2 μg benzo(a)pyrene is simultaneously applied. The spots containing the benzo(a)pyrene are separated under illumination with UV-light, scratched out and extracted with 100 ml pure ether. The extinction is spectroscopically measured according to the base-line method of Commins [48] and Cooper [49] at 375, 382 and 390 mμ. When the determination is repeated several times, the relative standard deviation is $\pm$ 7%.

Since the hydrocarbons to be isolated occur in very small amounts, only the purest (free of fluorescence) solvents, filter papers, adsorption agents, etc., should be used.

It is necessary to check the analytical procedure selected with the aid of a test solution (prepared by the chemist himself) of the hydrocarbons to be isolated. Pure substances of this group can be obtained from the "Gesellschaft für Teerverwertung" Duisburg–Meidrich. They can also be used as standard solutions for photometric analysis.

The principal separation method involves chromatographic columns with Al_2O_3; however, also here the individual components are not completely separated but eluted in groups of 2 or 3. In these groups, however, the individual components can be determined by UV-spectrophotometry via their absorption maxima. Figure 6.15.1a shows the UV-spectrum of benzo(a)pyrene. Some authors use the fluorescence spectra, which are about ten times more sensitive (Figure 6.15.1b). The disadvantage of the spectra, however, is that they are not additive in mixtures of various fluorescent hydrocarbons.

Groups of polycyclic aromatics exhibiting a similar behavior can also be separated by paper or thin-layer chromatography into the individual substances; however, these methods are principally intended for the separation of the polycyclic aromatics from other substance groups.

Since the volatility of the most important members of this group is high, the individual substances can be separated and quantitatively determined by gas chromatography. This method requires complicated instruments, but the ultimate aim of the analysis is rapidly achieved.

In the following section we shall describe some of the relevant working procedures; for the numerous other variants the reader is referred to the literature.

6.15.3 Procedure of Grimmer and Hildebrandt

Grimmer and Hildebrandt [84] developed a procedure for the separation and determination of 13 polycyclic aromatics which is not only limited to dust but also comprises other substances with which humans come into contact, for example, foodstuffs. This procedure thus involves several successive separation runs. The authors used model mixtures and achieved a yield of about 90% with a very good reproducibility (mean deviation ± 5%).

Extraction: 2 g dust are extracted in the Soxhlet device 3 times with benzene and 2 times with methanol; each extraction process takes 30 min.

Preliminary separation: The extract evaporated to 0.5 ml is separated by shaking between 50 ml cyclohexane and 55 ml 90% (aqueous) methanol; the methanol phase is again shaken with 50 ml cyclohexane.

Preliminary column chromatography: A 400 × 8 mm column is filled with 20 g silica gel (0.125–0.16 mm). The silica gel was previously treated with 5 *N* HCl in the cold to extract traces of heavy metals, washed to neutral reaction and its moisture adjusted to 9–10% by drying in vacuo. To determine its activity 5 μg 3,4-benzopyrene dissolved in a small amount of cyclohexane are introduced into the column and eluted with cyclohexane. The aromatic should appear after 90 ml at the end of the column and should be completely eluted after 140 ml. During the actual analysis the cyclohexane extract from the previous separation, evaporated to 0.5 ml, is introduced into the column and eluted with cyclohexane; the 35–300 ml fraction contains the polycyclic aromatics in question.

Paper chromatographic separation of the group: Descending chromatography is used on 50 × 8 cm strips of Schleicher and Schüll paper No. 2043 a or b. The paper is immersed into a 40% solution of dimethylformamide in acetone and squeezed for a short time between dry filter paper. The cyclohexane solution (evaporated to 0.5 ml) from the silica gel column is applied in addition to 1,2-benzopyrene as the tracer substance. Semisaturated decalin (decahydronaphthalene) serves as the mobile phase (decalin is saturated with dimethylformamide in acetone and mixed with the same volume of untreated decalin). The R_F-values of the 13 hydrocarbons in question lie between 0.38 and 0.57. This zone is marked under the quartz lamp, cut out and extracted still wet (before decalin is evaporated) with cyclohexane.

Final separation on the Al_2O_3 column: A column (200 × 8 mm) is packed with 15 g Al_2O_3. The absorbing agent was cleaned with acid (same procedure as for silica gel), washed and dried to a water level of 4%. A test elution with 3,4-benzopyrene in cyclohexane should give this hydrocarbon in 210–320 ml eluate. In accordance with the data in the last row of Table 6.15, 7 fractions are trapped. Each fraction is thoroughly evaporated and made up to 3.00 ml with cyclohexane.

Spectrophotometric determination: Since the individual fractions still constitute mixtures of 2–3 components, the individual compounds must be determined by

Table 6.15.3
SPECTROPHOTOMETRIC DETERMINATION
ACCORDING TO GRIMMER AND HILDEBRANDT

Compound	Maximum		Minimum		Difference
	mμ	log I_o/I 1.0 μg/ml	mμ	log I_o/I 1.0 μg/ml	log I_o/I 1.0 μg/ml
Anthracene	254	1.280	260	0.073	1.207
Phenanthrene	295	0.094	300	0.016	0.078
Pyrene	338	0.280	345	0.012	0.268
Fluoranthene	289	0.276	295	0.041	0.235
1,2-Benzanthracene	290	0.434	296	0.081	0.353
Chrysene	270	0.666	276	0.127	0.539
3,4-Benzopyrene	387	0.123	394	0.034	0.089
1,2-Benzopyrene	334	0.160	340	0.013	0.146
Perylene	441	0.159	448	0.030	0.129
Anthanthrene	435	0.294	440	0.035	0.259
1,12-Benzoperylene	388	0.099	394	0.017	0.082
1,2,5,6-Dibenzanthracene	299	0.695	305	0.145	0.550
Coronene	304	0.943	308	0.192	0.751

extinction measurements at wavelengths characteristic of the absorption bands of the individual substances. For this purpose the Zeiss PMQ II can be used. In accordance with Table 6.15.3 for each substance the maximum and the next minimum toward higher wavelengths is measured. The difference between the extinctions corresponds to the concentration of the substance to be measured. The values listed in the last column apply to concentrations of 1 μg/ml.

6.15.4 Determination of Pailer, Begutter, Baumann and Schedling

A highly simplified procedure for determining the content of benzo(a)pyrene in road dust was recently developed by Pailer, Begutter, Baumann and Schedling [195].

The dust sample is collected on a glass fiber filter (No. 8 of Schleicher and Schüll), moistened with acetone and extracted with benzene in the Soxhlet device for 2 hr. Benzene is then evaporated and the residue is dissolved in acetone. Aliquots of the solution are directly, or with known amounts of benzo(a)pyrene, evaporated to about 0.3 ml as the tracer substance and as the basis for the quantitative evaluation, and applied to Kieselgel-G-Dünnschichtplatten (silica gel-G-thin-layer disks). The substances are developed 45 min with cyclohexane, containing 5% pyridine. Then cyclohexane is removed by drying (5–10 min) and the disk is again immersed into the ascending liquid, dried and the entire procedure is repeated a third time. The benzopyrene zone observed under the UV-lamp (350 mμ) is scraped off and extracted on a suction frit with acetone until fluorescence ceases. The extract is evaporated in vacuo to dryness and dissolved in 5 ml cyclohexane. The fluorescence spectrum between 350 and 500 mμ is taken with an excitation wavelength of 381 mμ. The height of the maximum at 405 mμ is proportional to a benzo(a)pyrene concentration between 0.1 and 1.0 μg/ml. The results are evaluated on the basis of a calibration curve.

Variants of the separation method

Column chromatography with Al_2O_3: Fischer [68] uses basic aluminium oxide, activity degree I of the firm Woelm, Eschwege, and elutes with a cyclohexane–benzene mixture.

Cleary [42] uses a 1-m Al_2O_3 column and elutes with cyclohexane, to which increasing amounts of ethyl ether are added.

Sawicky, Elbert, Stanley, Hauser and Fox [222] elute their Al_2O_3 column with pentane; here, as well, increasing amounts of ethyl ether are added.

See also Cooper [49], Commins [48] and Moore, Thomas and Monkman [181a].

The following method of thin-layer chromatography is recommended by Köhler, Golder and Schiesser [137], who use the two-dimensional procedure. One part of acetyl cellulose and 2 parts of Al_2O_3 brand G serve as the thin layer. The mobile phase in one direction is 90:5:5 n-hexane–toluene–n-pentane and in the other

4:4:1 methanol–ether–water. See also Kucharczyk, Fohl and Vymetal [141] as well as Fischer [68].

Schmeltz, Stedman and Chamberlain [238] used silicic acid as a packing for the separating column and Silica Gel G as the thin layer with 19:1 n-pentane–ether as the mobile phase (the eluent is hexane with the addition of increasing amounts of benzene).

For a comparison of the methods for the isolation of benzo(a)pyrene see Sawicky, Stanley, Elbert, Meeker and McPherson [226a], and Thomas, Moore, Dubois, Monkman and Katz [251a].

Chemical identification reactions

Various color reactions have been proposed for the identification of the individual polycyclic aromatics. Sawicky and his co-workers produced characteristic colors, for example, with benzal chloride and trifluoroacetic acid (Sawicky [221]) or with piperonal chloride and trifluoroacetic acid (Sawicky, Miller, Stanley and Hauser [226]). From these color reactions we can expect unequivocal results only in the case when a single polycyclic aromatic has to be determined. Thus, these reactions cannot serve as a substitute for chromatographic separation and, consequently, are no longer used in modern analytical procedures.

Gas-chromatographic separation

Since the polycyclic aromatics in question possess an appreciable vapor pressure at temperatures as low as 200°C, these compounds can be separated by gas chromatography. Here, first a rough separation for compounds of other types should be carried out. The task to determine the most important representatives of this substance class (especially benzo(a)pyrene) as isolated peaks, which are quantitatively measurable, requires a higher separation efficiency and, hence, additional equipment (capillary column, automatically controlled temperature increase, sensitive and selective detectors). So far, no completely satisfactory procedures are available, but the problem is being intensively studied at numerous institutes. The advantage of the above procedure is that it is considerably less time-consuming than the other chromatographic methods.

Lijinsky, Domsky, Mason, Ramahi and Safavi [164] describe the gas-chromatographic separation of the polycyclic hydrocarbons in coal tar. For the preliminary concentration the corresponding fraction is dissolved in 25 ml cyclohexane, shaken with 25 ml 9:1 (vol.) methanol–water in a separating funnel and the methanol phase washed three times with 25-ml portions of cyclohexane. Cyclohexane is evaporated, the residue dissolved in cyclohexane and the solution shaken 5 times with 25-ml portions of nitromethane. Nitromethane is removed by evaporation in vacuo and the residue dissolved in 0.5 ml benzene. One μl of this solution is used for the gas-chromatographic separation.

The 2400-cm separating column has an inside diameter of 6 mm. The fixed phase consists of glass beads (diameter 0.18–0.25 mm) wetted with 0.25% Silicone SE 30. The carrier gas is argon and a strontium-90-ionization detector is used.

Retention times at 200°C (in min).

Pyrene 2	Chrysene 7	Dibenz(a,h)anthracene 40
Benz(a)anthracene 6.5	Benzo(a)pyrene 16	Benzoperylene 44

De Maio and Corn [56] analyzed a dust sample, whose benzene extract is directly fed into the column. They used a double column (6 m × 3 mm) to eliminate the drift caused by volatilization of the fixed phase (Silicone rubber SE 30 on the carrier "Gas Chrom 2"). The temperature increase from 220 to 270°C is automatically controlled. The carrier gas is helium and a flame ionization detector is used. However, the results obtained are not sufficient for the isolation and determination of, for example, benzo(a)pyrene.

The determination of the polycyclic hydrocarbons in dust is also being studied in Italy. Liberti, Cartoni and Cantuti [162] preliminarily extracted the dust with cyclohexane. Further processing by shaking with aqueous methanol and nitromethane is carried out in accordance with the procedure of Lijinsky et al. (see above). The nitromethane extracts are evaporated and the residue is dissolved in small amounts of ether.

Further advances in gas-chromatographic processing are described by Cantuti, Cartoni, Liberti and Torri [38]. The authors used 50-m capillary columns packed with Silicone rubber SE 52 (methylphenylsilicone of the firm Applied Scientific Laboratories). The temperature is automatically controlled between 150 and 230°C. The carrier gas is nitrogen. Both the flame ionization detector and the electron-capture detector (see page 93) of the firm Erba, Milan, are used. The latter is specifically intended for the individual components and thus permits differentiating the peaks obtained by the FID. Unfortunately, it was not possible to achieve an adequate separation of both isomeric benzopyrenes, of which only the 3,4-derivative is considered as a strong "carcinogenous" substance. We hope, however, that these difficulties will be overcome in the near future (see also Carugno and Rossi [39]).

Appendix

Table I

CONVERSION OF TOXICANT CONCENTRATION
(ppm in mg/m^3 and vice versa)
AT TEMPERATURES OF 0 AND 20°C, 760 mm Hg,
MOLECULAR WEIGHT 1–200

$ppm \times A_0 = mg/m^3$ (0°, 760 mm Hg)
$ppm \times A_{20} = mg/m^3$ (20°, 760 mm Hg)
mg/m^3 (STP) $\times B_0 = ppm$
mg/m^3 (20°) $\times B_{20} - ppm$

Examples: 5 ppm SO_2 (Mol. wt. = 64) = 5 × 2.86 = 14.3 mg/Nm^3
5 ppm SO_2 (Mol. wt. = 64) = 5 × 2.66 = 13.3 mg/m^3 (20°)
0.75 mg SO_2/Nm^3 = 0.75 × 0.350 = 0.26 ppm
0.75 mg SO_2 (20°) = 0.75 × 0.376 = 0.28 ppm

Mol. wt.	A_o	B_o	A_{20}	B_{20}
1	0.045	22.4	0.042	24.1
2	0.089	11.2	0.083	12.0
3	0.134	7.13	0.125	8.0
4	0.178	5.62	0.166	6.0
5	0.223	4.48	0.208	4.71
6	0.268	3.73	0.250	4.01
7	0.312	3.21	0.291	3.44
8	0.357	2.80	0.333	3.01
9	0.401	2.49	0.374	2.67
10	0.446	2.24	0.416	2.41
11	0.491	2.04	0.458	2.19
12	0.53	1.87	0.50	2.00
13	0.58	1.72	0.54	1.85
14	0.62	1.60	0.58	1.72
15	0.67	1.49	0.62	1.60
16	0.71	1.40	0.66	1.50
17	0.76	1.32	0.71	1.41
18	0.80	1.25	0.75	1.34
19	0.85	1.18	0.79	1.27
20	0.89	1.12	0.83	1.20
21	0.94	1.09	0.87	1.15
22	0.98	1.02	0.91	1.09
23	1.03	0.97	0.96	1.05
24	1.07	0.94	1.00	1.00
25	1.12	0.90	1.04	0.96
26	1.16	0.86	1.08	0.93
27	1.20	0.83	1.12	0.89
28	1.25	0.80	1.16	0.86
29	1.29	0.78	1.21	0.83
30	1.34	0.75	1.25	0.80
31	1.38	0.73	1.29	0.78
32	1.43	0.70	1.33	0.75

Table I (continued)

Mol. wt.	A_o	B_o	A_{20}	B_{20}
33	1.47	0.68	1.37	0.73
34	1.52	0.66	1.41	0.71
35	1.56	0.64	1.45	0.69
36	1.61	0.63	1.50	0.67
37	1.65	0.61	1.54	0.65
38	1.69	0.59	1.58	0.63
39	1.74	0.58	1.62	0.62
40	1.78	0.56	1.66	0.60
41	1.83	0.55	1.70	0.59
42	1.87	0.54	1.75	0.57
43	1.92	0.52	1.79	0.56
44	1.96	0.51	1.83	0.55
45	2.01	0.50	1.87	0.53
46	2.05	0.488	1.91	0.52
47	2.10	0.476	1.95	0.51
48	2.14	0.467	1.99	0.50
49	2.18	0.459	2.04	0.491
50	2.23	0.448	2.08	0.481
51	2.27	0.441	2.12	0.472
52	2.32	0.431	2.16	0.463
53	2.36	0.424	2.20	0.454
54	2.41	0.415	2.24	0.446
55	2.45	0.408	2.29	0.437
56	2.50	0.400	2.33	0.429
57	2.54	0.394	2.37	0.422
58	2.58	0.388	2.41	0.415
59	2.63	0.380	2.45	0.408
60	2.68	0.373	2.49	0.401
61	2.72	0.368	2.54	0.394
62	2.76	0.362	2.58	0.388
63	2.81	0.356	2.62	0.382
64	2.86	0.350	2.66	0.376
65	2.90	0.345	2.70	0.370
66	2.94	0.340	2.74	0.364
67	2.98	0.336	2.78	0.359
68	3.03	0.330	2.83	0.354
69	3.08	0.325	2.87	0.349
70	3.12	0.321	2.91	0.344
71	3.17	0.315	2.95	0.339

Mol. wt.	A_o	B_o	A_{20}	B_{20}
72	3.21	0.312	2.99	0.334
73	3.26	0.307	3.03	0.329
74	3.30	0.303	3.08	0.325
75	3.35	0.299	3.12	0.321
76	3.39	0.295	3.16	0.316
77	3.43	0.292	3.20	0.312
78	3.48	0.287	3.24	0.308
79	3.52	0.284	3.28	0.304
80	3.57	0.280	3.32	0.301
81	3.61	0.277	3.37	0.297
82	3.66	0.273	3.41	0.293
83	3.70	0.270	3.45	0.290
84	3.74	0.267	3.49	0.286
85	3.79	0.264	3.53	0.283
86	3.84	0.260	3.57	0.280
87	3.88	0.258	3.62	0.277
88	3.92	0.255	3.66	0.273
89	3.97	0.252	3.70	0.270
90	4.01	0.249	3.74	0.267
91	4.06	0.246	3.78	0.264
92	4.10	0.244	3.82	0.261
93	4.15	0.241	3.87	0.259
94	4.20	0.238	3.91	0.256
95	4.24	0.236	3.95	0.253
96	4.28	0.234	3.99	0.251
97	4.33	0.231	4.03	0.248
98	4.37	0.229	4.07	0.245
99	4.42	0.227	4.12	0.243
100	4.46	0.224	4.16	0.241
110	4.91	0.204	4.57	0.219
120	5.4	0.187	4.99	0.201
130	5.8	0.172	5.40	0.185
140	6.2	0.160	5.81	0.173
150	6.7	0.149	6.24	0.160
160	7.1	0.140	6.65	0.150
170	7.6	0.132	7.04	0.141
180	8.0	0.125	7.49	0.134
190	8.5	0.118	7.91	0.127
200	8.9	0.112	8.32	0.120

Table II
MAXIMUM GROUND-LEVEL CONCENTRATIONS OF ORGANIC COMPOUNDS IN ACCORDANCE WITH VDI-RICHTLINIE 2306

Substance	MIK_D		MIK_K		For comparison MAK (1963)	
	cm^3/m^3	mg/m^3	cm^3/m^3	mg/m^3	cm^3/m^3	mg/m^3
1. *Hydrocarbons* (mixtures)						
*Gasoline (< 10% aromatics; solvent benzine)	20	80	60	240	500	2000
Turpentine oil	5	25	15	75	100	560
2. *Aromatics*						
*Benzene	1	3	3	10	25	80
Toluene	5	20	15	60	200	750
Xylene (all isomers)	5	20	15	60	200	870
Higher alkylated benzenes	5		15			
Styrene	5	20	15	65	100	420
*Naphthalene	0.5	2.5	1.5	7.5	10	50
3. *Alcohols*						
*Methanol	10	15	30	40	200	260
*Ethanol	50	100	150	300	1000	1900
*Propanol (all isomers)	20	50	60	150	400	980
Butanol (all isomers)	5	15	15	45	100	300
Amyl alcohol (all isomers)	5	20	15	60	100	360
4. *Chlorinated hydrocarbons*						
Methylene chloride	5	20	15	55	500	1750
Carbon tetrachloride	0.5	3	1.5	10	10	65
Chloroform	2	10	6	30	50	240
1,1- and 1,2-Dichloroethane	2	8	6	25	100	400
1,1,1-Trichloroethane	5	30	15	90	200	1080
Trichloroethylene	5	30	15	90	100	520
Tetrachloroethylene	5	35	15	110	100	670
Chlorobenzene	1	5	3	15	75	350
5. *Aldehydes*						
Formaldehyde	0.02	0.03	0.06	0.07	5	6
Acetaldehyde	2	4	6	12	200	360
Acrolein	0.005	0.01	0.01	0.025	0.1	0.25
Furfural	0.02	0.08	0.06	0.25	5	20
6. *Ketones*						
*Acetone	50	120	150	360	1000	2400
*Methyl ethyl ketone	10	30	30	90	200	590
Cyclohexanone	2	10	6	30	50	200
Methyl isobutyl ketone	5	20	15	65	100	410

Table II (continued)

Substance	MIK_D		MIK_K		For comparison MAK (1963)	
	cm^3/m^3	mg/m^3	cm^3/m^3	mg/m^3	cm^3/m^3	mg/m^3
7. *Phenols*						
Phenol	0.05	0.2	0.15	0.6	5	19
Cresol (all isomers)	0.05	0.2	0.15	0.6	5	22
8. *Isocyanates*						
2,4- and 2,6-Toluylene diisocyanate	0.001	0.007	0.003	0.021	0.02	0.14
9. *Acids*						
*Acetic acid	2	5	6	15	25	65
10. *Esters*						
*Methyl acetate	5	15	15	45	200	610
*Ethyl acetate	20	75	60	225	400	1400
*n-Butyl acetate	5	25	15	75	200	950
*Amyl acetate	5	30	15	90	100	525
*Vinyl acetate	5	20	15	60		
11. *Ethers*						
*Diethyl ether	20	65	60	195	400	1200
Ethylene oxide	2	4	6	12	50	90
Dioxane	5	20	15	60	100	360
*Tetrahydrofuran	10	30	30	90	200	590
12. *Amines*						
Monomethylamine	0.01	0.02	0.03	0.06	25	31
Dimethylamine	0.01	0.02	0.03	0.06		
Trimethylamine	0.01	0.02	0.03	0.06		
Monoethylamine	0.01	0.02	0.03	0.06	25	45
Diethylamine	0.01	0.03	0.03	0.09	25	75
Triethylamine	0.01	0.04	0.03	0.12	25	100
Aniline	0.2	0.8	0.6	0.24	5	19
Pyridine	0.2	0.7	0.6	2.1	5	15
13. *Nitro compounds*						
Nitrobenzene	0.05	0.30	0.15	0.85	1	5
Dinitrobenzene (all isomers)	0.005	0.035	0.015	0.10		

*Substances whose MIK-values are currently being investigated. The MIK_K-value is permitted once (for 30 min) in 4 hr.

Table III

MAXIMUM CONCENTRATION AT THE WORK SITE, 1966

No MAK-values are given for the compounds marked with an asterisk since even the lowest concentrations are dangerous

Substance	Formula	MAK	
		ppm	mg/m^3
Acetaldehyde	$CH_3 \cdot CHO$	200	360
Acetic acid	CH_3COOH	25	65
Acetic anhydride	$(CH_3 \cdot CO)_2O$	5	20
Acetone	$CH_3 \cdot CO \cdot CH_3$	1000	2400
Acetonitrile	$CH_3 \cdot CN$	40	70
Acrolein	$CH_2{:}CH \cdot CHO$	0.1	0.25
Acrylonitrile	$CH_2{:}CH \cdot CN$	20	45
Aldrin		—	0.25
Allyl alcohol	$CH_2{:}CH \cdot CH_2OH$	2	5
Allyl chloride	$CH_2{:}CH \cdot CH_2 \cdot Cl$	1	3
Allylglycide ether	$CH_2{:}CH \cdot CH_2 \cdot O \cdot CH_2 \cdot$ $\cdot CH \cdot CH_2 \cdot O$	10	45
Allyl propyl disulfide	$CH_2{:}CH \cdot CH_2 \cdot S_2 \cdot C_3H_7$	2	12
2-Aminoethanol	$CH_2 \cdot NH_2 \cdot CH_2 \cdot OH$	3	3
2-Aminopyridine	$NC_5H_4 \cdot NH_2$	0.5	2
Ammate		—	15
Ammonia	NH_3	50	35
Amyl acetate	$CH_3 \cdot COO \cdot C_5H_{11}$	100	525
iso-Amyl alcohol	$(CH_3)_2CH \cdot CH_2 \cdot CH_2 \cdot OH$	100	360
Aniline	$C_6H_5 \cdot NH_2$	5	19
Antimony		—	0.5
ANTU		—	0.3
Arsenic trioxide	As_2O_3	—	0.5
Arsine	AsH_3	0.05	0.2
Azinphos–methyl		—	0.2
Barium compounds (soluble)		—	0.5
Benzene	C_2H_6	25	80
Benzidine*		—	—
Benzoyl peroxide	$(C_6H_5 \cdot CO)_2O_2$	—	5
Benzyl chloride	$C_6H_5 \cdot CH_2 \cdot Cl$	1	5
Beryllium		—	0.002
Boron trifluoride	BF_3	1	3
Boron trioxide	B_2O_3	—	15
Bromine	Br_2	0.1	0.7
Bromochloromethane	$CH_2 \cdot Cl \cdot Br$	200	1050
1,3-Butadiene	$CH_2{:}CH \cdot CH{:}CH_2$	1000	2200

Table III (continued)

Substance	Formula	MAK	
		ppm	mg/m³
Butane	C_4H_{10}	1000	2350
n-Butanol	$CH_3(CH_2)_2 \cdot CH_2 \cdot OH$	100	300
tert. Butanol	$(CH_3)_3C \cdot OH$	100	300
Butan-2-one	$CH_3 \cdot CO \cdot C_2H_5$	200	590
n-Butyl acetate	$CH_3 \cdot COO \cdot C_4H_9$	200	950
Butylamine	$C_4H_9 \cdot NH_2$	5	15
n-Butylglycid ether	$C_4H_9 \cdot O \cdot CH_2 \cdot CH \cdot CH_2 \cdot O$ (CH and O joined in ring)	50	270
Butyl mercaptan	$C_4H_9 \cdot SH$	10	35
p-tert. Butyltoluene	$C_4H_9 \cdot C_6H_4 \cdot CH_3$	10	60
Cadmium oxide (smoke)	CdO	—	0.1
Calcium arsenate	$Ca_3(AsO_4)_2$	—	0.1
Calcium oxide	CaO	—	5
Camphor		—	2
Carbaryl		—	5
Carbon dioxide	CO_2	5000	9000
Carbon monoxide	CO	50	55
Carbon tetrachloride	CCl_4	10	65
Carbon tetrasulfide	CS_2	20	60
Chlordane		—	0.5
Chlorine	Cl_2	0.5	2
Chloroacetaldehyde	$Cl \cdot CH_2 \cdot CHO$	1	3
2-Chloroethanol	$Cl \cdot CH_2 \cdot CH_2 \cdot OH$	5	16
Chlorine dioxide	ClO_2	0.1	0.3
Chlorinated camphene (chlorine content 60%)		—	0.5
Chlorinated diphenyl (chlorine content 42%)		—	1
Chlorinated diphenyl (chlorine content 54%)		—	0.5
Chlorinated diphenyl oxide		—	0.5
Chlorobenzene	$C_6H_5 \cdot Cl$	75	350
1-Chloro-1-nitropropane	$C_2H_5 \cdot CH \cdot (NO_2) \cdot Cl$	20	100
Chloroform	$CHCl_3$	50	240
2-Chloroprene	$CH_2{:}C \cdot Cl \cdot CH{:}CH_2$	25	90
Chloropicrin	$CCl_3 \cdot NO_2$	0.1	0.7
Chlorine trifluoride	ClF_3	0.1	0.4
Chromic acid and chromates (calculated as CrO_3)		—	0.1
Cobalt		—	0.5
Copper (dust)		—	1
Copper (smoke)		—	0.1

Table III (continued)

Substance	Formula	MAK ppm	MAK mg/m^3
Cresol (all isomers)	$CH_3 \cdot C_6H_4 \cdot OH$	5	22
Crotonaldehyde	$CH_3 \cdot CH:CH \cdot CHO$	2	6
Cumene	$C_6H_5 \cdot CH(CH_3)_2$	50	245
Cyanides (calculated as CN)		—	5
Cyclohexane	C_6H_{12}	300	1050
Cyclohexanol	$C_6H_{11} \cdot OH$	50	200
Cyclohexanone	$C_6H_{10}O$	50	200
Cyclohexene	C_6H_{10}	300	1015
Cyclopentadiene	C_5H_6	75	200
2,4-D		—	10
DDT		—	1
Decaborane	$B_{10}H_{14}$	0.05	0.3
Demeton		—	0.1
Demeton–methyl		—	5
Diacetone alcohol	$(CH_3)_2 \cdot C \cdot (OH) \cdot CH_2 \cdot \cdot CO \cdot CH_3$	50	240
Diethylamine	$(C_2H_5)_2NH$	25	75
β-Diethylaminoethanol	$(C_2H_5)_2N \cdot C_2H_4 \cdot OH$	10	50
Diborane	B_2H_6	0.1	0.1
Dibrom		—	3
1,2-Dibromoethane	$CH_2Br \cdot CH_2Br$	25	190
Dibromodifluoromethane	CF_2Br_2	100	860
1,1-Dichloroethane	$CHCl_2 \cdot CH_3$	100	400
1,2-Dichloroethane	$CH_2Cl \cdot CH_2Cl$	100	400
o-Dichlorobenzene	$C_6H_4Cl_2$	50	300
p-Dichlorobenzene	$C_6H_4Cl_2$	75	450
β,β'-Dichloroethyl ether	$C_2H_4Cl \cdot O \cdot C_2H_4Cl$	15	90
Dichlorodifluoromethane	CF_2Cl_2	1000	4950
1,2-Dichloroethylene	$CHCl:CHCl$	200	790
Dichloromethane	CH_2Cl_2	500	1750
Dichloromonofluoromethane	$CHFCl_2$	1000	4200
1,1-Dichloro-1-nitroethane	$CH_3 \cdot C(NO_2)Cl_2$	10	60
Dichlorphos		—	1
1,2-Dichloropropane	$CH_3 \cdot CHCl \cdot CH_2Cl$	75	350
Dichlorotetrafluoroethane	$CF_2Cl \cdot CF_2Cl$	1000	7000
Dieldrin		—	0.25
Diglycid ether	$(O \cdot CH_2 \cdot CH \cdot CH_2)_2O$ (ring: O bonded to CH)	0.5	2.8
Diisobutyl ketone	$((CH_3)_2CH \cdot CH_2)_2CO$	50	290
Dimethylacetamide	$CH_3 \cdot CO \cdot N(CH_3)_2$	10	35
Dimethylamine	$(CH_3)_2NH$	10	18
Dimethylaniline	$C_6H_5 \cdot N(CH_3)_2$	5	25

Table III (continued)

Substance	Formula	MAK	
		ppm	mg/m³
Dimethylformamide	$H \cdot CO \cdot N(CH_3)_2$	20	60
1,1-Dimethylhydrazine	$NH_2 \cdot N(CH_3)_2$	0.5	1
Dimethylnitrosamine*		—	—
Dimethyl sulfate	$(CH_3)_2SO_4$	1	5
Dinitrobenzene	$C_6H_4(NO_2)_2$	—	1
Dinitroorthocresol	$CH_3 \cdot C_6H_2(OH)(NO_2)_2$	—	0.2
Dinitrotoluene	$CH_3 \cdot C_6H_3(NO_2)_2$	—	1.5
Dioxane	$\underline{O \cdot CH_2CH_2 \cdot O \cdot CH_2CH_2}$	100	360
Diphenyl ether (vapor)	$C_6H_5 \cdot O \cdot C_6H_5$	1	7
Diphenyl ether/diphenyl mixture (vapor)		1	7
Dipropylene glycol methyl ether	$(CH_3CH(OCH_3)CH_2)_2O$	100	600
Endrin		—	0.25
Epichlorohydrin	$\underline{O \cdot CH_2 \cdot CH} \cdot CH_2Cl$	5	18
EPN		—	0.5
Ethanol	$C_2H_5 \cdot OH$	1000	1900
Ethyl acetate	$CH_3 \cdot COO \cdot C_2H_5$	400	1400
Ethyl acrylate	$CH_2{:}CH \cdot COO \cdot C_2H_5$	25	100
Ethylamine	$C_2H_5 \cdot NH_2$	10	18
Ethylbenzene	$C_6H_5 \cdot C_2H_5$	100	435
Ethyl bromide	$C_2H_5 \cdot Br$	200	890
Ethyl chloride	$C_2H_5 \cdot Cl$	1000	2600
Ethylenediamine	$NH_2 \cdot C_2H_4 \cdot NH_2$	10	25
Ethylene glycol monoethyl ether	$C_2H_5 \cdot O \cdot C_2H_4 \cdot OH$	200	740
Ethylene glycol monoethyl ether acetate	$C_2H_5 \cdot O \cdot C_2H_4 \cdot O \cdot CO \cdot CH_3$	100	540
Ethylene glycol monobutyl ether	$C_4H_9 \cdot O \cdot C_2H_4 \cdot OH$	50	240
Ethylene glycol monomethyl ether	$CH_3 \cdot O \cdot C_2H_4 \cdot OH$	25	80
Ethylene glycol monomethyl ether acetate	$CH_3 \cdot O \cdot C_2H_4 \cdot O \cdot CO \cdot CH_3$	25	120
Ethyleneimine	$\underline{CH_2 \cdot CH_2 \cdot NH}$	0.5	1
Ethylene oxide	$\underline{CH_2 \cdot CH_2 \cdot O}$	50	90
Ethyl ether	$C_2H_5 \cdot O \cdot C_2H_5$	400	1200
Ethyl formate	$HCOO \cdot C_2H_5$	100	300
Ethyl mercaptan	$C_2H_5 \cdot SH$	10	25
Ethyl silicate	$Si(O \cdot C_2H_5)_4$	100	850
Ferbam		—	15
Ferrovanadium (dust)		—	1
Fluorine	F_2	0.1	0.2
Fluoride (calculated as F)		—	2.5

Table III (continued)

Substance	Formula	MAK	
		ppm	mg/m^3
Fluorotrichloromethane	$CFCl_3$	1000	5600
Formaldehyde	$HCHO$	5	6
Formic acid	$HCOOH$	5	9
Furfural	$O \cdot CH{:}CH \cdot CH{:}C \cdot CHO$	5	20
Furfuryl alcohol	$O \cdot CH{:}CH \cdot CH{:}C \cdot CH_2OH$	50	200
Gasoline		500	2000
Glycidol	$O \cdot CH_2 \cdot CH \cdot CH_2 \cdot OH$	50	150
Hafnium		—	0.5
Heptachlor		—	0.5
n-Heptane	C_7H_{16}	500	2000
n-Hexane	C_6H_{14}	500	1800
Hexan-2-one	$CH_3(CH_2)_3 \cdot CO \cdot CH_3$	100	410
sec.-Hexyl acetate	$CH_3 \cdot COO \cdot C_6H_{13}$	50	300
Hydrazine	$NH_2 \cdot NH_2$	1	1.3
Hydrogen bromide	HBr	5	17
Hydrogen chloride	HCl	5	7
Hydrogen cyanide	HCN	10	11
Hydrogen fluoride		—	2
Hydrogen peroxide	H_2O_2	1	1.4
Hydrogen selenide	H_2Se	0.05	0.2
Hydrogen sulfide	H_2S	10	15
Hydroquinone	$C_6H_4(OH)_2$	—	2
Iodine	O_2	0.1	1
Iron oxide (smoke)		—	15
Isophorone	$C_9H_{14}I$	25	140
Ketene	$CH_2{:}CO$	0.5	0.9
Lead		—	0.2
Lead arsenate	$Pb_3(AsO_4)_2$	—	0.15
Lindane		—	0.5
Lithium hydride	LiH	—	0.025
Magnesium oxide (smoke)	MgO	—	15
Malathion		—	15
Manganese		—	5
Mercury		—	0.1

Table III (continued)

Substance	Formula	MAK	
		ppm	mg/m³
Mercury compounds, organic (calculated as Hg)		—	0.01
Methanol	$CH_3 \cdot OH$	200	260
Methoxychlor		—	15
Methyl acetate	$CH_3 \cdot COO \cdot CH_3$	200	610
Methylacetylene	$CH_3 \cdot C{:}CH$	1000	1650
Methyl acrylate	$CH_2{:}CH \cdot COO \cdot CH_3$	10	35
Methylal	$CH_2(O \cdot CH_3)_2$	1000	3100
Methylamine	$CH_3 \cdot NH_2$	10	12
Methyl bromide	$CH_3 \cdot Br$	20	80
Methyl chloride	$CH_3 \cdot Cl$	50	105
Methylcyclohexane	$CH_3 \cdot C_6H_{11}$	500	2000
Methylcyclohexanol	$CH_3 \cdot C_6H_{10} \cdot OH$	100	470
o-Methylcyclohexanol	$CH_3 \cdot CH \cdot CO(CH_2)_3CH_2$ (ring)	100	460
Methyl formate	$HCOO \cdot CH_3$	100	250
Methylisobutylcarbinol	$CH_3 \cdot CH \cdot OH \cdot CH_2 \cdot \cdot CH(CH_3)_2$	25	100
Methyl isobutyl ketone	$CH_3 \cdot CO \cdot CH_2 \cdot CH(CH_3)_2$	100	410
Methyl mercaptan	$CH_3 \cdot SH$	10	20
Methyl methacrylate	$CH_2 \cdot C(CH_3) \cdot COO \cdot CH_3$	100	410
2-Methyl-2-penten-4-one	$(CH_3)_2C{:}CH \cdot CO \cdot CH_3$	25	100
α-Methylstyrene	$C_6H_5 \cdot C \cdot (CH_3){:}CH_2$	100	480
Molybdenum compounds (insoluble) (calculated as Mo)		—	15
Molybdenum compounds (soluble) (calculated as Mo)		—	5
Monomethylaniline	$C_6H_5 \cdot NH \cdot CH_3$	2	9
Naphtha (coal tar)		200	800
Naphtha (petroleum)		500	2000
Naphthalene	$C_{10}H_8$	10	50
β-Naphthylamine		—	—
Nickel carbonyl	$Ni(CO)_4$	0.1	0.7
Nicotine		—	0.5
Nitric acid	HNO_3	10	25
p-Nitroaniline	$C_6H_4 \cdot (NO_2) \cdot NH_2$	1	6
Nitrobenzene	$C_6H_5 \cdot NO_2$	1	5
Nitroethane	$C_2H_5 \cdot NO_2$	100	310
Nitrogen dioxide	NO_2	5	9
Nitroglycerin	$C_3H_5(ONO_2)_3$	0.5	5
Nitroglycol	$C_2H_4(ONO_2)_2$	0.25	1.6
Nitromethane	$CH_3 \cdot NO_2$	100	250

Table III (continued)

Substance	Formula	MAK ppm	MAK mg/m^3
1-Nitropropane	$CH_2 \cdot (NO_2) \cdot CH_2 \cdot CH_3$	25	90
2-Nitropropane	$CH_3 \cdot CH \cdot (NO_2) \cdot CH_3$	25	90
Nitrotoluene (all isomers)	$CH_3 \cdot C_6H_4 \cdot NO_2$	5	30
Octane	C_8H_{18}	500	2350
Oil mists (mineral oil)		—	5
Osmium tetroxide		—	0.002
Ozone	O_3	0.1	0.2
Parathion		—	0.1
Pentaborane	B_5H_9	0.005	0.01
Pentachloroethane	$CCl_3 \cdot CHCl_2$	5	40
Pentachloronaphthalene	$C_{10}H_3Cl_5$	—	0.5
Pentachlorophenol	C_6Cl_5OH	—	0.5
Pentane	C_5H_{12}	1000	2950
Pentan-2-one	$CH_3(CH_2)_2 \cdot CO \cdot CH_3$	200	700
Perchloromethylmercaptan	$CCl_3 \cdot S \cdot Cl$	0.1	0.8
Phenol	C_6H_5OH	5	19
p-Phenylenediamine	$C_6H_4(NH_2)_2$	—	0.1
Phenylglycide ether	$C_6H_5 \cdot O \cdot CH_2 \cdot \underline{CH \cdot CH_2 \cdot O}$	50	310
Phenylhydrazine	$C_6H_5 \cdot NH \cdot NH_2$	5	22
Phosdrin		—	0.1
Phosgene	$COCl_2$	0.1	0.4
Phosphine	PH_3	0.1	0.15
Phosphorus (yellow)		—	0.1
Phosphorus oxychloride	$POCl_3$	0.5	3
Phosphorus pentachloride	PCl_5	—	1
Phosphorus pentasulfide	P_2S_5	—	1
Phosphorus pentoxide	P_2O_5	—	1
Phosphorus trichloride	PCl_3	0.5	3
Picric acid	$C_6H_2(OH)(NO_2)_3$	—	0.1
Platinum compounds (calculated as Pt)		—	0.002
Propane	C_3H_8	1000	1800
β-Propiolacetone			
Propyl acetate	$CH_3 \cdot COO \cdot C_3H_7$	200	840
iso-Propyl alcohol	$(CH_3)_2CH \cdot OH$	400	980
iso-Propylamine	$(CH_3)_2CH \cdot NH_2$	5	12
1-Propyl ether	$([CH_3]_2CH)_2O$	500	2100
Propyleneimine	$\underline{NH \cdot CH_2 \cdot CH_2 \cdot CH_2}$	2	5
Propylene oxide	$CH_3 \cdot CH \cdot CH_2 \cdot O$	100	240

Table III (continued)

Substance	Formula	MAK	
		ppm	mg/m³
iso-Propylglycide ether	$C_3H_7 \cdot O \cdot CH_2 \cdot \underline{CH \cdot CH_2 \cdot O}$	50	240
n-Propyl nitrate	$C_3H_7 \cdot ONO_2$	25	110
Pyrethrum		—	5
Pyridine	C_5H_5N	5	15
Quinone	$C_6H_4O_2$	0.1	0.4
Rotenone		—	5
Selenium compounds (calculated as Se)		—	0.1
Silver		—	0.05
Sodium hydroxide	$NaOH$	—	2
Sodium fluoroacetate	$CH_2F \cdot COONa$	—	0.05
Stibine	SbH_3	0.1	0.5
Strychnine		—	0.15
Styrene	$C_6H_5 \cdot CH:CH_2$	100	420
Sulfuric acid	H_2SO_4	—	1
Sulfur dioxide	SO_2	5	13
Sulfur hexafluoride	SF_6	1000	6000
Sulfur monochloride	S_2Cl_2	1	6
Sulfur pentafluoride	SF_5	0.025	0.25
Tantalum		—	5
TEDP		—	0.2
TEPP		—	0.05
Tellurium and its compounds (calculated as Te)		—	0.1
1,1,2,2-Tetrabromoethane	$CHBr_2 \cdot CHBr_2$	1	14
1,1,2,2-Tetrachloroethane	$CHCl_2 \cdot CHCl_2$	1	7
Tetrachloroethylene	$CCl_2:CCl_2$	100	670
1,1,1,2-Tetrachloro-2,2-difluoroethane	$CF_2Cl \cdot CCl_3$	1000	8340
1,1,2,2-Tetrachloro-1,2-difluoroethane	$CFCL_2 \cdot CFCl_2$	500	4170
Tetraethyllead (calculated as Pb)	$Pb(C_2H_5)_4$	—	0.075
Tetrahydrofuran	$\underline{CH_2(CH_2)_3 \cdot O}$	200	590
Tetramethyllead	$Pb(CH_3)_4$	—	0.075
Tetramethylsuccinonitrile		0.5	3
Tetranitromethane	$C(NO_2)_4$	1	8
Tetryl		—	1.5
Thallium compounds (soluble) (calculated as Tl)		—	0.1
Thiram		—	5

Table III (continued)

Substance	Formula	MAK	
		ppm	mg/m^3
Tin compounds, inorganic			
(calculated as Sn)		—	2
organic (calculated as Sn)		—	0.1
Titanium dioxide	TiO_2	—	15
Toluene	$C_6H_5 \cdot CH_3$	200	750
o-Toluidine	$CH_3 \cdot C_6H_4 \cdot NH_2$	5	22
2,4-Toluylene diisocyanate	$CH_3 \cdot C_6H_3(NCO)_2$	0.02	0.14
2,6-Toluylene diisocyanate	$CH_3 \cdot C_6H_3(NCO)_2$	0.02	0.14
Triethylamine	$(C_2H_5)_3N$	25	100
1,1,1-Trichloroethane (Methylchloroform)	$CCl_3 \cdot CH_3$	200	1080
1,1,2-Trichloroethane	$CHCl_2 \cdot CH_2Cl$	10	45
Trichloroethylene	$CCl_2:CHCl$	100	520
Trichloronaphthalene	$C_{10}H_5Cl_3$	—	5
2,4,5-Trichlorophenoxyacetic acid	$Cl_3C_6H_2 \cdot OCH_2 \cdot COOH$	—	10
1,2,3-Trichloropropane	$CH_2Cl \cdot CHCl \cdot CH_2Cl$	50	300
1,1,2-Trichloro-1,2,2-trifluoroethane	$CF_2Cl \cdot CFCl_2$	1000	7600
Trifluoromonobromomethane	CF_3Br	1000	6100
Trinitrotoluene	$CH_3 \cdot C_6H_2(NO_2)_3$	—	1.5
Turpentine		100	560
Uranium compounds (soluble)			
(calculated as U)		—	0.05
Uranium compounds (insoluble)			
(calculated as U)		—	0.25
Vanadium (V_2O_5-dust)		—	0.5
Vanadium (V_2O_5-smoke)		—	0.1
Vinyl chloride	$CH_2:CHCl$	500	1300
Vinyltoluene	$CH_3 \cdot C_6H_4 \cdot CH:CH_2$	100	480
Warfarin		—	0.5
Xylene (all isomers)	$(CH_3)_2C_6H_4$	200	870
Xylidine (all isomers)	$(CH_3)_2C_6H_3 \cdot NH_2$	5	25
Yttrium and its organic compounds		—	5
Zinc oxide (smoke)		—	5
Zirconium compounds (calculated as Zr)		—	5

Bibliography

1 ABEL, E. and H. BARTH: *Arch. Eisenhüttenw.* **29** (1958) 683; *Z. anal. Ch.* **169** (1959) 210

1a ADAMS, D. F. and R. K. KOPPE: *Anal. Chemistry* **31** (1959) 1249; **33** (1961) 117

2 ALTSHULLER, A. P.: *J. Gas-Chrom.* **1** (1963) 6

2a ALTSHULLER, A. P. AND C. A. CLEMONS: *Anal. Chemistry* **34** (1962) 466

3 ALTSHULLER, A. P. and J. R. COHEN: *Anal. Chemistry* **32** (1960) 802

4 ALTSHULLER, A. P., C. M. SCHWAB and M. BARE: *Anal. Chemistry* **31** (1959) 1987

5 ALTSHULLER, A. P., A. F. WARTBURG, I. R. COHEN and S. F. SLEVA: *Int. Journ. Air Water Poll.* **6** (1962) 75

6 ANDREATCH, A. J. and R. FEINLAND: *Anal. Chemistry* **32** (1960) 1021

7 ANDREW, T. R. and P. N. R. NICHOLS: *Analyst* **90** (1965) 367

8 ANGLERAUD, O. and F. BORELLI: *Gas-Wasserfach* **107** (1966) 573

9 ARNOLD, C. and C. MENZEL: *Ber.* **35** (1902) 1324

10 AUBEAU, R., L. CHAMPEIX and J. REISS: *J. Chromatogr.* **16** (1964) 7

10a AURAND, K. and J. BOSCH: *Staub* **27** (1967) 445

11 BAKER, R. A. and R. C. DOERR: *Int. J. Air Poll.* **2** (1959) 142

12 BARRY, G., J. W. COOK, G. A. HASLEWOOD, C. L. HEWETT, J. HIEGLER and E. L. KENNAWAY: *Proc. Roy. Soc. London* **117** (1935) 318

12a BASSETT, P. M. and J. DAVIES: *Air Water Poll.* **10** (1966) 633

13 BAUER, H. D.: *Staub* **24** (1964) 290

13a BAUM, F., L. HERMANN and W. STEINBACH: *Gesundh.-Ing.* **86** (1965) 319; *Staub* **26** (1966) 131

13b BAUM, F.: *Gesundh.-Ing.* **87** (1966) 357

14 BAYER, F. and G. WAGNER: *Gasanalyse (Gas Analysis)* 3rd ed., Verlag Enke, Stuttgart 1960

15 BECKMAN, A. O., J. D. CULLOUGH and R. A. CRANE: *Anal. Chemistry* **20** (1948) 674

16 BELCHER, R., M. A. LEONARD and T. S. WEST: *J. Chem. Soc. London* 1959, 3577

17 BELLAR, T. A., M. F. BROWN and J. E. SIGBY: *Anal. Chemistry* **35** (1963) 1924

18 BELLAR, T., J. E. SIGBY, C. A. CLEMONS and A. P. ALTSHULLER: *Anal. Chemistry* **34** (1962) 763

19 BENDER, D. F. and A. W. BREIDENBACH: *Anal. Chemistry* **35** (1963) 417

19a BERSIN, L. R., F. J. BRONSAIDES and C. O. HOMMEL: *J. Air Poll. Control Assoc.* **12** (1962) 129

19b BENSON, F. B., V. A. NEVILL, J. E. THOMPSON, M. TERABE and S. OMICHI: *J. Air Poll. Contr. Assoc.* **17** (1967) 673
20 BIENSTOK, D., J. H. FIELD, S. KATELL and K. D. PLANTS: *J. Air Poll. Control Assoc.* **15** (1965) 459; *Staub* **26** (1966) 505
21 BLOM, L. and L. EDELHAUSER: *Anal. chim. acta* **13** (1955) 120; *Z. anal. Ch.* **150**, 233
22 BOBER, H.: Beckman Report 2/66, 10
23 BOKHOVEN, C. and H. J. NIESSEN: *Nature* (London) **192** (1961) 458
24 BOKHOVEN, C. and H. J. NIESSEN: *Int. J. Air Water Poll.* **10** (1966) 233; *Staub* **26** (1966) 502
25 BOKHOVEN, C. and P. H. TOMMASSEN: *Nature* (London) **190** (1961) 435
26 BOVEE and R. J. ROBINSON: *Anal. Chemistry* **33** (1961) 1115
27 BRAVO, H. A. and J. P. LODGE: *Anal. Chemistry* **36** (1964) 671
27a BREUER, W.: *VDI-Ztschr.* **107** (1965) 1434
28 BREUER, H.: *Staub* **20** (1960) 290
29 BREUER, H.: *Staub* **24** (1964) 324
30 BREUER, H.: *Staub* **25** (1965) 57
30a BUCK, R. P. and R. W. ELDRIDGE: *Anal. Chemistry* **37** (1965) 1242
31 BUCK, M. and H. STRATMANN: *Staub* **24** (1964) 241
32 BUCK, M. and H. STRATMANN: *Z. anal. Ch.* **213** (1965) 241
33 BUCK, M. and H. STRATMANN: *Brennst.-Chem.* **46** (1965) 231
34 BUCK, M. and H. STRATMANN: *Staub* **27** (1967) 265
35 BUCK, M. and H. GIES: *Staub* **26** (1966) 379
35a BURKE, D. E., G. C. WILLIAMS and C. A. PLANK: *Anal. Chemistry* **39** (1967) 544
36 BUTENANDT, A.: *Ch. Ztg.* **74** (1950) 7
37 BYERS, D. H. and B. E. SALTZMAN: *Amer. Ind. Hyg. Assoc. J.* **19** (1958) 251
38 CANTUTI, V., G. P. CARTONI, A. LIBERTI and A. G. TORRI: *J. Chromatogr.* **17** (1965) 60
39 CARUGNO, N. and S. ROSSI: *J. Gaschrom.* **5** (1967) 103
39a CASTELLO, G. and S. MUNARI: *J. Chromatogr.* **31** (1967) 103
40 CAUER: *Z. anal. Ch.* **103** (1935) 166
40a CHERNIAK, J. and R. J. BRYAN: *J. Air Poll. Contr. Assoc.* **15** (1965) 351; *Staub* **26** (1966) 173
41 CIACCIO, L. L. and TH. COTSIS: *Anal. Chemistry* **39** (1967) 260
42 CLEARY, G. J.: *J. Chromat.* **9** (1962) 204
43 CLEMONS, C. A.:Private communication
44 CLEMONS, C. A. and A. P. ALTSHULLER: *Anal. Chemistry* **38** (1966) 133
45 CLIPSON, J. L. and L. C. THOMAS: *Analyst* **88** (1963) 971
46 COENEN, W.: *Staub* **23** (1963) 119
46a COENEN, W.:*Staub* **26** (1966) 415
47 COHEN, J. R. and A. P. ALTSHULLER: *Anal. Chemistry* **33** (1961) 726
48 COMMINS, B. T.: *Analyst* **83** (1958) 386
49 COOPER, R. L.: *Analyst.* **79** (1958) 573
49a CRIDER, W. L.: *Anal. Chemistry* **37** (1965) 1770
50 CROPPER, F. R. and S. KAMINSKY: *Anal. Chemistry* **35** (1963) 735
51 DAMASCHKE, K. and M. LÜBKE: *Ch. Ztg.* **88** (1964) 547

52 DANIEL, J. W. and J. C. GAGE: *Analyst* **81** (1956) 594
53 DARLEY, E. F., K. A. KETTNER and E. R. STEPHENS: *Anal. Chemistry* **35** (1963) 589
54 DECKERT, W.: *Z. anal. Ch.* **176** (1960) 163
55 DECRISTOFORO, P. J.: *Dechema-Monogr.* **54** (1965) 251
56 DE MAIO, L. and M. CORN: *Anal. Chemistry* **38** (1966) 131
56a DIEM, M.: *Staub* **21** (1961) 345
57 DIXON, B. E. and P. METSON: *Analyst* **85** (1960) 122
58 DORTA-SCHÄPPI, Y. and W. D. TREADWELL: *Helv. chim. acta* **32** (1949) 356
58a EAVES, A. and R. C. MACAULAY: *Int. J. Air Water Poll.* **8** (1964) 645
59 EGEROW, M. S.: *Z. Lebensm.* **56** (1928) 355
60 EGGERTSEN, F. T. and F. M. NELSON: *Anal. Chemistry* **30** (1958) 1040
61 EHMERT, A.: *Z. Naturforschg.* **4 B** (1949) 321
62 EHMERT, A.: *Meteorolog. Rundsch.* **4** (1954) 65
63 ELLIS, C. F.: *Intern. J. Air Water Poll.* **8** (1963) 297
64 ENGELHARDT, H.: *Elektrotechn. Ztschr.* **18** (1966) 38
65 FARRINGTON, P. S., R. L. PECSOK, R. L. MEEKER and T. J. OLSON: *Anal. Chemistry* **31** (1959) 1512
66 FEINLAND, R., A. J. ANDREATCH and D. P. COTRUPE: *Anal. Chemistry* **33** (1961) 991
67 FIELDER, R. S. and C. H. MORGAN: *Anal. chim. acta* **23** (1960) 538; cited in *Z. anal. Chem.* **183** (1961) 455
68 FISCHER, F.: Centenary volume 100 *Jahre BASF*
69 FORTUIN, J. M. H.: *Anal. chim. acta* **15** (1956) 521
69a FRAUST, Ch. L. and E. R. HERMANN: *Amer. Ind. Hyg. Assoc. J.* **27** (1966) 68; *Staub* **26** (1966) 501
70 FREIER, R. K.: *Wasseranalyse (Water Analysis)*, Verlag De Gruyter, Berlin 1964
70a FRIEDRICHS, K. H.: *Staub* **26** (1966) 240
71 FUHRMANN, H.: *Staub* **25** (1965) 266
71a FUHRMANN, H. and H. WINTER: *Wasser-Luft-Betrieb* **8** (1964) 260
72 GAGE, J. C.: *Analyst* **82** (1957) 587
73 GAGE, J. C.: *Analyst* **85** (1960) 196
74 GALSTER, H.: *Z. anal. Chem.* **186** (1962) 359
75 GAST, T.: *Staub* **21** (1961) 136
76 GAST, T.: *Chem. Ing. Techn.* **29** (1957) 262
77 GERICKE, S. and B. KURMIES: *Z. anal. Chem.* **132** (1951) 335
78 GESSNER, H.: *Staub* **24** (1964) 314
79 GILL, W. E.: *Amer. Ind. Hyg. Assoc. Journ.* **21** (1960) 87
80 GOLDMAN, F. H. and H. YAGODA: *Ind. Eng. Ch. Anal. Ed.* **15** (1943) 377
81 GREENE, S. A.: *Anal. Chemistry* **31** (1959) 480
82 GRIGORESCU, I and G. TOBA: *Rev. Chem.* (Bucharest) **15** (1964) 572; cited in *Z. anal. Ch.* **215** (1966) 64
83 GRIMMER, G.: *Erdöl und Kohle* **19** (1966) 578
84 GRIMMER, G. and A. HILDEBRAND: *J. Chromatogr.* **20** (1965) 89
85 GROSSKOPF, K.: *Chem. Ztg.* **87** (1963) 270
86 GRUPINSKY, L.: *Wasser-Luft-Betr.* **9** (1965) 38
86a GRUPINSKY, L.: *Wasser-Luft-Betr.* **10** (1966) 77

87 Guerrant, G. O.: *Anal. Chemistry* **37** (1965) 516

88 Gunther, F. A., J. H. Barkley, M. J. Kolbezen, R. C. Blinn and E. A. Staggs: *Anal. Chemistry* **28** (1956) 1985

89 Guthmann, K.: *Stahl u. Eisen* **79** (1959) 1129; **85** (1962) 296

90 Guthmann, K. and H. Kahnwald: *Stahl u. Eisen* **82** (1962) 296

91 Haagen-Smit, A. J. and M. J. Brunelle: *Intern. J. Air Poll.* **1** (1959) 51

92 Haagen-Smit, A. J., M. F. Brunelle and J. Hara: *Arch. Ind. Health* **20** (1959) 399; *Ch. Abstr.* **54**, 2667

93 Häntzsch, S. and K. E. Prescher: *Z. anal. Ch.* **213** (1965) 408; *Staub* **26** (1966) 332

94 Hands, G. C. and A. F. F. Bartlett: *Analyst.* **85** (1960) 147

95 Hanst, P. L., E. R. Stephens, W. E. Scott and R. L. Doerr: *Anal. Chemistry* **33** (1961) 1113

96 Hasenclever, D.: *Ch. Ing. Techn.* **26** (1954) 180

97 Hasenclever, D.: *Staub* **21** (1961) 131

98 Hasenclever, D.: *Staub* **22** (1962) 99

99 Hasenclever, D. and C. Siegmann: *Staub* **20** (1960) 212

100 Hauser, Th. R. and D. W. Bradley: *Anal. Chemistry* **38** (1966) 1529

101 Hauser, Th. R. and R. L. Cummins: *Anal. Chemistry* **36** (1964) 679

102 Helwig, H. L. and Ch. L. Gordon: *Anal. Chemistry* **30** (1958) 1810

102a Hendricks, R. H. and L. B. Larsen: *Amer. Ind. Hyg. Assoc. J.* **27** (1966) 81

103 Herrmann, G.: *Chem. Technik* **15** (1963) 342

104 Herrmann, G.: *Chem. Technik* **17** (1965) 97

105 Herrmann, E.: *Chemie-Ing. Techn.* **37** (1965) 905

106 Herrmann, E.: *Staub* **26** (1966) 403

106a Hersch, P. and R. Deuringer: *Anal. Chemistry* **35** (1963) 897

107 Hochheiser, S., J. Santer and W. F. Ludmann: *J. Air Poll. Contr. Assoc.* **16** (1966) 266; *Staub* **27** (1967) 108

108 Hoeschele, K.: *Staub* **24** (1964) 140

108a Hodges, C. T. and R. F. Matson: *Anal. Chemistry* **37** (1965) 1065

109 Hollis, O. L.: *Anal. Chemistry* **38** (1966) 309

110 Holm-Jensen, J.: *Anal. chim. acta* **23** (1960) 13; *Z. anal. Ch.* **183** (1961) 303

111 Hughes, E. E. and S. G. Lias: *Anal. Chemistry* **32** (1960) 707

112 Huit, H. A. and J. P. Lodge: *Anal. Chemistry* **36** (1964) 1305

112a Hummel, H.: *Chem. Ing. Techn.* **32** (1960) 455

113 Hunold, G. A. and W. Pietrulla: *Z. anal. Chem.* **178** (1961) 271

114 Hunold, G. A. and B. Schühlein: *Z. anal. Chem.* **179** (1961) 81

115 Huygen, C.: *Anal. chim. acta* **28** (1963) 349

116 Iwasaki, J., S. Utsumi, K. Hagino and T. Ozawa: *Bull. Chem. Soc. Japan* **29** (1956) 860; cited in *Z. anal. Chem.* **157** (1957) 293

116a Jacobs, M. B.: *The Chemical Analysis of Air Pollutants*, New York, 1960

117 Jacobs, M. B., M. M. Braverman and S. Hochheiser: *Anal. Chemistry* **29** (1957) 1349

118 Jacobs, M. B. and S. Hochheiser: *Anal. Chemistry* **30** (1958) 426

119 Janak, J. and M. Rusek: *Chem. Listy* **48** (1954) 397; cited in *Z. anal. Ch.* **143** (1954) 442

120 Jeffery, P. G. and P. J. Kipping: *Gas Analysis by Gas Chromatography*, Pergamon Press, Oxford, 1964

121 JOHSWICH, F.: *Brennst.-Wärme-Kraft* **17** (1965) 238
122 JONES, K. and R. GREEN: *Nature* (London) **205** (1965) 67
123 JONES, K. and P. HALFORD: *Nature* (London) **202** (1964) 1003
124 JÜNTGEN, H.: *Chemie-Ing. Techn.* **38** (1966) 734
125 JUNGE, CH. E.: *Air Chemistry and Radioactivity*, Academic Press, New York and London, 1963
126 KAHLE, H. and B. KARLIK: *Fresenius–Jander Handb. d. Anal. Chem.* (*Handbook of Analytical Chemistry*) **VIIIa**, Berlin, 1949
126a KAISER, E. R.: in A. C. STERN, *Air Pollution*, **1**, p. 552, Acad. Press, New York and London, 1962
127 KAISER, H. and H. SPECKER: *Z. anal. Chem.* **149** (1956) 46
127a KAMPF, W. D. and B. SCHMIDT: *Staub* **27** (1967) 395
128 KARTHAUS, H.: *Gaswärme* 1957, 148
129 KAST, W.: *Staub* **21** (1961) 215
130 KEIDEL, F. A.: *Anal. Chemistry* **31** (1959) 2043
131 KETTNER, H.: *Gesundh.-Ing.* **87** (1966) 105; *Staub* **26** (1966) 449
131a KIRKLAND, J.: *Anal. Chemistry* **35** (1963) 2003
131b KIRSTE, H.: Report. Landesanst. f. Bodennutzungsschutz Bochum **61** (1961)
132 KIYOURA, R.: *Staub* **26** (1966) 524
133 KLEIN, G.: Linde-Reports No. 17 (1964) 24
134 KNIGHT, H. S. and F. T. WEISS: *Anal. Chemistry* **34** (1962) 749
135 KOCH, O. G. and G. A. KOCH-DEDIC: *Handb. d. Spurenanalyse (Handbook of Trace Analysis)*, Springer-Verlag, Berlin, 1964
135a KÖHLER, A.: *Beiträge z. Physik. d. Atmosphäre* **36** (1962)
136 KÖHLER, A. and W. FLECK: *Staub* **26** (1966) 105
137 KÖHLER, M., H. GOLDER and R. SCHIESSER: *Z. anal. Chem.* **206** (1964) 430
138 KOHLSCHÜTTER, H. W. and W. HOPPE: *Z. anal. Chem.* **197** (1963) 133
139 KOLBEZEN, M. J., J. W. ECKERT and C. W. WILSON: *Anal. Chemistry* **36** (1964) 593
140 KOLTHOFF, J. M. and C. S. MILLER: *J. Amer. Soc.* **63** (1941) 2818
141 KUCHARCZYK, N., J. FOHL and J. VYMETAL: *J. Chromat.* **11** (1963) 55
142 KÜNDIG, S.: *Chem. Rundschau* **18** (1965) 123
143 KÜNDIG, S. and D. HÖGGER: *Staub* **24** (1964) 408
144 KULEY, C. J.: *Anal. Chemistry* **35** (1963) 1472
145 KUNZ, J.: *Monatsh. Chem.* **83** (1952) 946
146 LAHMANN, E.: *Staub* **25** (1965) 346
146a LAHMANN, E.: *Gesundh.-Ing.* **86** (1965) 69
147 LAHMANN, E.: *Staub* **26** (1966) 530
147a LAHMANN, E. and K. PRESCHER: *Staub* **25** (1965) 527
147b LAHMANN, E. and M. MÖLLER: *Gesundh.-Ing.* **88** (1967) 182
148 LANDWEHR, M.: *Staub* **24** (1964) 329: **26** (1966) 359
149 LARD, W. and R. C. HORN: *Anal. Chemistry* **32** (1960) 878
150 LEICHNITZ, K.: *Ch. Ztg.* **91** (1967) 141
151 LEITHE, W.: *Die Chemie* **56** (1943) 235
152 LEITHE, W.: *Mikrochem.-Microchim. acta* **33** (1946) 42
153 LEITHE, W.: *Mikrochem.-Michrochim. acta* **33** (1947) 16

154 LEITHE, W.: *Fresenius–Jander Handb. d. Anal. Chemie (Manual of Analytical Chemistry)*, Springer-Verlag, Berlin, 1957, vol. Va
155 LEITHE, W.: *Michrochim. acta* 1962, 166
156 LEITHE, W.: *Angew. Chem.* **73** (1961) 488
157 LEITHE, W.: *Z. anal. Chem.* **193** (1963) 24
158 LEITHE, W.: *Chem. Ing. Techn.* **36** (1964) 112
159 LEITHE, W. and G. PETSCHL: *Z. anal. Chem.* **226** (1967) 352
160 LEITHE, W. and G. PETSCHL: *Z. anal. Chem.*, in press
160a LEITHE, W. and A. HOFER: *Allgem. Prakt. Chem.* (Wien), in press
161 LEY, H. and H. DIRKING: *Ber.* **67** (1934) 1331.
162 LIBERTI, A., G. P. CARTONI and V. CANTUTI: *J. Chromat.* **15** (1964) 141
163 LIESEGANG, W.: *Gesundh.-Ing.* **54** (1931) 705
164 LIJINSKY, W., I. DOMSKY, G. MASON, H. Y. RAMAHI and T. SAFAVI: *Anal. Chemistry* **35** (1963) 952
165 LIJINSKY W., I. DOMSKY and J. WARD: *J. Gaschromat.* **3** (1965) 152
165a LINCH, A. L. and M. CORN: *Amer. Ind. Hyg. Assoc. J.* **26** (1965) 601; *Staub* **26** (1966) 346
165b LINCH, A. L., S. S. LORD, K. A. KUBITZ and M. R. DE BRUNNER: *Amer. Ind. Hyg. Assoc. J.* **26** (1965) 465; *Staub* **26** (1966) 348
166 LINDSLEY, Ch. H. and J. H. YOE: *Anal. Chemistry* **21** (1949) 513; *Z. anal. Chem.* **131** (1950) 308
167 LINSER, H.: *Mitt. D. Landw. Gesellschaft.* 1965, issue 6
167a LING, C.: *Anal. Chemistry* **39** (1967) 798
168 LITTMANN, F. E. and R. W. BENOLIEL: *Anal. Chemistry* **25** (1953) 1480
169 LOHS, K. H.: *Chem. Technik* **17** (1965) 38
170 LOVELOCK, J. E.: *Anal. Chemistry* **33** (1961) 162
170a LUFT, K. F., G. KESSELER and K. H. ZÖRNER: *Chem Ing. Techn.* **39** (1967) 937
171 LUGG, G. A.: *Anal. Chemistry* **38** (1966) 1532
172 LYLES, G. R., F. B. DOWLING and V. J. BLANCHARD: *J. Air Poll. Contr. Assoc.* **15** (1965) 106; cited in *Chem. Abstr.* **63**, 2303
173 LYSYI, J. and P. R. NEWTON: *J. Chromat.* **11** (1963) 173
174 MCCULLOUGH, J. D., R. A. CRANE and A. O. BECKMAN: *Anal. Chemistry* **19** (1947) 999
175 MCEVEN, D. J.: *Anal. Chemistry* **38** (1966) 1047
176 MCKELVEY, J. M. and H. E. HOELSCHER: *Anal. Chemistry* **29** (1957) 123
176a MCQUAIN, R. H.: see M. B. JACOBS, *Chemical Analysis of Air Pollutants* p. 223
177 MADER, P. P., W. J. HAMMING and A. BELLIN: *Anal. Chemistry* **22** (1950) 1181
177a MALANCHUK, M.: *Amer. Ind. Hyg. Assoc. J.* **28** (1967) 76
178 MALISSA, H. and G. WAGNER: *Microchim. acta* 1962, 332
179 MAY, J.: *Staub* **25** (1965) 153
180 MAY, J.: *Staub* **26** (1966) 385
181 MEDICUS, L. and W. POETHKE: *Kurze Anleitung zur Massanalyse (Short Guide for Routine Analysis)*, 17th ed. Verlag Steinkopf, Leipzig, 1962
181a MOORE, G. E., R. S. THOMAS and J. L. MONKMAN: *J. Chromatogr.* **26** (1967) 456
182 MORRISON, M. E. and W. H. CORCORAN: *Anal. Chemistry* **39** (1967) 255
183 MOSS, R. and E. V. BROWETT: *Analyst* **91** (1966) 428
184 MOUNDREA, V.: Thesis, University of Vienna, June, 1966: see also *Staub* **26** (1966) 30

185 MÜLLER, E. and A. THAER: *Staub* **25** (1965) 251
186 MURRAY, J. N. and J. B. DOE: *Anal. Chemistry* **37** (1965) 941
187 NAUMANN, R. V., P. W. WEST, F. TRON and G. C. GAEKE: *Anal. Chemistry* **32** (1960) 1307
187a NEERMAN, J. C. and F. R. BRYAN: *Anal. Chemistry* **31** (1959) 532
188 NIEDERMAIR, T.: *Z. anal. Chem.* **223** (1966) 336
189 NIETRUCH, F. and K. E. PRESCHER: *Z. anal. Chem.* **226** (1967) 259
189a NORMAN, V. and C. H. KEITH: *Nature* **205** (1965) 915
190 NOVAK, J., V. VASAK and J. JANAK: *Anal. Chemistry* **37** (1965) 660
191 OBERMILLER, E. L. and R. W. FREEDMAN: *J. Gaschr.* **3** (1965) 242; *Anal. Abstr.* **13**, 6566
192 ÖDELYKE, P.: *Staub* **26** (1966) 526
193 OEHME, F. and H. WYDEN: *Glas- und Instrumententechn.* **9** (1965) 107; *Staub* **26** (1966) 252
194 OETTEL, H.: *Angew. Ch.* **70** (1958) 532; see also *Ullmann Encycl. d. Techn. Chem.* vol. 10, p. 762, section *"Krebs" (Cancer)*
194a O'KEEFFE, A. E. and G. ORTMAN: *Anal. Chemistry* **38** (1966) 761
194b OST, K. and G. MIRISCH: *Mitt. Verein Grosskesselbes.* **37** (1955)
195 PAILER, M., H. BEGUTTER, R. BAUMANN and J. A. SCHEDLING: *Österr. Ch.-Ztg.* **67** (1966) 222
196 PANETH, F. A. and E. GLÜCKHAUF: *Nature* (London) **147** (1941) 61
197 PATE, J. B., B. E. AMMONS, G. A. SWANSON and J. P. LODGE: *Anal. Chemistry* **37** (1965) 942
198 PATE, J. B., J. P. LODGE and M. P. NEARY: *Anal. chim. acta* **28** (1963) 341
199 PATE, J. B., J. P. LODGE and A. F. WARTBURG: *Anal. Chemistry* **34** (1962) 1660
199a PAVELKA, F.: *Microchim. acta* 1964, 1121
199b PERSSON, G. A.: *Intern J. Air Water Poll.* **10** (1966) 845
200 PETERS, K.: *Chem. Fabrik* **10** (1937) 371
201 PETERS, K.: *Microchim. acta* 1956, 1023
202 PETERS, K. and H. STRASCHIL: *Angew. Chem.* **68** (1956) 291
203 PILZ, W.: *Microchim. acta* 1958, 383
204 PILZ, W. and E. STELZL: *Z. anal. Chem.* **219** (1966) 416
204a PORTER, K. and D. H. VOLMAN: *Anal. Chemistry* **34** (1962) 748
205 PRIESTLY, L. J., F. E. CRITCHFELD, N. H. KETCHAM and J. D. CAVENDER: *Anal. Chemistry* **37** (1965) 70
205a PRESCHER, K. E. and E. LAHMANN: *Gesundh.-Ing.* **87** (1966) 351
206 PROCHAZKA, R.: *Staub* **26** (966) 202
207 QUIRAM, E. R. and W. F. BILLER: *Anal. Chemistry* **30** (1958) 1166
208 QUIRAM, E. R., S. J. METRO and J. B. LEWIS, *Anal. Chemistry* **26** (1954) 352
209 RAYNER, A. C.: *J. Air Poll. Control Assoc.* **16** (1966) 418; *Staub* **27** (1967) 108
210 RAYNER, A. C. and C. M. JEPHCOTT: *Anal. Chemistry* **33** (1961) 627
211 REGENER, V. H.: *Naturwiss.* **26** (1938) 155; *Chem. Zbl.* 1938 **II** 677
212 REGENER, E. and V. H. REGENER: *Physikal. Z.* **35** (1935) 788; *Chem. Zbl.* 1935 **I** 355
213 RENZETTI, N. A.: *Anal. Chemistry* **29** (1957) 869
214 RIPLEY, D. L., J. M. CLINGENPEEL and R. W. HURN: *Intern. J. Air Water Poll.* **8** (1964) 455
214a RONDIA, D.: *Intern. J. Air Water Poll.* **9** (1965) 113
215 ROTH, F. and A. SCHULZ: *Brennst.-Chem.* **206** (1939) 317
216 SALSBURY, J. M., J. W. COLE and J. H. YOE: *Anal. Chemistry* **19** (1947) 66

217 SALTZMAN, B. E.: *Anal. Chemistry* **26** (1954) 1949
218 SALTZMAN, B. E.: *Anal. Chemistry* **32** (1960) 135
219 SALTZMAN, B. E.: *Anal. Chemistry* **33** (1961) 1100
219a SALTZMAN, B. E., A. S. COLEMAN and C. A. CLEMONS: *Anal. Chemistry* **38** (1966) 753
220 SALTZMAN, B. E. and A. F. WARTBURG: *Anal. Chemistry* **37** (1965) 1261
221 SAWICKY, E.: *Chemist-Analyst* **46** (1957) 67
222 SAWICKY, E., W. ELBERT, T. W. STANLEY, T. R. HAUSER and F. T. FOX: *Anal. Chemistry* **32** (1960) 810
223 SAWICKY, E. and Th. R. HAUSER: *Anal. Chemistry* **34** (1962) 1460
224 SAWICKY, E., T. R. HAUSER and G. MCPHERSON: *Anal. Chemistry* **34** (1962) 1460
225 SAWICKY, E., T. R. HAUSER, T. W. STANLEY and W. ELBERT: *Anal. Chemistry* **33** (1961) 93
226 SAWICKY, E., R. MILLER, T. STANLEY and T. HAUSER: *Anal. Chemistry* **30** (1958) 1130
226a SAWICKY, E., T. W. STANLEY, W. C. ELBERT, J. MEEKER and G. MCPHERSON: *Atmosph. Environment* **1** (1967) 131
227 SAWICKY, E., T. W. STANLEY, W. C. ELBERT and J. D. PFAFF: *Anal. Chemistry* **36** (1964) 497
228 SAWICKY, E., R. MILLER, T. STANLEY, J. PFAFF and A. D'AMICO: *Talanta* **10** (1963) 641
228a SCARINGELLI, F. P., S. A. FREY and B. E. SALTZMAN: *Amer. Ind. Hyg. Assoc. J.* **28** (1967) 260
228b SELUCKY, M., J. NOVAK and J. JANAK: *J. Chromatogr.* **28** (1967) 285
229 SERGEANT, G. A., B. E. DIXON and R. G. LIDZEY: *Analyst* **82** (1957) 27
230 SHEPHERD, M.: *Anal. Chemistry* **19** (1947) 77
231 SHEPHERD, M., S. M. ROCK, R. HOWARD and J. STORMES: *Anal. Chemistry* **23** (1951) 1431
231a SIMECEK, J.: *Staub* **26** (1966) 372
231b SIETH, J., *Mitt. Verein Grosskesselbes.* **108** (1967) 160
232 SMITH, R. G. and P. DIAMOND: *Am. Ind. Hyg. Assoc. Quarterly* **13** (1952) 235
233 SMITH, R. G., J. D. MCEWEN and R. E. BARROW: *J. Amer. Ind. Hyg. Assoc.* **20** (1959) 142
234 SPECTOR, N. A. and B. F. DODGE: *Anal. Chemistry* **19** (1947) 55
235 SPENGLER, G.: *Die Schwefeloxide in Rauchgasen und in der Atmosphäre (The Sulfur Oxides in Fume Gases and in the Atmosphere)*, 2nd ed. VDI-Verlag Düsseldorf, 1965
236 SWINNERTON, J. W., V. J. LINNENBOM and C. H. CHEEK: *Anal. Chemistry* **36** (1964) 1671
237 SCHEDLING, J. A.: *Staub* **22** (1962) 96
238 SCHMELTZ, J., R. L. STEDMAN and W. J. CHAMBERLAIN: *Anal. Chemistry* **36** (1964) 2499
239 SCHMIDT, K. G.: *Staub* **26** (1966) 190
240 SCHMIDTS, W. and W. BARTSCHER: *Z. anal. Chem.* **181** (1961) 54
240a SCHOLS, J. A.: *Anal. Chemistry* **33** (1961) 359
241 SCHNEIDER, W. and H. NAGEL: *Staub* **26** (1966) 389
242 SCHÜTZ, A.: *Staub* (1966) 198; 409
242a SCHULZE, F.: *Anal. Chemistry* **38** (1966) 748
242b STALKER, W. W., R. C. DICKERSON and G. D. KRAMER: *Amer. Ind. Hyg. Assoc. J.* **24** (1963) 68
242c STANLEY, T. W., M. J. MORGAN and J. E. MEEKER: *Anal. Chemistry* **39** (1967) 1327
243 STEPHENS, B. G. and F. LINDSTROM: *Anal. Chemistry* **36** (1964) 1308
244 STETTER, G.: *Anz. d. Math.-natw. Klasse d. Österr. Akad. d. Wiss.* No. 10 (1952) 131; *Z.f. Aerosolforschg.* **5** (1956) 361

244a Storp, N.: *Staub* **26** (1966) 340

245 Stratmann, H.: *Microchim. acta* 1954, 668

245a Stratmann, H. and M. Buck: *Intern. J. Air Water Poll.* **9** (1965) 211

246 Stratmann, H. and M. Buck: *Intern. J. Air Water Poll.* **10** (1966) 313

246a Stratmann, H. and M. Buck: *Intern. J. Air Water Poll.* **9** (1965) 199

247 Stuke, J. and J. Rzeznik: *Staub* **24** (1964) 366

248 Theer, J.: *Chem. Technik* **14** (1962) 164

249 Thilliez, G.: *Anal. Chemistry* **39** (1967) 427

250 Thoenes, H. W.: *Brennst.-Chem.* **42** (1961) 116

251 Thomas, M. D., J. A. McLeod, et al.: *Anal. Chemistry* **28** (1956) 1810

251a Thomas, R., G. E. Moore, L. Dubois, J. L. Monkman and M. Katz: *Air Water Poll.* **9** (1965) 833

252 Todd, G. W.: *Anal. Chemistry* (1955) 1490

253 Tomberg, V.: *Experientia* (Basel) **10** (1954) 388; *Z. anal. Chem.* **147** (1955) 306

254 Treadwell, F. P.: *Lehrbuch d. Anal. Chemie (Textbook of Analytical Chemistry)*, 11th ed. Verlag Deuticke Wien, 1949

255 Vasak, V. and V. Sedivec: *Chem. Listy* **46** (1952) 341

256 Vizard, G. S. and A. Wynne: *Chem. and Ind.* 1959, 196

258 Wadelin, C. W.: *Anal. Chemistry* **29** (1957) 441

259 Wagner, G.: *Österr. Ch. Ztg.* **54** (1953); *Z. Aerosolforschg.* **2** (1953) 422

259a Wahnschaffe, E.: *Mitt. Verein Grosskesselbes.* **108** (1967) 168

260 Walter, E.: *Staub* **22** (1962) 162

261 Walter, E.: *Staub* **23** (1963) 112

262 Wartburg, A. F., A. W. Brewer and J. P. Lodge: *Intern. J. Air Water Poll.* **8** (1964) 21; *Staub* **24** (1964) 377

263 Werth, H.: *Dechema-Monographie* **52** (1964) 219; *Wasser-Luft-Betr.* **9** (1965) 161

264 West, P. W. and G. C. Gaeke: *Anal. Chemistry* **28** (1956) 1816

265 West, P. W. and F. Ordoveza: *Anal. Chemistry* **34** (1962) 1324

266 West, P. W. and B. Sen: *Z. anal. Chem.* **153** (1956) 177

267 West, P. W., B. Sen and N. A. Gibson: *Anal. Chemistry* **30** (1958) 1390

268 Wieland, Th. and W. Kracht: *Angew. Ch.* **69** (1957) 172

269 Wiethaupt, H.: *Staub* **24** (1964) 161

270 Willard, H. H. and O. B. Winter: *Ind. Eng. Ch. Anal. Ed.* **5** (1933) 710

271 Williams, D. D. and R. R. Miller: *Anal. Chemistry* **34** (1962) 225

272 Williams, J. H.: *Anal. Chemistry* **37** (1965) 1723

273 Wilsdon, B. H.: *Chem. Abstr.* **32** (1938) 2263

274 Wilson, K. W.: *Anal. Chemistry* **30** (1958) 1127

275 Wilson, H. N. and W. Hutchinson: *Analyst.* **72** (1947) 149

276 Winkel, A.: *Staub* **19** (1959) 253

277 Winkel, A.: *Staub* **22** (1962) 77

278 Winkel, A.: *Staub* **24** (1964) 1

279 Winkel, A. and W. Coenen: *Staub* **26** (1966) 9

280 Winnaker, K.: *Chem. Ing. Techn.* **36** (1964) 1

280a Wolf, A.: *Staub* **27** (1967) 190

281 Zepf, K. and F. Vetter: *Mikrochemie* 1930, 280

Author's Addenda to the English Edition

During the period of two years which has elapsed since the publication of the German edition, fresh experience has expanded our knowledge of the control of air pollution. In this topical field, many new methods have been devised and analyses of pollutants have been accumulated. These developments are reflected by the numerous additions and bibliographical references in the book.

The German edition was intended mainly for German speaking readers, therefore regulations operative in the German-speaking countries predominated. The present English edition was supplemented by additions and bibliographical references that apply in particular to British and American conditions.

The former data, pertaining specifically to Central Europe, were not, however, specifically abridged. Thus, those readers familiar with the subject are enabled to compare regulations valid in Europe and elsewhere. Despite common aims and assumptions, problems of air pollution and its analysis were unfortunately often tackled quite differently in different countries. This edition may therefore help to establish better cooperation in this important field.

W. Leithe

p. 2, below:

The ASTM (Designation D2356-67a) also defines Air Pollution as "the presence of unwanted material in the air." The term "unwanted material" refers to material in sufficient concentration, present for a sufficient time and under circumstances in which they interfere significantly with the comfort, health, or welfare of persons or the full use and enjoyment of property.

p. 6, after third paragraph:

In the USA, due to the widely varying economic, social and meteorological factors governing its life, air pollution control is largely delegated to local governments and state authorities. The Clean Air Act of 1963 broadened the role of the Federal Government and strengthened its ability to assist and participate in a comprehensive national effort, by Federal financial grants to State and local control programs, and to take enforcement action against interstate air pollution. It includes a directive for the development of air quality criteria for the guidance of State, regional, and local air pollution control agencies.

Amendments in 1965 provided for Federal regulations of air pollution from new motor vehicles. National standards for control of emissions from new cars have been set (see page 11).

The US Public Health Service, Division of Air Pollution Control spent 18 million dollars in 1967 on research in this field. The Air Pollution Association with 3000 members, 13 local sections and 26 technical committees has been in existence in the US for over 60 years. It also publishes a journal on the subject.

Great Britain was the first country to provide a legal basis for official control of air pollution in industrial regions, by passing the *Alcali Acts* of 1863. Construction and operation of installations (plants) that may cause air pollution are subject to official approval and their registration is renewed every year. A special authority, the Alcali Inspectorate, is in charge of industrial air pollution control. In particular instances, this authority prescribes the "best practical means" and, among others, also the height of chimneys for sulphuric acid plants according to their production capacity, as well as limiting the content of noxious gases and dust in flue gases.

The Clean Air Act 1956 (see p. 10) contains prohibitions concerning the emission of smoke liable to cause soot damage. Strict regulations concerning air pollution are applicable in some city sections which are declared "smoke control areas" or "smokeless areas." Local authorities are empowered to take steps against exceedingly heavy smoke issuing from domestic chimneys. Householders in smoke control areas who have been using smoky fuel must change to a solid smokeless fuel, or electricity, gas or oil. To help this changeover, a 70% grant was given.

The Warren Spring Laboratory in Stevenage, Hertfordshire, which is subject to the Ministry of Technology, acts as a research institute on all problems of air pollution.

p. 9, line 2 from above:

In 1967 this author published a new edition of his book "Analytical Toxicology of Industrial Inorganic Poisons".

p. 36, line 4 from above:

See also F. J. Schuette [228d].

p. 45, last line:

An automatic device for taking samples with 9 wash bottles which can be inserted as desired within a period of 1–24 hours, has been developed by the Warren Spring Laboratory (manufacturer Charles Austen Pumps Ltd, Byfleet, Surrey) (see P. Beddoes [15a]). A similar device has been described by W. R. Parker and M. A. Huey [196b].

p. 51, addition to second paragraph:

Directly applicable toxicant standards (SO_2, NO_2, H_2S, hydrocarbons, fused into small Teflon tubes) are offered by Poly Science Corp., Evanston, Ill.

p. 51, before 3.2:

See Saltzman [219] and Hersch [106b].

p. 60, before 4.4:

The firm Unico, USA manufactures 65 tubes for a range of various toxicants.

p. 78, before 4.4.6:

A description of a micro-coulometer for determining SO_2, H_2S and mercaptans simultaneously by selective scrubbing processes is given by D. F. Adams, W. L. Bamesberger and T. J. Robertson [1c].

p. 80, top of page:

Concerning the use of a micro-coulometer for determining halogenated hydrocarbons in the air after their enrichment on Porapak Q and S, see D. M. Coulson, F. W. Williams, M. E. Umstead and L. A. Cavanagh.

. . . and Feinland, also S. Hantzsch and K. E. Prescher [93c], F. Baum, J. Reichardt and W. Steinbach [13c] as well as S. Häntzsch and E. Lahmann [93a].

p. 80, after second paragraph:

Even more sensitive than the FID, according to J. G. Price, D. C. Fenimore, G. P. Simmonds and A. Zlatkis [204b] is the photo-ionization detector.

p. 81, end of sixth paragraph:

. . . purified helium (see also H. Hackenberg and J. Gutbier [92b]).

p. 93, end of page:

For the determination of ethyleneglycol dinitrate, nitroglycerine, benzene and nitrotoluene compounds with the ECD (see E. Camera and D. Pravisani [37a]). Organochlor pesticides in the air are also determined with the ECD by D. C. Abbott, R. B. Harrison, J. O. G. Tatton and J. Thomson [1b].

p. 94, last paragraph:

. . . see also p. 120. The separation by gas chromatography of Ar, O_2 and N_2 after hydrogenation of O_2 to H_2O is described by E. J. Havlena and K. A. Hutchinson [101a], the separation of N_2, O_2, Ar, CO, CO_2, H_2S and SO_2 over Porapak by means of a heated wire detector is described by E. L. Obermiller and G. O. Charlier [191b].

p. 97, before 4.7:

According to E. E. Hughes and W. D. Darko [111b], direct mass-spectrometric determination of atmospheric CO_2 is possible if most of the oxygen present is removed by means of white phosphorus. Errors are less than 1 %. Assuming the argon content to be constant, the CO_2/Ar ratio may also be selected as the value to be determined.

p. 98, after Table 4.7:

On the determination of the odor threshold values of 53 chemical see also G. Leonardos, D. Kendall and N. Barnar [160c].

p. 99, addition to chapter:

For further references on the technique of quantitative determination of odors see W. C. L. Hemeon [102a], H. C. Wohlers [280a], R. A. Duffee [58d], F. V. Wilby [269a] and D. M. Benforado, W. J. Rotella and D. L. Horton [19c].

p. 103, add:

8. Gravimetric measurements of dust with the American Staplex instrument (air throughput 40–100 m^3/hr, detection limit 0.14 mg/m^3 for hourly sampling, 0.006 mg/m^3 for 24-hour sampling) are reported by E. Lahmann and K. E. Prescher [147d].

p. 103, line 12:

. . . Landwehr [148] and K. Kimura [131e].

p. 106, top:

. . . workshops (see K. Aurand, E. Lahmann and H. Rühle [10b]).

p. 110, before 5.1.7:

For comparative measurements at the work place with various measuring instruments (thermal precipitator — Konimeter, Porticon) see A. Winkel [279a], also VDI Directives 2266 and 2267.

p. 114, second line:

. . . Fleck 136, also A. Deuber, A. Gilgen and E. Grandjean [566] as well as A. F. Fisher [64b].

For measurements with the dust-deposit measuring instrument CERL see H. H. Macey [176b].

p. 123, middle of fourth paragraph:

. . . titration (W. Leithe and A. Hofer [160b]).

p. 125, second paragraph:

(e.g. the firm Beckman and by A. H. Thomas and Co., Philadelphia).

p. 127, last paragraph but one:

For the chemical mechanisms of atmospheric oxidation processes, see also E. R. Stephens [243a].

p. 129, end of 5.4.2.4:

For the determination of oxidation processes with acid KI solution, see S. Deutsch [56c].

p. 130, after first paragraph:

To measure ozone in the presence of excess NO_2 J. C. Cohen, A. F. Smith and R. Wood [47a] first determined the amounts with a neutral buffered KI solution and in a second sample they determined ozone with a cotton plug.

p. 133, before 5.4.2.6:

For continual ozone measurements with a bromometric ozone analyzer see also E. Lahmann, J. Westphal, K. Damaschke and M. Lübke [147f].

p. 138, before 5.4.3.4:

T. Asami [9a] describes a small test tube of cobalt (II) chloride on silica gel.

p. 142, fourth paragraph:
This type of dew point hygrometer is manufactured by K. Schafer, Frankfurt/Main.

p. 147, before 5.5.1.5:
For automatic sampling with 12 standard impingers see E. Lahmann and K. E. Prescher [147e].

p. 153, addition to 5.5.2.4:
Sulfur from flue gases is deposited in the soil in the form of sulfates, mainly by precipitation. Moreover, N. Faller [611c] of the Linser Institute used radioactive indicator SO_2 at sublethal doses to prove that SO_2 can be absorbed by plants within their normal sulfur metabolism and can be found again as protein materials, mustard oil etc.

p. 162, line 17:
... Prescher 189; see also F. P. Scaringelli, B. E. Saltzman and S. A. Frey [228c], H. G. C. King and G. Pruden [131f] and N. M. Trieff, H. C. Wohlers, J. A. O'Malley and H. Newstein [254a].
For SO_2 determinations at outside temperatures below 0°C, J. Huhn, J. Wagner and R. Fahnert [111c] added 50% glycerine or diglycol to the absorption solution.

p. 164, before 5.5.2.13:
Further experimental details were reported by R. K. Stevens, A. E. O'Keefe and G. C. Ortman [244a].

p. 165, second paragraph:
... Buck [245a], also G. Booras and C. E. Zimmer [25a].

p. 165, fourth paragraph:
A. Lyshkow [172a] also avoids the use of tetrachloromercurate and inserts a rotating cuvette instead.

p. 165, sixth paragraph:
J. P. Terry [262a] as well as G. B. Morgan, C. Golden and E. C. Tabor [181c] report on the use of auto-analyzers (see p. 71) for automation of the method by West and Ordoveza.

p. 167, line 4 from below:
For the automatic titration of the sulfate with thoron see G. A. Persson [199b]. Further simplifications of the lead dioxide method are suggested by N. A. Huey [111a].

p. 168, third paragraph from bottom:
For further information on the analysis of sulfuric acid aerosols see F. P. Scaringelli and K. A. Rehme [228b] as well as L. Dubois, C. J. Baker, T. Teichman, A. Zdrojewski and J. L. Monkman [58a].

p. 174, third paragraph:
For the determination of diethanolamine and 2-methylethanolamine after oxidation to

formaldehyde with periodic acid and color measurement with chromotropic acid, see F. A. Miller [181b].

p. 179, third paragraph should read:

According to previous information (H. Czech and N. Northdurft [99c]), cultures suffered slight or no damage at 30 ppm NO_2 after gassing for one hour. However, H. van Haut and H. Stratmann [101b] have recently observed a damage limit of 0.4 ppm NO_2 for medium-sensitive one-year old cultures after long-term gassing.

p. 186, second paragraph:

For further details see D. Wilson and St. L. Kopczynski [237a]. W. Forweg and H. J. Crecelius [69b] used the Drägerwerk oxidation-gravimetric method for the oxidation of NO.

p. 187, fifth paragraph:

... Wartburg [220], and J. T. Shaw [229a].

p. 188, before 5.6.5.9:

On rechecking the "Saltzman-factor" F. Nietruch and K. E. Prescher [189b], in agreement with Stratmann and Buck, found the factor 1 for a NO_2 concentration of 0.15 mg NO_2/m^3 = = 0.07 ppm; for higher concentrations (1.6 and 4.9 mg NO_2/m^3), however, they found the factors 0.8–0.9.

Tests made in the main laboratory of the Österreichischen Stickstoffwerke indicate that in reactions where the Saltzmann factor is smaller than 1, NO is formed again according to the equation $3NO_2 + H_2O = NO + 2HNO_3$.

p. 188, end of 5.6.5.9:

S. Häntzsch, F. Nietruch and K. E. Prescher [93b] describe the use of auto-analyzers with the Saltzman reagent, whereas G. B. Morgan, C. Golden and E. C. Tabor [181e] report on the application of auto-analyzers to the method of Jacobs and Hochheiser.

An automatic colorimeter (manufactured by Atlas Electric Devices Comp., Chicago) that can be used with sulfanilic acid instead of 2-aminobenzene disulfonic acid is now available.

p. 196, before 5.7.2.4:

(see also R. C. Robbins, K. M. Borg and E. Robinson [214b]).

p. 210, addition to fourth paragraph:

For the determination of elementary fluorine by the coulometric method see S. Kaye and M. Griggs [129a].

Simplifications for the determination of F^- have been described by B. S. Marshall and R. Wood [178a].

Recently, F^--specific electrodes are frequently used for the potentiometric determination of low F^--concentrations. L. A. Elfers and C. E. Decker [62a] used electrodes made by the firm Orion (lanthanum fluoride crystals covered with europium); M. S. Frant did likewise (see also J. J. Lingane and G. H. Farrah [64a] who used $Bi{-}BiF_3$ electrodes).

For separate sampling of fluorine compounds in the form of gas or dust, see J. A. Dorseyu and D. A. Kemnitz [57a] and K. Habel [92a].

Another instrument for automatic continual determination of fluorine is the Fluoride Recorder of the Stanford Research Institute, which has been modified by M. D. Thomas, G. A. St. John and S. W. Chaikin for concentrations of 0.2–10 ppb. It measures the fluorescence extinction of magnesium oxinate by HF. The instrument described by C. R. Thompson, L. F. Zielenski and J. O. Ivic is based on the same principle.

p. 213, after table 5.8.2:

Determination with methyl orange

A procedure for the determination of elementary chlorine based on the bleaching of a methyl orange solution containing KBr, was given by H. Kettner and W. Forweg [131d]. The NO_2 concentrations generally present in the atmosphere do not interfere.

The sensitivity limit is 0.3 μg Cl_2, the standard deviation is $\pm$ 0.05 mg Cl_2/m^3.

Indicator solution: 0.1 g methyl orange is dissolved in 100 ml water, together with 0.5 g $HgCl_2$ (as fungicide), 0.5 g KBr and 20 ml glycol. The mixture is then made up to 1 liter.

Absorption solution: 100 ml 1 *N* H_2SO_4, 50 ml of above solution and 0.1 ml of 30% H_2O_2 are made up with water to 1 liter.

At 510 mμ and a cell thickness of 1 cm the extinction should amount to 0.49–0.51.

Sampling: 20 liter air (500–700 ml/min) is passed through a fritted glass wash bottle. The results can be found on a calibration curve obtained with air mixtures containing a known amount of Cl_2. With an absorption solution of 25 ml, a Cl_2 amount of 4.7 μg produced an extinction reduction of 0.1.

The automatic measurement with the Mikometer (a variant of the Okometer described on p. 64, see H. Hümmel) as described by W. Forweg and P. Dopfer [69c] is based on the same reaction. At throughput amounts up to 38 liter/hr, the detection sensitivity (triple standard deviation of the zero value) is 0.01 ppm Cl_2.

p. 220, addition to fourth paragraph:

A detailed description of a method based on the same principle for the determination of lead in the gas and dust state, is the subject of ASTM-Designation D 2681-68 T.

p. 221, before 5.10.4:

L. J. Snyder [233a], adsorbed the organic lead compounds from the air with activated charcoal, dissolved the lead with aqua regia and determined it as the dithizonate.

p. 225, line 3 from above:

See also A. A. Cristie, A. J. Dunsdon and B. S. Marshall [496].

p. 230, second paragraph:

For the determination of formaldehyde with chromotropic acid (4,5-dihydroxy-2,7-naphthalenedisulfonic acid) after absorption in bisulfite solution see A. P. Altshuller and S. P. McPherson [2b], and J. W. Cares [38a].

p. 232, addition to paragraph on "Variants":

... 560 mμ (see also E. Lahmann and K. Jander [147c].

p. 237, before 6.5.4:

For the combustion of air samples containing halogenated hydrocarbons over a platinum catalyst, absorption of quantitatively formed elementary halogens in phenol red and colorimetric determination of the resulting halogenized dyes see R. Sidor [231b].

p. 239, addition to 6.5.5:

For the gas-chromatographic determination of halogenated hydrocarbons and aromatic hydrocarbons (benzene, toluene and Xylene) after enrichment over activated charcoal and desorption with carbon disulfide, see F. H. Reid and W. R. Halpin [212a].

p. 246, addition to 6.10.3:

For the determination of styrene and ethylbenzene by UV-measurement in iso-octane see R. K. Yamamoto and W. A. Cook [280b].

p. 261, addition to *Variants of the separation method*:

For further details on the separation procedure see G. B. Lindstedt [164a], L. Dubois, A. Zdrojewski and J. L. Monkman [58c] as well as A. Zdrojewski, L. Dubois, G. E. Moore, R. S. Thomas and J. L. Monkman [280c].

A detailed method for isolating and determining polynuclear aromatic hydrocarbons in air forms the subject of ASTM-Designation D 2682-68 T.

p. 263, addition to last chapter:

For further details on gas chromatographic separation methods of polycyclic hydrocarbons see H. J. Davis [53a], and H. Arito, R. Soda and H. Matsushita [8a].

Additional Bibliography

1b. ABBOTT, D. C., R. B. HARRISON, J. O. G. TATTON and J. THOMSON: *Nature* **211** (1966) 259

1c. ADAMS, D. F., W. L. BAMESBERGER and T. J. ROBERTSON: *J. Air Poll. Contr. Assoc.* **18** (1968) 145

2b. ALTSHULLER, A. P. and S. P. MCPHERSON: *J. Air Poll. Contr. Assoc.* **13** (1963) 109

8a. ARITO, H., R. SODA and H. MATUSHITA: *Industr. Health* **5** (1967) 243; cited in *Staub* **29** (1969) 37

9a. ASAMI, T.: *Anal. Chemistry* **40** (1968) 648

10b. AURAND, K., E. LAHMANN and H. RÜBLE: *Gesundh.-Ing.* **89** (1968) 360

13c. BAUM, F., J. REICHARDT and W. STEINBACH: *Staub* **27** (1967) 269

15a. BEDDOES, P.: *Atmosph. Envir.* **1** (1967) 595

19c. BENFORADO, D. M., W. J. ROTELLA and D. L. HORTON: *J. Air Poll. Contr. Assoc.* **19** (1969) 101

25a. BOORAS, G. and C. E. ZIMMER: *J. Air Poll. Contr. Assoc.* **18** (1968) 612

37a. CAMERA, E. and D. PRAVISANI: *Anal. Chemistry* **39** (1967) 1645

38a. CARES, J. W.: *Amer. Ind. Hyg. Assoc. J.* **29** (1968) 405

47a. COHEN, J. C., A. F. SMITH and R. WOOD: *Analyst* **93** (1968) 507

47b. COULSON, D. M. and L. A. CAVANAGH: *Anal. Chemistry* **32** (1960) 1245

49b. CRISTIE, A. A., A. J. DUNSDON and B. C. MARSHALL: *Analyst* **92** (1967) 185

49c. CZECH, M. and N. NOTHDURFT: *Landw. Forschung* **IV**, No. 1, 1 (1952)

53a. DAVIS, H. J.: *Anal. Chemistry* **40** (1968) 1583

56b. DEUBER, A., A. GILGEN and E. GRANDJEAN: *Staub* **29** (1969) 148

56c. DEUTSCH, S.: *J. Air Poll. Contr. Assoc.* **18** (1968) 78

57a. DORSEY, J. A. and D. A. KEMNITZ: *J. Air Poll. Contr. Assoc.* **18** (1968) 12

58a. DUBOIS, L., C. J. BAKER, T. TEICHMAN, A. ZDROJEWSKI and J. L. MONKMAN: *Microchim. acta* 1969, 269

58b. DUBOIS, L., A. ZDROJEWSKI, C. BAKER and J. L. MONKMAN: *J. Air Poll. Contr. Assoc.* **17** (1967) 818

58c. DUBOIS, L., A. ZDROJEWSKI and J. L. MONKMAN: *Microchim. acta* 1967, 834, 903

58d. DUFFEE, R. A.: *J. Air Poll. Contr. Assoc.* **18** (1968) 472

62a. ELFERS, L. A. and C. E. DECKER: *Anal. Chemistry* **40** (1968) 1658

64a. FARRAH, G. H.: *J. Air Poll. Contr. Assoc.* **17** (1967) 738

64b. FISHER, A. F.: *J. Air Poll. Contr. Assoc.* **7** (1957) 47

64c. FALLER, N.: Thesis, Giessen, 1968

69b. FORWEG, W. and H. J. CRECELIUS: *Staub* **28** (1968) 514

69c. FORWEG, W. and P. DOPFER: *Wasser/Luft/Betr.* **13** (1969) 92

69d. FRANT, M. S.: *Science* **154** (1966) 1553

92a. HABEL, K.: *Staub* **28** (1968) 283

92b. HACKENBERG, H. and J. GUTBIER: *Brennst.-Chem.* **49** (1968) 242

93a. HÄNTZSCH, S. and E. LAHMANN: *Erdöl u. Kohle* **20** (1967) 642

93b. HÄNTZSCH, S., F. NIETRUCH and K. E. PRESCHER: *Mikrochim. acta* 1969, 550

93c. HÄNTZSCH, S. and K. E. PRESCHER: *Wasser/Luft/Betr.* **12** (1968) 68

101a. HAVLENA, E. J. and K. A. HUTCHINSON: *J. Gas-Chrom.* **6** (1968) 419

101b. VAN HAUT, H. and H. STRATMANN: *Schriftenreihe, Landesanst. f. Immiss.-u. Bodennutzungsschutz Essen*, No. 7, p. 50 (1967)

102a. HEMEON, W. C. L.: *J. Air Poll. Contr. Assoc.* **18** (1968) 166

106b. HERSCH, P. A.: *J. Air Poll. Contr. Assoc.* **19** (1969) 164

111a. HUEY, N. A.: *J. Air Poll. Contr. Assoc.* **18** (1968) 610

111b. HUGHES, E. H. and W. D. DORKO: *Anal. Chemistry* **40** (1968) 866

111c. HUHN, J., J. WAGNER and R. FAHNERT: *Chem. Techn.* **19** (1967) 564

111d. HUMMEL, H.: *Chem. Ing. Techn.* **34** (1962) 704

129a. KAYE, S. and M. GRIGGS: *Anal. Chemistry* **40** (1968) 2217

131c. KETTNER, H.: *Staub* **29** (1969) 153

131d. KETTNER, H. and W. FORWEG: *Atmosph. Envir.* **3** (1969) 215

131e. KIMURA, K.: *Staub* **28** (1968) 163

131f. KING, H. G. C. and G. PRUDEN: *Analyst* **94** (1969) 43

147c. LAHMANN, E. and K. JANDER: *Gesundh.-Ing.* **89** (1968) 18

147d, LAHMANN, E. and K. E. PRESCHER: *Wasser/Luft/Betr.* **11** (1967) 677

147e. LAHMANN, E. and K. E. PRESCHER: *Wasser/Luft/Betr.* **12** (1968) 529

147f. LAHMANN, E., J. WESTPHAL, K. DAMASCHKE and M. LÜBKE: *Gesundh.-Ing.* **89** (1968) 144

160b. LEITHE, W. and A. HOFER: *Microch. acta* 1968, 1066

160c. LEONARDOS, G., D. KENDALL and N. BARNARD: *J. Air Poll. Contr. Assoc.* **19** (1969) 91

160d. LINGANE, J. J.: *Anal. Chemistry* **39** (1967) 881

164a. LINSTEAD, G.: *Atmosph. Envir.* **2** (1968) 1

169a. LONEMAN, W. A., T. A. BELLAR and A. P. ALTSHULLER: *Envir. Sci. Techn.* **2** (1968) 1017

172a. LYSHKOW, N. A.: *J. Air Poll. Contr. Assoc.* **17** (1967) 687

176b. MACEY, H. H.: *Atmosph. Envir.* **1** (1967) 637

178a. MARSHALL, B. S. and R. WOOD: *Analyst* **94** (1969) 821

181b. MILLER, F. A.: *Amer. Ind. Hyg. Assoc. J.* **29** (1968) 411

181c. MORGAN, G. B., C. GOLDEN and E. C. TABOR: *J. Air Poll. Contr. Assoc.* **17** (1967) 300

189b. NIETRUCH, F. and K. E. PRESCHER: *Z. anal. Chem.* **234** (1968) 118; **244** (1969) 294

191b. OBERMILLER, E. L. and G. O. CHARLIER: *J. Gas-Chrom.* **6** (1968) 446

196b. PARKER, W. R. and N. A. HUEY: *J. Air Poll. Contr. Assoc.* **17** (1967) 388

204b. PRICE, J. G., D. C. FENIMORE, G. P. SIMMONDS and A. ZLATKIS: *Anal. Chemistry* **40** (1968) 541

212a. REID, F. H. and W. R. HALPIN: *Amer. Ind. Hyg. Assoc. J.* **29** (1968) 390

214b. ROBBINS, R. C., K. M. BORG and E. ROBINSON: *J. Air Poll. Contr. Assoc.* **18** (1968) 106
228b. SCARINGELLI, F. P. and K. A. REHME: *Anal. Chemistry* **41** (1969) 707
228c. SCARINGELLI, F. P., B. E. SALTZMAN and S. A. FREY: *Anal. Chemistry* **39** (1967) 1709
228d. SCHUETTE, F. J.: *Atmosph. Envir.* **1** (1967) 515
229a. SHAW, J. T.: *Atmosph. Envir.* **1** (1967) 81
231b. SIDOR, R.: *Amer. Ind. Hyg. Assoc. J.* **30** (1969) 188
233a. SNYDER, L. J.: *Anal. Chemistry* **39** (1967) 591
243a. STEPHENS, E. R.: *J. Air Poll. Contr. Assoc.* **19** (1969) 181
244a. STEVENS, R. K., A. E. O'KEEFE and G. C. ORTMAN: *Envir. Sci. Techn.* **3** (1969) 652
251b. THOMAS, M. D., G. A. ST. JOHN and S. W. CHAIKIN: *ASTM Special Techn. Publ.* No. 250
251c. THOMPSON, C. R., L. F. ZIELENSKI and J. O. IVIC: *Atmosph. Envir.* **1** (1967) 253
254a. TRIEFF, N. M., H. C. WOHLERS, J. A. O'MALLEY and H. NEWSTEIN: *J. Air Poll. Contr. Assoc.* **18** (1968) 329
262a. WELCH, A. F., and J. P. TERRY: *J. Amer. Ind. Hyg. Assoc.* **21** (1960) 316
269a. WILBY, F. V.: *J. Air Poll. Contr. Assoc.* **19** (1969) 96
272a. WILLIAMS, F. W. and M. E. UMSTEAD: *Anal. Chemistry* **40** (1968) 2232
273a. WILSON, D. and ST. L. KOPCZYNSKI: *J. Air Poll. Contr. Assoc.* **18** (1968) 161
279a. WINKEL, A.: *Staub* **28** (1968) 1
280a. WOHLERS, H. C.: *J. Air Poll. Contr. Assoc.* **17** (1967) 609
280b. YAMAMOTO, R. K. and W. A. COOK: *Amer. Ind. Hyg. Assoc. J.* **29** (1968) 238
280c. ZDROJEWSKI, A., L. DUBOIS, G. E. MOORE, R. S. THOMAS and J. L. MONKMAN: *J. Chromat.* **28** (1967) 317

Subject Index